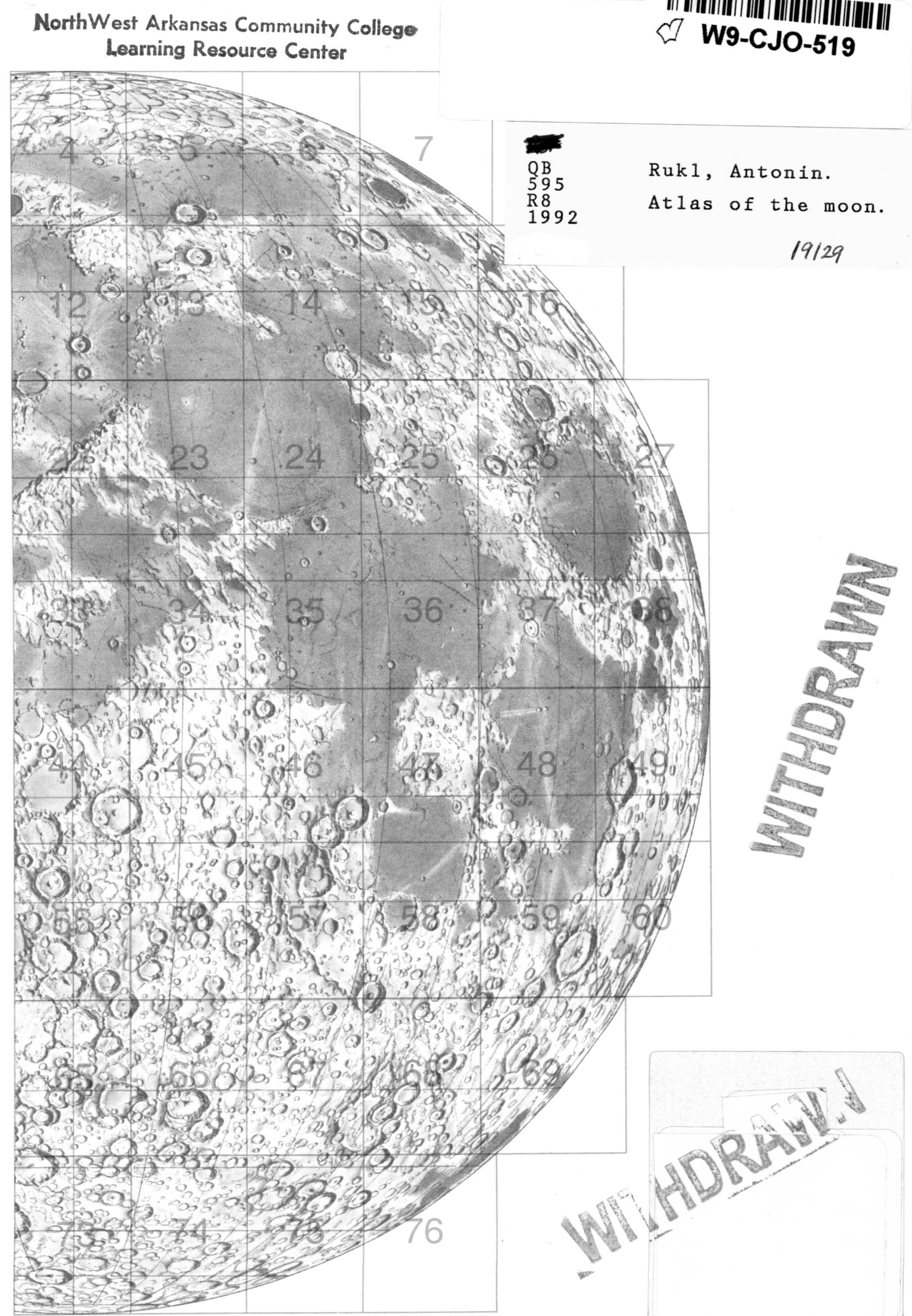
4
5
6
7
12
13
14
15
16
23
24
25
26
27
34
35
36
37
44
45
46
47
48
49
56
57
58
59
60
65
66
67
68
69
73
74
75
76

ATLAS OF THE MOON

ATLAS OF THE MOON

ANTONÍN RÜKL

EDITED BY

DR. T.W. RACKHAM

KALMBACH BOOKS®

Front endpaper:
Key for rapid location of individual sections of the detailed map of the near-side of the Moon (pp. 29–179).

Back endpaper:
Map showing the locations of the 50 regions on the near-side of the Moon, illustrated and described in the section 'Fifty views of the Moon' (pp. 194–207).

Frontispiece:
The medieval astronomical clock of the city of Prague, which dates from 1410 and shows for every day the Moon's phase, its position against the Zodiacal constellations and the time of its rising and setting. Note the Moon shortly after First Quarter as shown on the clock and on the inset photo above.

Photo by A. Rükl

Text and illustrations by Antonín Rükl
English version edited by T. W. Rackham
With thanks to Peter B. J. Gill
Graphic design by Stanislav Seifert
Designed and produced by Aventinum, Prague

This edition published in the U.S.A. in 1992
by Kalmbach Publishing Co., 21027 Crossroads Circle,
P. O. Box 1612, Waukesha, Wisconsin 53187,
by arrangement with Reed Consumer Books, part of
Reed International Books Ltd.

ISBN 0-913135-17-8

For copyright reasons this edition may not be sold outside the
United States of America

Printed in Czechoslovakia by Neografia, Martin
3/21/05/51–02
10 9 8 7 6 5 4 3 2 1

Contents

Introduction 6
The Moon: a satellite of the Earth 7
Phases of the Moon 9
Lunar phases 1989–2000 10
The Moon in the sky 11
The surface of the Moon 12
Origin and history of the Moon 15
Lunar cartography 16
Numerical data on the Moon 19

Becoming acquainted with the Moon 20
The Moon at First Quarter 20
The Moon at Last Quarter 22
The Full Moon 24
Atlas of the near-side of the Moon 26
Maps of libration zones 180
General map of the Moon 190
Flights to the Moon 192
Fifty views of the Moon 194

Observation of the Moon 208
Tables for the calculation of co-longitude 212
Lunar eclipses 213
Glossary 215
Bibliography 218
Sources used for compilation of the map and pictorial sections 219
Acknowledgements 219
Index of named formations 219

Introduction

The Moon, our closest celestial neighbour, has been attracting the attention of the inhabitants of the Earth since ancient times. This is hardly surprising, for apart from the Sun it is the most conspicuous astronomical object seen from the surface of our own planet. Its changing phases were used by ancient scholars to compile the first calendars, and even now remind us of the passing of time; and its surface features may be observed in detail using even quite a small telescope.

This close proximity and ease of observation had resulted by the late 1950s in a vast accumulation of knowledge, which made the Moon the first target of space flights into the solar system. It was the first celestial body to be reached by space probes and the first – and until now the only natural extraterrestrial body to bear the imprint of man. Intense exploration of the Moon during the 1960s and '70s has greatly improved our knowledge of the earliest phases of development of the solar system and has shed light on the early history of the Earth.

The science and technology of the second half of the twentieth century have brought the Moon so close to us that we now regard it in much the same way as fifteenth century navigators contemplated the exploration of distant continents. Like them, we know that there are distant lands on the Earth that we may never see, yet our object of discovery, the Moon, is often plainly inviting us to explore its surface with the aid of maps and modest telescopes.

The present atlas is intended for observers, amateur astronomers and all who would like to familiarize themselves with lunar features as seen from the Earth. The main part of the atlas consists of a detailed map of the near-side of the Moon, subdivided into 76 sections and with complete nomenclature as authorized by the International Astronomical Union. Each section is accompanied by a 'Who's Who on the Moon' summary containing details of the eminent people whose names have been given to lunar formations and a brief description of the formations themselves. Additional text describes how the maps may be used when observing the Moon. A map of this kind was previously published by the author in his book *Moon–Mars–Venus* (published by Hamlyn/Artia, Prague 1976). The present map contains more detail and the nomenclature has been brought up to date.

The Moon: a satellite of the Earth

Just as the planets orbit the Sun, taking with them their families of orbiting moons, so our Moon describes an elliptical path around the Earth (Fig. 1). That is why its distance from our planet is constantly changing. When it occupies the orbital point at which it is closest to the Earth, the Moon is said to be at *perigee*. *Apogee* occurs when it reaches its most distant point. At perigee the Moon's distance from the Earth is 356 400 km and the angular diameter of its disk is about 33.5 minutes of arc (33.5′)*. At apogee its distance increases to 406 700 km, while its angular diameter shrinks to 29.4′.

Such significant changes in angular diameter can be detected easily by comparing photographs of the Moon taken at perigee and apogee using the same telescope. Coincidentally, the apparent angular diameter of the Moon is similar to that of the Sun as seen from the Earth – about half a degree. Theoretically, when the Moon is at perigee finer details should be observable than when it is at apogee, but in practice telescopic visibility is governed by the ever-changing and often turbulent state of the Earth's atmosphere, not forgetting the experience and skill of the observer.

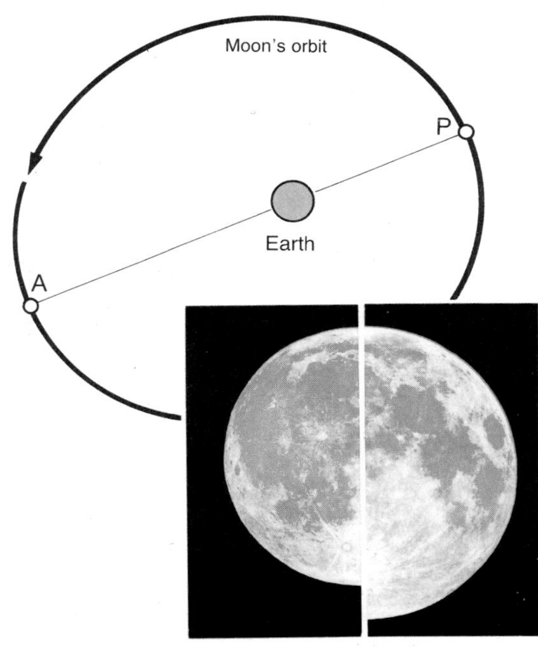

Fig. 1. The Moon's orbit. Comparison of the Moon's angular diameter at perigee P (to the right) and at apogee A (to the left).

Our star, the Sun, is four-hundred times further away from the Moon than the Earth, yet the Sun and the Earth produce the principal gravitational effects that constantly influence the Moon's orbit. Not only does the orbital velocity of the Moon vary, but changes also occur in the shape, dimensions and orientation of the plane of the orbit as measured against the stars. A precise description of the Moon's motion is one of the most difficult tasks of theoretical astronomy. The plane of the Moon's orbit around the Earth is within about five degrees of the plane of the Earth's orbit around the Sun. It is perhaps surprising that the resulting trajectory of the Moon around the Sun is an orbit similar to that of the Earth, which is at all times concave towards the Sun (see Fig. 6).

Since the distance of the Moon from the Earth is a mere thirty Earth diameters, it is little wonder that the Moon exerts a considerable gravitational effect on the Earth: this is clearly observed in the alternating ebb and flow of the ocean tides. The Earth–Moon system is, in fact, regarded as a double planet, for both bodies are planetary sized, orbiting around a common centre of gravity which lies some 4700 km from the Earth's centre, along a line joining the centres of the two bodies.

Among the many dynamical properties of moons in general is *synchronous rotation*. This is shown by our Moon, which completes one rotation on its axis in the same time as it takes to orbit the Earth, which is why it always presents the same hemisphere to the Earth. This well-known fact has often been explained erroneously by asserting that the Moon does not rotate on its axis at all! If this were so, the Moon would orbit the Earth as shown in Fig. 2a. A defined lunar crater A, which in position 1 would occupy the centre of the near-side, would in position 2 be at the edge of the visible hemisphere, and in position 3 be out of sight in the middle of the averted hemisphere. In fact, the Moon spins on its axis of rotation as shown in Fig. 2b: from position 1 to position 2 it completes one quarter of its orbital revolution while simultaneously turning one quarter of a rotation (90°) on its axis, so that crater A stays in the middle of the side facing the Earth. In the course of one complete orbital revolution one lunar hemisphere (indicated by hatching in Fig. 2b) remains forever hidden from the Earth, although the whole Moon is successively illuminated by the Sun.

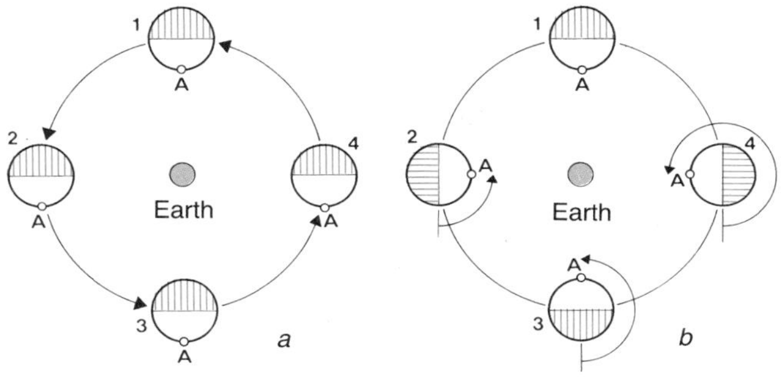

Fig. 2a, b. The Moon's rotation.

* For numerical data on the Moon see p. 19

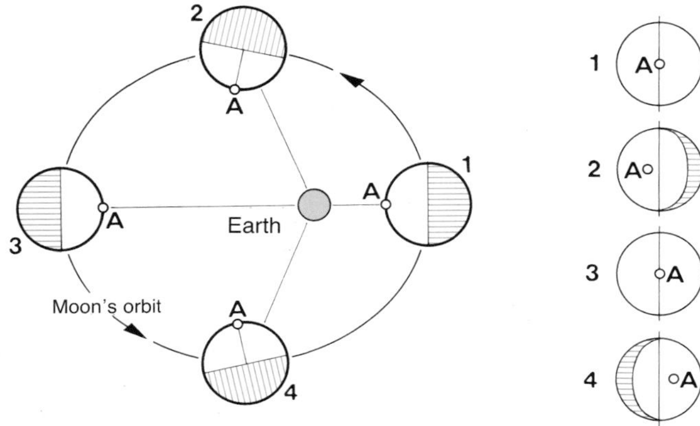

Fig. 3. Libration in longitude.

If what has been considered so far about synchronous rotation is true, we would expect just 50% of the lunar surface to be visible from the Earth. In reality the complexities of lunar dynamics favour terrestrial observers. While we can never see more than one hemisphere at a time, slender 'crescents' of the averted hemisphere are brought into view by swinging or oscillating motions called *librations,* which enable us to see some 59% of the total surface.

The maximum effect, and the most beneficial for the terrestrial observer, is caused by the optical librations in longitude and latitude.

Libration in longitude (Fig. 3) is caused by the fact that the axial rotation of the Moon is constant, while its orbital velocity around the Earth is perpetually changing. The latter reaches a maximum at perigee and then slows down to a minimum at apogee, after which it accelerates again, and so on... For example, from perigee (position 1), the Moon completes one quarter of its orbit around the Earth in less time that it requires for an axial rotation of 90°. This manifests itself by an apparent 'swinging' to the left of the lunar globe to reveal an area that normally lies beyond the right-hand edge of the Moon. Similarly, from apogee (position 3), the Moon takes longer to reach position 4 than it does to make a quarter of an axial rotation; therefore features that are usually beyond its left-hand edge are brought into view. Libration in longitude can cause an east–west displacement of ±7°54′.

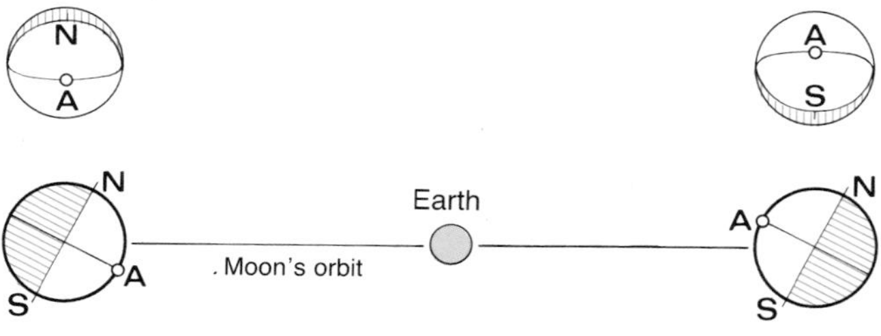

Fig. 4. Libration in latitude.

Libration in latitude (Fig. 4) is caused by the fact that the Moon's equator is inclined to its orbital plane. This tilt, of 6.41°, inclines first one pole and then, when the Moon is on the opposite side of its orbit, the other towards the Earth, as the rotational axis of the Moon (like that of the Earth and any other massive 'gyroscope') maintains its spatial orientation and points to the same position 'on the sky' no matter where the Moon is in its orbit. The apparent displacements in latitude amount to ± 6°50′.

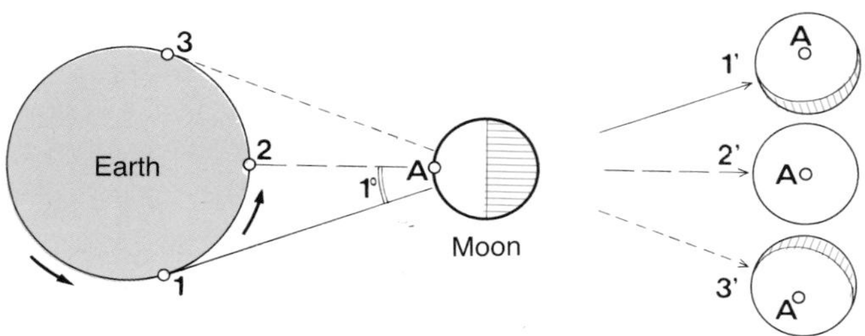

Fig. 5. Diurnal libration.

Librations in latitude and longitude occur simultaneously and continuously and their combined effects bring into view peripheral areas called *libration zones.* An additional *diurnal libration* (Fig. 5), amounting to about one degree, arises from the fact that from different points on the Earth's surface the Moon is seen from slightly different angles. Librations in latitude and longitude are of fundamental importance to anyone observing the Moon and so their values are tabulated in most astronomical almanacs.

Before passing on to other topics it would be as well to mention *physical libration,* which is caused by gravitational irregularities in the rotation of the Moon. Although the angles involved are small (several minutes of arc), they are of importance to astronomers studying the shape and internal structure of the Moon.

Fig. 6. Orbital motion and phases of the Moon. The Moon's orbits around the Earth and around the Sun in column **a** *are not shown to scale. The section* **s** *of the Moon's orbit around the Sun is, in fact, equal to about 12 times the diameter of the Moon's orbit around the Earth.*

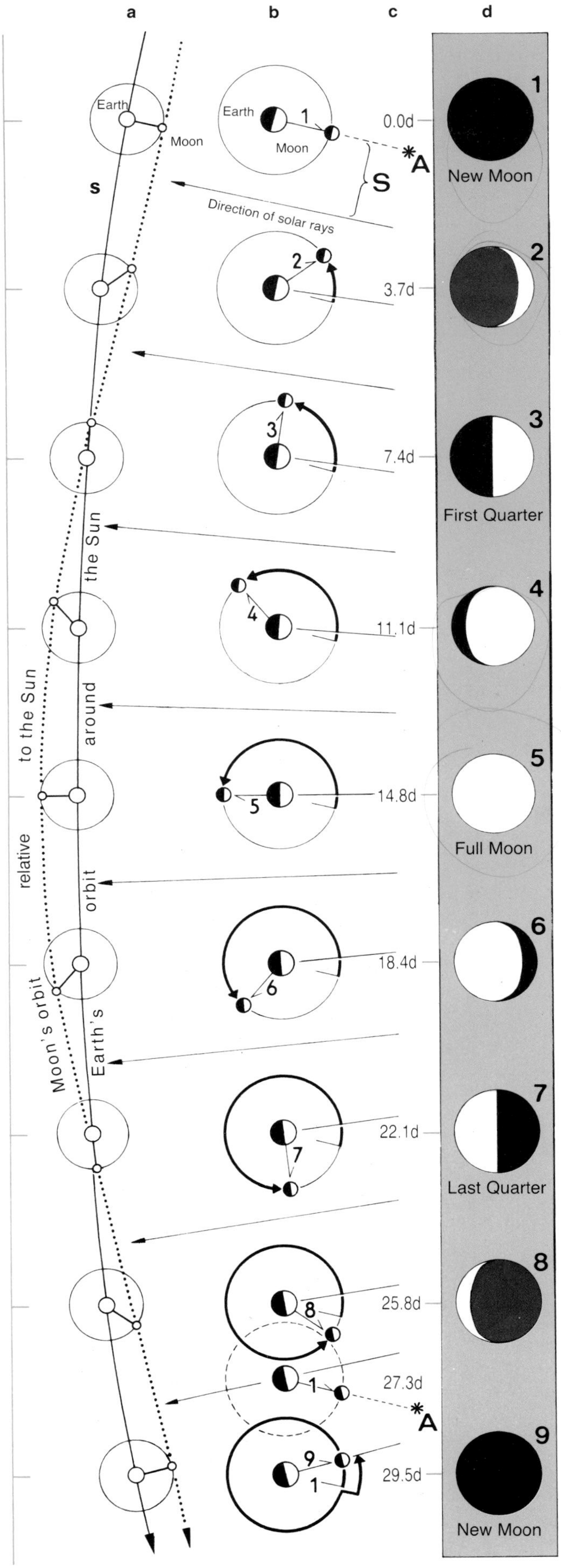

Phases of the Moon

The most striking phenomenon arising from the Moon's revolution around the Earth is its changing phases. The light of the Sun falls upon the Moon and those parts that are illuminated can be observed from the Earth, depending on the angle separating the Sun and Moon as seen from the Earth. In Fig. 6 successive values of this angle are shown in column **b** and the corresponding appearance or phase of the Moon is illustrated in column **d**. In position 1 the Moon lies in the same direction **S** as the Sun as seen from the Earth and what we call the *New Moon* is not observable. Starting from this instant the age of the Moon in Earth-days is given in column **c**. As the Moon 'waxes' it gradually becomes visible. After about 3.7 days it has completed one-eighth of its orbit (45°), having arrived at position 2, and from the Earth a part of its illuminated hemisphere is seen as a narrow crescent. At this time the remainder of the lunar disk is also faintly visible, because its 'night-time' side is lit by 'earthshine', which is sunlight reflected from the Earth. The boundary between day and night on the Moon is called the *terminator*.

In position 3 the age of the Moon is about 7.4 days and precisely half of the illuminated hemisphere is observable: the Moon is at *First Quarter*. The whole of the illuminated hemisphere can be seen at position 5, when the Moon lies in the opposite direction to the Sun. This is *Full Moon*. After the 'full' phase the Moon begins to wane and the illuminated area, as seen from the Earth, diminishes through *Last Quarter* (position 7) and the ever-narrowing crescent phases to New Moon.

Consider a straight line passing through the centres of the Earth and Sun and projected beyond the Sun to a point 'on the sky'. As the Earth and Moon orbit the Sun, this line will change its direction in space and its projected 'end' will move through the so-called 'fixed' stars. Returning to Fig. 6, let us imagine that when the Moon is in position 1 a fixed star **A** is exactly at the end of this imaginary projected line. In other words, the centres of the Earth, Moon and Sun are in line with star **A**. From the instant of this celestial 'line-up' 27.3 days will elapse before the Moon, having apparently encircled the sky in the meantime, will return to be in line with the position of star **A**. This period is called one *sidereal month*. However, the Moon will not now be aligned with the Sun because the Sun has apparently moved to a different position against the background stars. It takes the Moon an additional 2 days to 'catch up' with the Sun and again show the phase of New Moon. The period of 29.5 days between successive New Moons is known as a *synodic month* or *lunation* (lunar month). The numbering of lunations was introduced in 1923, with lunation no.1 starting at Full Moon on 17 January 1923. Most astronomical almanacs contain the lunation numbers, together with the dates of the main lunar phases, and the age of the Moon at 0 hours UT for each day of the year.

LUNAR PHASES 1989 – 2000

This table contains the dates of New Moon, First Quarter, Full Moon and Last Quarter. Each vertical column represents a particular year. Months are represented by the numerals I–XII in the left-hand margin. All dates are based on UT with a precision of 0.1 day. The date of Full Moon is printed in **bold type.**

Example

When will New Moon, First Quarter, Full Moon and Last Quarter occur in May 1991? At midnight on 6/7 May the Moon will be at Last Quarter. New Moon will occur during the morning of 14 May. This will be followed by First Quarter on the evening of 20 May (0.8 × 24 hours = 19 h UT). Full Moon occurs at noon on 28 May.

Month \ Year	1989	1990	1991	1992	1993	1994	1995	1996	1997	1998	1999	2000
I	7.8 14.6 **21.9** 30.1	4.4 **11.2** 18.9 26.8	7.8 16.0 23.6 **30.3**	4.9 13.1 **19.9** 26.6	1.2 **8.5** 15.2 22.8 31.0	5.0 12.0 19.8 **27.6**	1.5 8.7 **16.8** 24.2 31.0	**5.9** 13.9 20.5 27.5	2.1 9.2 15.8 **23.6** 31.8	5.6 **12.7** 20.8 28.2	**2.1** 9.6 17.7 24.8 **31.7**	6.8 14.6 **21.2** 28.3
II	6.3 13.0 **20.6** 28.8	2.8 **9.8** 17.8 25.4	6.6 14.7 22.0 **28.8**	3.8 11.7 **18.3** 25.3	**7.0** 13.6 21.6	3.3 10.6 18.7 **26.0**	7.5 **15.5** 22.6	**4.7** 12.4 19.0 26.2	7.6 14.4 **22.4**	4.0 **11.4** 19.6 26.7	8.5 16.3 23.1	5.5 13.0 **19.7** 27.2
III	7.8 14.4 **22.4** 30.4	4.1 **11.5** 19.6 26.8	8.4 16.3 23.2 **30.3**	4.6 12.1 **18.8** 26.1	1.7 **8.4** 15.2 23.3 31.2	4.7 12.3 20.5 **27.5**	1.5 9.4 **17.1** 23.8 31.1	**5.4** 12.7 19.4 27.1	2.4 9.0 16.0 **24.2** 31.8	5.4 **13.2** 21.3 28.1	**2.3** 10.4 17.8 24.4 **31.9**	6.2 13.3 **20.2** 28.0
IV	6.2 13.0 **21.1** 28.9	2.4 **10.1** 18.3 25.2	7.3 14.8 21.5 **28.9**	3.2 10.4 **17.2** 24.9	**6.8** 13.8 22.0 29.5	3.1 11.0 19.1 **25.8**	8.2 **15.5** 22.1 29.7	**4.0** 11.0 17.9 25.9	7.5 14.7 **22.9** 30.1	3.8 **11.9** 19.8 26.5	9.1 16.2 22.8 **30.6**	4.8 11.6 **18.7** 26.8
V	5.5 12.6 **20.8** 28.2	1.8 **9.8** 17.8 24.5 31.3	7.0 14.2 20.8 **28.5**	2.7 9.7 **16.7** 24.7	**6.2** 13.5 21.6 28.8	2.1 10.7 18.5 **25.2**	7.9 **14.9** 21.5 29.4	**3.5** 10.2 17.5 25.6	6.9 14.5 **22.4** 29.3	3.4 **11.6** 19.2 25.8	8.7 15.5 22.2 **30.3**	4.2 10.8 **18.3** 26.5
VI	3.8 11.3 **19.3** 26.4	**8.5** 16.2 22.8 29.9	5.6 12.5 19.2 **27.1**	1.2 7.9 **15.2** 23.3 30.5	**4.5** 12.2 20.1 26.9	1.2 9.4 16.8 **24.0** 30.8	6.4 **13.2** 19.9 28.0	**1.9** 8.5 16.1 24.2	5.3 13.2 **20.8** 27.5	2.1 **10.2** 17.4 24.2	7.2 13.8 20.8 **28.9**	2.5 9.2 **16.9** 25.0
VII	3.2 11.0 **18.7** 25.6	**8.1** 15.6 22.1 29.6	5.1 11.8 18.6 **26.8**	7.1 **14.8** 22.9 29.8	**4.0** 12.0 19.5 26.1	8.9 16.0 **22.8** 30.5	5.8 **12.4** 19.5 27.6	**1.2** 7.8 15.7 23.7 **30.4**	4.8 12.9 **20.1** 26.8	1.8 **9.7** 16.6 23.6 31.5	6.5 13.1 20.4 **28.5**	1.8 8.5 **16.6** 24.5 31.1
VIII	1.7 9.7 **17.1** 23.8 31.2	**6.6** 13.7 20.5 28.3	3.5 10.1 17.2 **25.4**	5.5 **13.4** 21.4 28.1	**2.5** 10.6 17.8 24.4	7.4 14.2 **21.3** 29.3	4.1 **10.8** 18.1 26.2	6.2 14.3 22.2 **28.8**	3.3 11.5 **18.5** 25.1	**8.1** 14.8 22.1 30.2	4.7 11.5 19.1 **27.0**	7.0 **15.2** 22.8 29.4
IX	8.4 **15.5** 22.1 29.9	**5.1** 11.9 19.0 27.1	1.8 8.5 15.9 **23.9**	3.9 **12.1** 19.8 26.4	**1.1** 9.3 16.1 22.8 **30.8**	5.8 12.5 **19.8** 28.0	2.4 **9.2** 16.9 24.7	4.8 13.0 20.5 **27.1**	2.0 10.1 **16.8** 23.6	**6.5** 13.1 20.7 28.9	2.9 9.9 17.8 **25.4**	5.7 **13.8** 21.1 27.8
X	8.0 **14.9** 21.6 29.6	**4.5** 11.2 18.6 26.8	1.0 7.9 15.7 **23.5** 30.3	3.6 **11.8** 19.2 25.9	8.8 15.5 22.4 **30.5**	5.2 11.8 **19.5** 27.7	1.6 **8.7** 16.7 24.2 30.9	4.5 12.6 19.8 **26.6**	1.7 9.5 **16.1** 23.2 31.4	**5.8** 12.5 20.4 28.5	2.2 9.5 17.6 **24.9** 31.5	5.5 **13.4** 20.3 27.3
XI	6.6 **13.6** 20.2 28.4	**2.9** 9.5 17.4 25.6	6.5 14.6 **22.0** 28.6	2.4 **10.4** 17.5 24.4	7.3 13.9 21.1 **29.3**	3.6 10.3 **18.3** 26.3	**7.3** 15.5 22.7 29.3	3.3 11.2 18.0 **25.2**	7.9 **14.6** 22.0 30.1	**4.2** 11.0 19.2 27.0	8.2 16.4 **23.3** 30.0	4.3 **11.9** 18.6 26.0
XII	6.1 **12.7** 20.0 28.1	**2.3** 9.1 17.2 25.1 **31.8**	6.2 14.4 **21.4** 28.1	2.3 **10.0** 16.8 24.0	6.7 13.4 20.9 **29.0**	3.0 9.9 **18.1** 25.8	**7.1** 15.2 22.1 28.8	3.2 10.7 17.4 **24.9**	7.3 **14.1** 21.9 29.7	**3.6** 10.8 18.9 26.4	7.9 16.0 **22.7** 29.6	4.2 **11.4** 18.0 25.7

After Meeus, J.: *Phases of the Moon 1801–2010. Memoirs* 2, Vereniging voor Sterrenkunde, Brussels, 1976.

The Moon in the sky

Day after day substantial changes occur in the times of the Moon's rising and setting, as well as in the arcs (sometimes called diurnal arcs) it describes above the physical horizon. As a consequence the position of the Moon relative to the background stars is constantly changing. Data on the Moon's position, expressed in astronomical co-ordinates, are available in most astronomical almanacs. However, a quick estimation of the visibility of the Moon for a particular date can quite easily be made.

The path of the Moon against the background stars is always in the vicinity of the *ecliptic,* which is the apparent path of the Sun against the sky during the course of the Earth's revolution around it. The Moon's maximum deviation north or south of the ecliptic amounts to about 5°. The direction of the orbital motion of the Moon around the Earth is from west to east, which is opposite to the apparent diurnal (or daily) rotation of the sky, and the Moon appears to move eastwards from the Sun by approximately 12° per day. With these facts in mind let us start from a convenient phase, the date of which can be found from any calendar or from the table in this book (p. 10). A 90° eastward motion of the Moon from the Sun brings it to First Quarter phase, at which time it is visible in the evening sky. At Full Moon it is on the opposite side of the sky to the Sun and, rising at about sunset, it shines throughout the night. The Moon's continuing motion relative to the Sun results in its being 90° west of the Sun at Last Quarter. At this time the Moon rises at midnight to dominate the early-morning sky. Further, we can estimate, relative to the Sun, the direction and altitude of the arc that represents the path of the Moon above the horizon. We can also obtain the time of the Moon's rising and setting.

The above relationships are illustrated in Fig. 7, where, for simplicity, the inclination of the Moon's orbit with respect to the ecliptic has been neglected. In the graph the annual variations of the Sun's altitude and the main phases of the Moon above the horizon are indicated. For example, in March, at the beginning of spring, the Sun and the Full Moon are both located on or very near the celestial equator and their paths attain mean or average altitudes above the horizon. Under these conditions both bodies remain above the horizon for about 12 hours. In spring the path of the waxing Moon at First Quarter attains the altitude of the path of the mid-summer Sun. At the other extreme, the path taken by the waning Moon at Last Quarter in spring is akin to that of the Sun in midwinter – very low in the sky. Therefore, when trying to determine the visibility of the Moon, we can commence by considering the apparent path of the Sun during the course of a year. To sum up, the Moon's altitude above the horizon varies like that of the Sun, the main difference being that the Sun completes one revolution along the ecliptic in a year whereas the Moon makes a similar journey around the sky in 27.3 days. In other words, the motion of the Moon around the sky in one lunation is nearly identical to that of the Sun in a whole year.

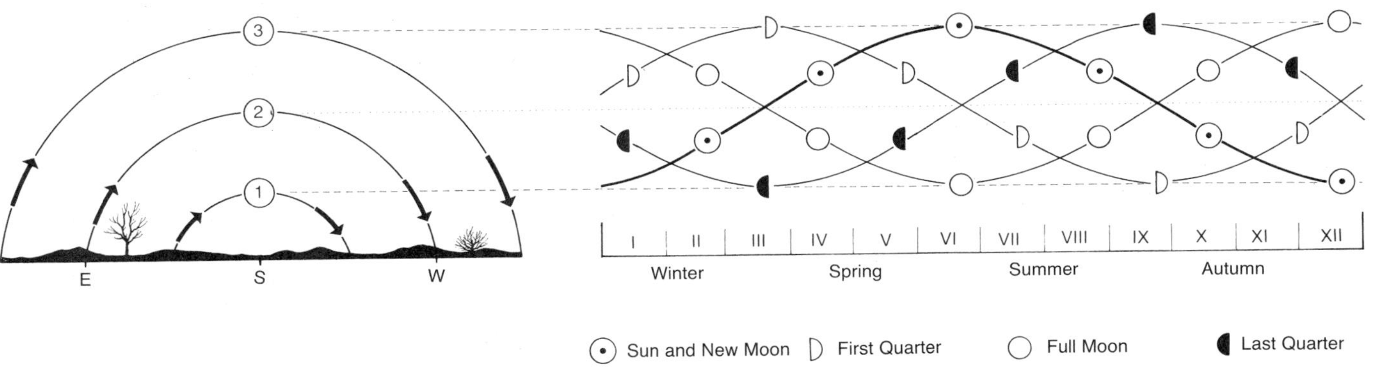

Fig. 7. How the Moon's height above the horizon changes. 1. The path of the Sun above the horizon in early winter, which is similar to the diurnal arc of the Full Moon in early summer, of the First Quarter Moon in early autumn, etc. 2. The diurnal arc of the Sun and of the Full Moon at around the time of the vernal or autumnal equinox, when each body stays above the horizon for 12 hours; it is similar to the path of the First or the Last Quarter Moon at around the time of the summer or winter solstice. 3. The diurnal arc of the Sun in early summer, of the Full Moon in early winter, of the First Quarter Moon in early spring, etc.

Note: *The diagram is valid for an observer in the northern hemisphere, and the inclination of the Moon's orbit with respect to ecliptic has been neglected.*

The surface of the Moon

In 1610 Galileo Galilei discovered that there are mountains and craters on the Moon. By the middle of the seventeenth century astronomers using improved telescopes were able to observe craters as small as 10 km in diameter. Three hundred years later, thanks to the continued developments in optical technology, astronomers are able to discern lunar crater pits as small as 300 m across, which is probably about the limit for Earth-based telescopes.

During the first half of the twentieth century astronomical interest in the Moon diminished, but the development of astronautics in the late 1950s opened up new horizons. In October 1959 the probe *Luna 3* reached the Moon and photographed its far-side. In July 1964 the *Ranger 7* photographs revealed pits and boulders that were about half-a-metre across and therefore some thousand times smaller than anything that could be seen from the Earth. Millimetre-sized grains and small fragments of the lunar surface were detected in the images sent back by *Luna 9* in January 1966, and so rapid was the progress that Earth scientists were examining lunar rocks under the microscope by the summer of 1969! In just one decade the Moon had forfeited its remoteness.

Intense research in the 1960s and '70s (see p. 192) has unveiled the complexities of the Moon as a physical body. The Moon is a spherical body with a diameter of 3476 km* and its surface, which does not enjoy the protection of an atmosphere, is immersed in the near-vacuum of space. The lunar 'soil' of regolith consists of an incoherent mixture of loose, but sticky, dust, small rock fragments and stones. This shallow layer hides the Moon's crust, which is formed of crushed layers of rock, extending to a depth of approximately 60 km. Below this is the lunar mantle, which extends for approximately 1000 km and encloses the plastic core where the temperature rises to about 1300 °C.

The Moon's magnetic field is extremely weak, its intensity being about one ten-thousandth that of the Earth. Seismic activity on the Moon is very slight, the 'moonquakes' ranging from 1 to 2 on the Richter scale; lunar explorers would hardly be aware of such events. Our satellite seems to be a practically 'dead' body and its manifestations of residual internal geological (or, more precisely, 'selenological') activity are both rare and inconspicuous (see p. 211). There is no water on the Moon – not even in the rocks – and there are no micro-organisms nor any traces of them. It is a dead, unfriendly world without any sign of life, yet a man in a space-suit can survive and even work on the lunar surface.

Following the cessation of the *Apollo* manned programme with *Apollo 17* in December 1972, three more automatic probes of the Luna series landed on the Moon and returned rock samples to Earth. Then came a long interval while scientists in terrestrial laboratories attempted, using all possible techniques, to analyse, evaluate and describe what had been discovered. This work will continue, but new projects for future flights are being planned, among them the installation of geophysical (or perhaps 'selenophysical') stations on the Moon. In our mind's eye, and as a result of this newly acquired information, we can be transported to the Moon to construct a detailed picture of its surface.

What can we observe on the lunar surface from the Earth, at a distance of 400 000 km? The first things that we notice are the darker areas, which can be seen even without the aid of a telescope. Early observers of the seventeenth century mistakenly believed them to be seas and named them accordingly, giving them the Latin name *maria* (see p. 22). Similarly the brighter areas were taken for continents (Latin: *terrae*) and this convention has been preserved up to the present time despite there being no surface water on the Moon at all!

With the optical aid of a small telescope it is easy to see that the 'terrae' are the lunar highlands and are covered with crater formations of all sizes (Fig. 9). By contrast, the 'maria' exhibit the character of smooth

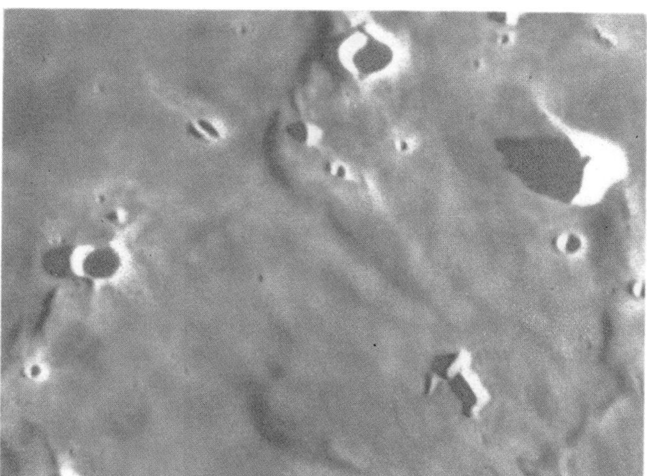

Fig. 8. The surface of a lunar mare.

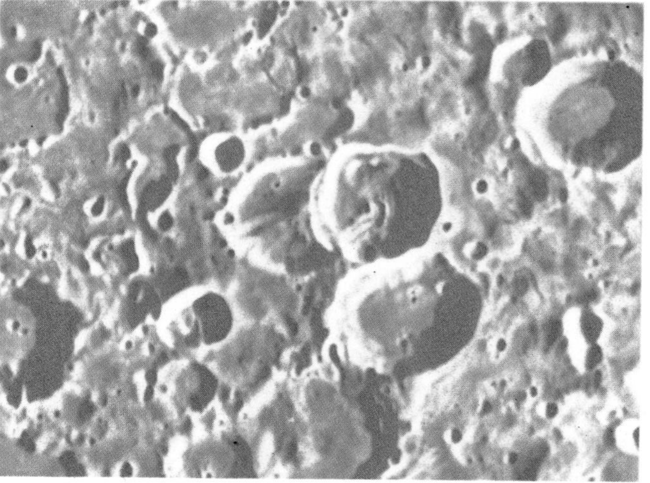

Fig. 9. The lunar highlands in the area of the crater Geber (see map 56).

* For numerical data on the Moon see p. 19.

plains, only here and there rippled by low mountain ridges and solidified flows of lava (Fig. 8). Limiting crater sizes to 1 km or more, it has been established statistically that a given highland area contains 30 times more craters than an equivalent mare area. At first sight it would seem that craters are comparatively rare in the maria. However, closer inspection by unmanned lunar probes and by astronauts during the *Apollo* lunar missions has yielded evidence that shows that both the maria and the terrae, and indeed the whole of the Moon's surface, are pockmarked with countless craters, ranging in size from a few tens of metres down to microscopic dimensions.

The maria represent 31.2% of the surface area of the near-side of the Moon, but, strangely, only 2.6% of the surface area of the far-side. Their darker hue is on account of their chemical composition. The brighter continental rocks are rich in calcium and aluminium, while the darker maria consist of basaltic lavas containing large amounts of magnesium, iron and titanium. It is worth noting also that the lunar maria do not conform to the shape of the Moon to produce what might be called 'horizontal' surfaces; instead, they are irregular and undulating.

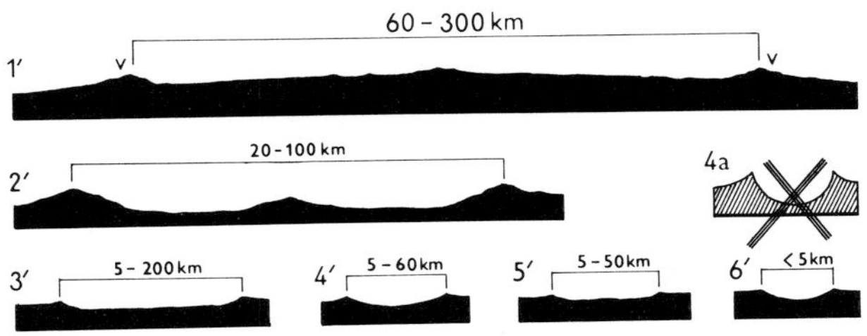

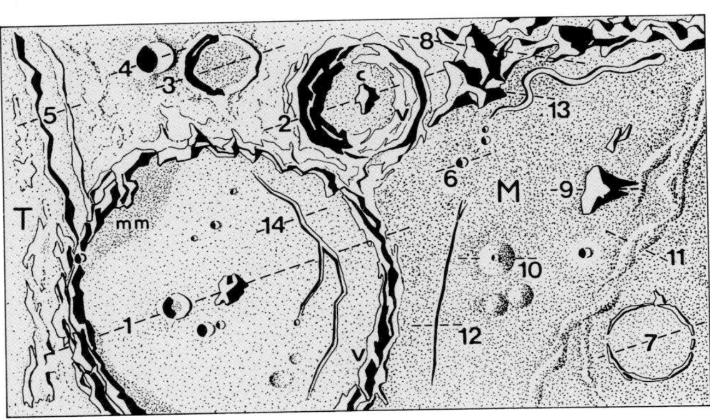

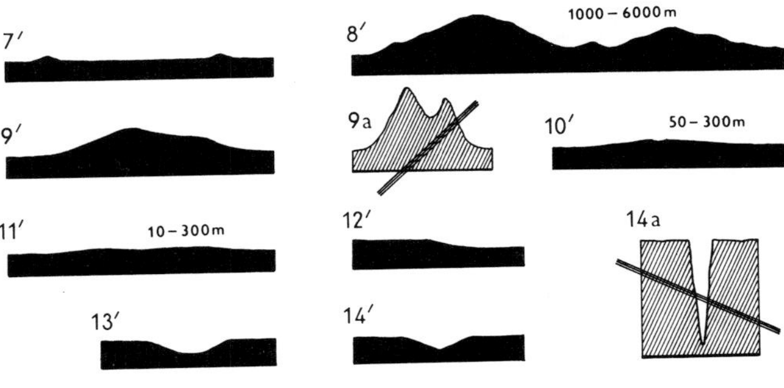

Fig. 10. *Ideal lunar landscape with typical formations. T = continent (terra), M = sea (mare), mm = mare material, c = central peak of a crater, v = wall of a crater, 1 = walled plain, 2 = ring mountain, 3 = crater with a flooded floor, 4 = crater with a sharp rim, 5 = valley (vallis), 6 = small crater, 7 = flooded crater (remains of a crater flooded with lava), 8 = mountain range (montes), 9 = mountain (mons), 10 = dome, 11 = wrinkle ridge, 12 = fault (rupes), 13 = sinuous rille (rima), 14 = rille, cleft (rima), 1′, 2′, ... 14′ = characteristic profiles of the preceding representative formations. Typical diameters of individual types of craters are given in kilometres. Typical heights are given in metres.*

Craters are the most common formations on the Moon's surface and on the surface of many other bodies in the solar system. On the near-side there are about 300 000 craters larger than 1 km, including 234 craters with diameters larger than 100 km. The term 'crater' has such a wide meaning that it is necessary to establish rough classifications according to shape and appearance, so here, in this atlas, we respect the classifications accepted in the selenographical literature.

Craters with diameters ranging from approximately 60 to 300 km are called *walled plains.* Their walls are usually massive, sometimes rugged and are often disturbed by numerous smaller craters, valleys and landslides. The original sharp ridges of the ramparts have been eroded by the continuous 'rain' of micrometeorites and by debris ejected during the cataclysmic formation of other craters, to say nothing of Moon tremors in the distant past.

The floor of a typical walled plain is covered with small craters, hills and rilles (see p. 15), and tends to follow the curved shape of the lunar sphere, as shown by profile 1 in Fig. 10. Good examples of walled plains are Clavius (p. 171), Schickard (p. 151), Posidonius (p. 55) and Ptolemaeus (p. 115). The diameter of the last one is 153 km and its depth (i.e. the height of the rim measured above the floor of the crater) is 2400 m, so that the ratio of the depth to the diameter is 1:64, which makes Ptolemaeus a surprisingly shallow formation!

Perhaps the most beautiful and symmetrical craters are the *ring mountains,* with diameters ranging from approximately 20 to 100 km. Typical examples of such craters are Copernicus (p. 89), Theophilus (p. 119), Arzachel (p. 137) and Tycho (p. 155). Characteristically, these possess regular, circular walls with sharp, well-defined summits; the inner slopes are terraced and, at close quarters, it can be seen that further terracing has been formed by successive landslides from the encircling walls. The inner slopes can be quite steep, with gradients of 20–30°, which contrast sharply with the more gentle gradients, ranging from 5° to 15°, that form the exterior slopes. Floors are usually lower in altitude than the surrounding terrain, and, according to the so-called Schröter's Law, the volume of the wall above the level of the surrounding terrain is equal to the spatial content of the crater depression. In some cases this may be true, but many ring mountains possess a central peak or, in some cases, a group of central peaks that rise to a considerable height above the centre of the depressed floor. From a typical profile, 2′ in Fig. 10, we can ascertain that ring mountains are relatively shallow depressions. For example, in the case of the crater Copernicus the ratio of the depth to the diameter of the outer enclosure is 1:25.

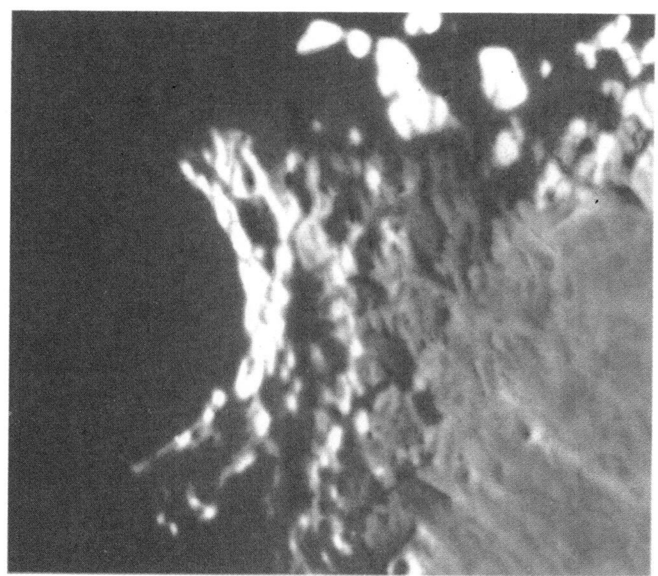

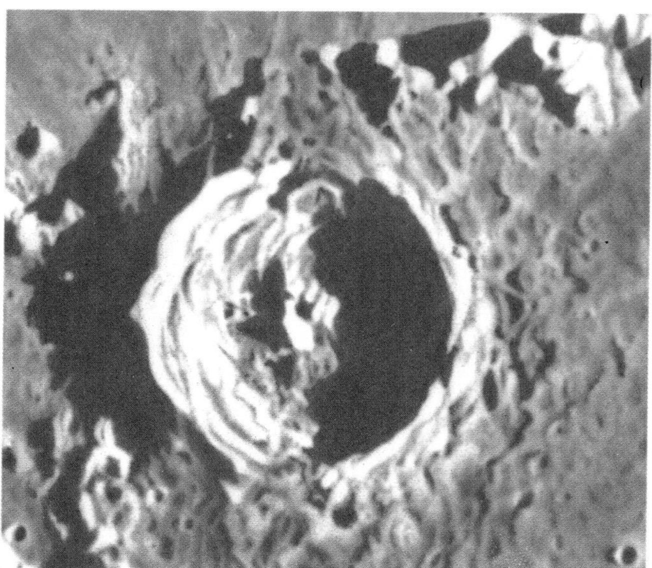

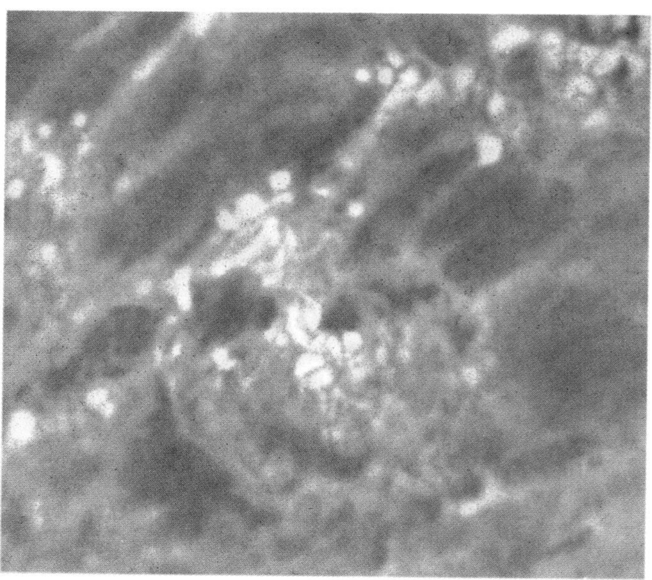

The circular depressions with diameters ranging from approximately 5 to 60 km are called simply *craters.* Such a crater has a single circular wall with a relatively sharp rim, and its floor has no central peak. From many examples let us consider the craters Kepler, which is 2570 m deep and 32 km in diameter (ratio 1:12), and Hortensius, which is 2860 m deep and 14.6 km in diameter (ratio 1:5); profile 4′ in Fig. 10 illustrates the cross-section of such a relatively deep crater as the latter.

The reader who comes across such data and profiles for the first time will probably find it hard to believe that craters are not deep hollows, as indicated by the exaggerated profile 4a in Fig. 10. The shadows in craters close to the lunar terminator exaggerate this impression of depth. The effect of solar illumination upon a crater's appearance is shown in Fig. 11.

The smallest craters visible from the Earth with telescopes are called *craterlets* or *crater pits.* These are less than 5 km in diameter. While they might be difficult to see from the Earth, a one-kilometre 'pit' would be a formidable obstacle in the path of a lunar astronaut!

The world of craters is one of almost infinite diversity. The wall of a crater need not always be circular: many craters have polygonal ramparts. On rare occasions double walls can be found: in these instances (e.g. Hesiodus A, Marth) a smaller crater is centrally located within a slightly larger one. From the point of view of crater depth, there are many craters that do not obey Schröter's Law. Some craters, for example, have been flooded with lava to the level of the surrounding terrain, and there are rare examples where the intruding magma has welled up from below to fill a crater up to its rim, thereby producing a floor level or plateau higher than the surrounding terrain. The best-known example of the latter type of crater is Wargentin (p. 167).

In many cases lunar craters are so closely packed that their walls merge together; Theophilus and Cyrillus (p. 119), for example, share common walls. In other areas a little searching will reveal composite craters; for example, a pair of slightly overlapping craters with no ramparts to separate them would appear elongated (e.g. Messier A), or perhaps pear-shaped (e.g. Torricelli). There are many variations on this theme and the combined shapes of overlapping craters can be complex. Nevertheless, the relative ages of craters can be estimated by studying how the more recent ones have distorted the earlier formations.

On the Moon craters far outnumber *mountain ranges.* The few mountain ranges that do exist were given terrestrial names, so within a few minutes we can explore the Alps, Caucasus and Apennines, without removing

Fig. 11. *Variations in the appearance of the crater Eratosthenes (diameter 58 km, depth 3570 m) owing to different illumination by the Sun. 1. The Sun is rising above the eastern wall of the crater; the inside of the crater is still immersed in shadow. 2. Sixteen hours after sunrise the central peak of the crater is illuminated. 3. On the days around Full Moon the Sun is high above Eratosthenes, the shadows have disappeared and the crater is hard to detect.*

our eye from the telescope! Lunar mountain ranges, however, bear little resemblance to their terrestrial counterparts – in character they are truly lunar. There are no networks of valleys akin to those that were carved by water and ice through the Earth's mountain ranges. Because the lunar mountain ranges seem to follow the boundaries of the lunar basins or maria (see below) it is widely believed that they are the remnants of walls of gigantic crater formations.

Profile 8′ in Fig. 10 shows a cross-section of a lunar mountain range, while profile 9′ illustrates the shape of an isolated mountain, such as Pico or Piton (pp. 49 and 51, respectively). These profiles emphasize the enormous differences between the gentle undulations of the lunar surface in reality and the steep and spiky lunar mountain ranges that were illustrated in the earlier literature – see exaggerated profile 9a. The lunar mountains actually possess gradual slopes of about 15–20°, with occasional steeper inclines of 30–35°, so there is nothing of interest to mountaineers on the Moon!

A completely different type of elevated formation is represented by curious structures called lunar *domes.* These objects have diameters in the region of 10–20 km and rise to a height of several hundred metres, so the slopes of these rounded bulges range from about 1° to 3°. Some of them have a single craterlet on top. The slopes and heights of lunar domes are very similar to those of the so-called *wrinkle ridges* or 'veins', as they are sometimes called because of their resemblance to protruding veins beneath the skin. Wrinkle ridges are abundant on the lunar maria.

The marial wrinkle ridges and domes bear witness to the endogenetic or internal forces that helped to sculpt the lunar surface: other examples include faults and depressions which, on the Moon, are known as *rilles* and *clefts* (see profiles 13 and 14 in Fig. 10). Even in the early years of the twentieth century, rilles were still described as perpendicular precipices, and clefts were thought to be deep canyons (see exaggerated profile 14a). Again this was an erroneous interpretation of the interplay of highlight and shadow in the terminator regions. *Sinuous rilles,* which resemble dry river beds, with numerous meanders extending for hundreds of kilometres, are of special interest, for they are probably the last vestiges of lava channels which may have been active when the maria were being formed.

Numerous examples of the various types of lunar formations described above are illustrated on pp. 194–207, and are shown on the map of the near-side of the Moon, which is subdivided into 76 sections on pp. 29–179.

Origin and history of the Moon

When we compare lunar surface features with those of the Earth it is plainly evident that they originated by completely different physical processes. Many scientists believe that the Moon and the Earth, together with the other bodies of the solar system, were formed at approximately the same time, about 4600 million years ago. So why these differences? On Earth we take for granted the protective shield of the atmosphere and the role of wind and water in the moulding of the landscape, but the Moon is utterly lacking an atmosphere and surface water. On the Moon there is no eroding 'weather' as we understand it, so the surface we observe is very ancient by terrestrial standards. Differences have also been found in the chemistry of lunar rocks that have been submitted to laboratory analyses after their successful return to Earth during the 1960s and '70s. The oldest of these were found to be 4200 million years old.

The immense energy of innumerable collisions of large and small high-velocity bodies on the primeval Moon melted its outer layers to depths of perhaps as much as 100 km, and in these oceans of molten rock chemical differentiation took place, with the heavier elements sinking towards the Moon's centre. As the 'heavy bombardment' eased off, cooling and solidification of the surface took place, but even so, the debris left over after the formation of the solar system continued to rain down

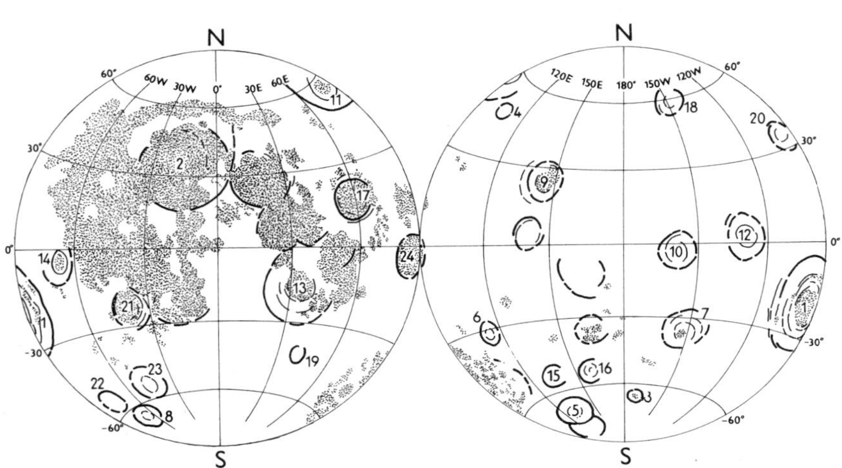

Fig. 12. Distribution of mare basins on the near-side (left) and far-side (right) of the Moon. The basins are numbered in order of increasing age, i.e. with the youngest basins first (according to Hartmann and Wood). 1 = *Orientale,* 2 = *Imbrium,* 3 = *Antoniadi,* 4 = Compton, 5 = *Schrödinger,* 6 = *Milne,* 7 = *Apollo,* 8 = *Bailly,* 9 = *Moscoviense,* 10 = *Korolev,* 11 = *Humboldtianum,* 12 = *Hertzsprung,* 13 = *Nectaris,* 14 = *Grimaldi,* 15 = *Planck,* 16 = *Poincaré,* 17 = *Crisium,* 18 = *Birkhoff,* 19 = *Janssen,* 20 = *Lorentz,* 21 = *Humorum,* 22 = *Pingré,* 23 = *a basin close to Schiller,* 24 = *Smythii.*

to produce craters of all shapes and sizes. Some 4000 million years ago a series of immense impacts created mare basins such as the vast Mare Imbrium and the Mare Orientale (Fig. 12).

From about 3800 million to 3100 million years ago radioactive heating at depths of 100–250 km started to melt the underlying layers and lava and magma burst through to the surface. The resulting lava flows each solidified within a few years to leave a modest thickness of basaltic lava of no more than a few tens of metres. superimposed upon existing lava flows overlying the mare basins. However, the radioactive heating period extended over hundreds of millions of years, so it is hardly surprising that sporadic eruptions continued to produce hundreds of lava flows. Many of these later lava flows submerged earlier ones, and built up layers which attained a considerable height – some more than a kilometre.

Volcanic and tectonic processes have also left their marks on the Moon in the shapes of sinuous rilles, faults, domes, wrinkle ridges, some craters and chains of craters. About 3000 million years ago the Moon's endogenetic activity practically ceased. However, unprotected by an atmosphere, the lunar surface continued to be eroded by countless small meteorites and micrometeorites, which gradually transformed the surface rocks into dust and fragmented material. Thus, over a period of more than 3000 million years, a layer of loose 'lunar soil' or *regolith* has been built up to a thickness of about 20 m. If this were the whole story, we would not have to mention the isolated impacts of larger meteoritic bodies, which excavated the youngest craters, centres of radial systems of bright *rays* composed of pulverized ejecta. Examples of such craters include the magnificent Copernicus, which was formed by an impact some 800 million years ago, and the bright Tycho, which is more than 100 million years old.

Nowadays, thanks to direct exploration, the history of the lunar surface is well understood, but the first 500 million years of the existence of the satellite as a whole is a mystery for which there is still no satisfactory explanation. The old question of the Moon's origin is still without an answer. Did it originate simultaneously with the Earth from a planetary cloud, or was the mass that formed the Moon once part of the Earth itself? A third possibility is that it came into being in some other part of the solar system and was captured by the gravitational attraction of the Earth. None of these three 'traditional' hypotheses is able to explain convincingly the facts we now know about the Moon, especially its unique chemical composition, which is so unlike that of the Earth. One of the more recent attempts to solve this puzzle suggests that a planetary body of Martian dimensions collided with the primeval Earth and that this cataclysmic event ejected a mixture of material from the mantles of both bodies into nearby space. This material consolidated to form a new celestial body – the Moon. Was this how it happened? We do not know, but it is to be hoped that future research will lead to more definite answers to these perplexing questions.

Lunar cartography

Mapping of the Moon's surface dates from the first half of the seventeenth century. Until 1959 this was limited to the near-side and the vast majority of maps presented the Moon in the same way as it was observed from the Earth. The Soviet probe *Luna 3* heralded a new epoch of global cartography, when it photographed the far-side in October 1959, and during the American *Lunar Orbiter* programme in 1966–67 the whole sphere was surveyed, with the exception of less than 1% of the south polar region. In addition, the *Apollo* lunar expeditions made considerable contributions to the precision of lunar mapping. Precise and detailed lunar maps are not only indispensable for observers, they also serve as a basis for scientific studies as well as for the planning of future flights to the Moon.

The mapping itself is preceded by the determination of the actual figure or shape, as well as the dimensions, of the Moon's globe. This task is known as *selenodesy.* Selenodetic measurements have indicated that the deviations of the Moon's actual surface from a perfect spherical shape lie within limits of ± 4 km. These differences are *absolute heights*. However, for mapping purposes the Moon is regarded as a sphere with a radius of 1738 km.

Just as geography is concerned with mapping of the Earth, so *selenography* deals with practical mapping of the Moon. This includes determination of the exact location of lunar formations, as well as estimation of the relative height of mountains above the surrounding terrain, the depth of craters, etc. Perhaps the primary task is the selection of a suitable cartographic projection, which in turn involves consideration of the scale of the map and of the number of sections into which it is to be divided. Last but not least comes the system of names or nomenclature to accompany each map.

Selenographic co-ordinates, namely selenographic latitude and longitude, define the precise and unambiguous position of individual points on the Moon's surface. They have terrestrial counterparts in the familiar geographical latitude and longitude. For the sake of completeness, a third dimension may be used to give the distance of a point from the Moon's centre. This is not usually given on ordinary maps.

The basic grid of selenographic co-ordinates is illustrated in Fig. 13. It is oriented in the same way as it would be seen from the Earth and the fundamental circle in the system is the lunar equator, **r,** which lies in a plane perpendicular to the Moon's axis of rotation, **o.** This axis intersects the surface of the Moon at two points: the north pole, N, and the south pole, S. When lunar observations

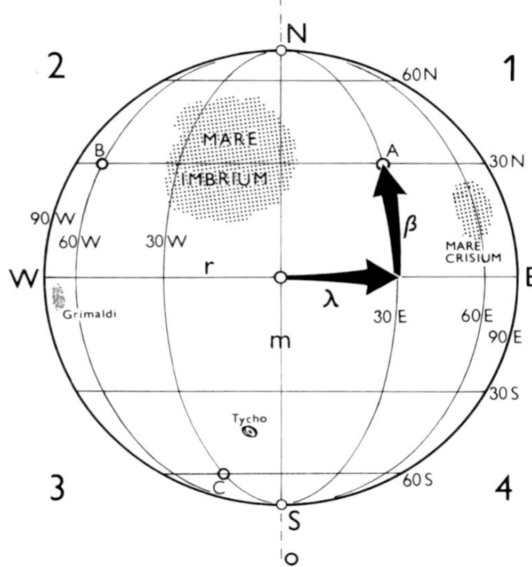

Fig. 13. Selenographic co-ordinates. The equator **r** *and prime meridian* **m** *divide the near-side of the Moon into four quadrants, designated 1, 2, 3 and 4. The co-ordinates of the three marked points are as follows:* A *(30°N, 30°E);* B *(30°N, 60°W);* C *(60°S, 30°W).*

are made with the naked eye (or through a non-inverting telescope) from the northern hemisphere of the Earth, lunar *north* is uppermost and well beyond the easily identified Mare Imbrium. *South,* on the lower, opposite edge of the lunar disk, is in the hemisphere containing the bright crater Tycho.

In 1961 the International Astronomical Union (IAU) decided to discard an older convention and to adopt the system used to determine the directions of east and west on the Earth. This, the *'astronautical' convention,* means that an observer on the Moon would see the Sun rising in the east and setting in the west. Consequently, when we look at the Moon with the naked eye from the northern hemisphere, east (E) is to the right and somewhat beyond Mare Crisium, while west is to the left of the borders of the Oceanus Procellarum. This is the orientation used in all modern maps of the Moon. For the sake of completeness, it should be mentioned that the formerly used reverse, or *'astronomical',* orientation assisted terrestrial observers, for east and west on the lunar disk coincided with east and west on the celestial sphere.

North and south of the lunar equator are the parallels of selenographic latitude. Other circles that intersect at the north and south poles and are perpendicular to the equator are termed 'meridians': these connect points of equal selenographic longitude. The central meridian is defined as that which passes exactly through the centre of the lunar disk at the moment when librations in latitude and longitude are zero, as seen from the Earth.

Selenographic longitude, λ (lambda), is measured along the equator from the central meridian to the meridian containing a given point. It is positive to the east and negative to the west of the central meridian, and always lies between 0° and 180°. The longitude of 180° is the meridian crossing the centre of the far-side of the Moon. In this atlas, both on the maps and throughout the text, the abbreviations E (for east) and W (for west) are used instead of the signs + and –. For example, 30°W signifies 30° of western selenographic longitude, which could also be written as $\lambda = -30°$.

Selenographic latitude, β (beta), is the angular distance of a given point from the lunar equator. It is measured along the appropriate meridian, and is positive to the north and negative to the south and always lies between 0° and 90°. The abbreviation N (for north) and S (for south) are often used instead of the corresponding signs + and –. For instance, 60°S means 60° of southern selenographic latitude, or $\beta = -60°$.

Lunar mapping is based on a network of numerous control points, usually small circular craters whose coordinates have been measured to a high degree of precision using selenographical measuring equipment. From this basic network the positions of other features to be included in the map can be ascertained.

The old problem of producing a two-dimensional map of a three-dimensional sphere is well known to cartographers, who, over the years, have experimented with several different map projections. The orthographic projection is commonly used for lunar charts and represents the Moon as it is observed on the celestial sphere at zero librations. It is the projection used in Fig. 13 and in the 76 sections comprising the detailed map of the Moon (pp. 29–179). For scientific purposes and flights to the Moon, maps drawn to the so-called 'conformal projections' are especially suitable, since they depict lunar formations without distortion: in these a ring crater is always circular, rather than elliptical (as seen from the Earth).

The cartographic representation of the lunar surface can be schematic or realistically detailed. In the first instance a crater may be indicated by a single circular outline, while in the second it could be made to resemble its photographic image. The air-brush technique of relief representation, which was developed in the early sixties in the USA for the production of the 1:1000000 LAC series of charts, has proved to be the best. Lunar slopes, which at certain periods during a lunation would be immersed in shadow, are depicted as inclined surfaces receiving grazing illumination from the Sun; in other words, the altitude of the Sun is made to equal the gradient of a given feature. Cast shadows are absent with this technique, so no details are hidden, and this is one of the advantages of this form of presentation over that of a photographic atlas. Moreover, a map can reflect the quality of the best photographs, to make the interpretation of depicted details easier. And there is one more thing: when one examines a drawing or photograph of the Moon's relief, the direction of solar illumination has to be taken into account, otherwise the eye can be deluded into accepting optical reversals, which result in concave, partly shadow-filled craters taking on the appearance of convex domes that bulge out of an 'upside-down' surface!

Numerical data must also be given to reinforce the cartographical interpretation, as should the relative heights of selected points. Determination of the heights of mountain ranges, by measurement of the lengths of cast shadows, dates from the eighteenth century. If we know the length of the cast shadow and the corresponding angle of the Sun above the horizon, we can calculate the height of the shadow-casting peak above the surrounding terrain. The method is fairly precise, since at sunrise and sunset a shadow is approximately 100 times longer than the height of the eminence from which it was cast. This technique has been adapted for use on lunar images returned to Earth from satellites in lunar orbit, and relative height measurements with a precision of a few metres have been obtained.

An essential component of every map is the system of names of selected formations – the *nomenclature.* This facilitates prompt identifications and aids description, as well as assisting in the search for a particular formation, without the need for lengthy positional specifications. It is, in fact, an easily remembered short-form coding method.

Following Galileo's first telescopic observations of the Moon, lunar charting started in earnest and by the middle of the seventeenth century several charts had been produced. Michel Floret van Langren, in his lunar map of 1645, was the first to assign names to features. Two years later Hevelius of Gdańsk published his chart, in which lunar mountain ranges were named after those on the Earth – e.g. the Alps, Carpathians, Apennines. It was, however, Giovanni Baptista Riccioli, a professor of philosophy, theology and astronomy in Bologna, who laid the foundations of the current system of lunar nomenclature. In 1651 he published a map of the Moon which included the names of 'seas', mountain ranges and craters; he divided these names into three categories: people's names, terrestrial names and symbolic names. Riccioli discarded most of the names proposed by Hevelius (who had disregarded those of van Langren) and gave the maria exotic names, which are perhaps more romantic than suitable – e.g. Mare Imbrium (Sea of Rains), Oceanus Procellarum (Ocean of Storms). He also preserved van Langren's idea of naming lunar craters after astronomers and other personalities and located them in historical order from north to south on the lunar disk.

Lunar nomenclature was further developed by the German selenographers J. H. Schröter (*Selenotopographical Fragments,* 1791 and 1802) and, in particular, W. Beer and J. H. Mädler. Included in Beer and Mädler's *Mappa Selenographica* (1837) were as many as 427 named features: 200 of these had been taken from Riccioli's map, 60 from Schröter's, and Mädler added another 145 names of seafarers and geographers. Beer and Mädler also introduced a system of naming small, secondary craters by the capital letters of the Latin alphabet, and mountain peaks, domes, etc. by lower-case Greek letters.

Succeeding selenographers, for the best part of the next century, increased the confusion by adding, deleting, replacing, duplicating and generally changing lunar names. The task of restoring order to this accumulated mass of discrepancies and of coming up with an acceptable nomenclature was taken on by the International Astronomical Union (IAU). The work was done by M. A. Blagg and K. Müller, with earlier contributions from an English scientific artist, W. H. Wesley. In 1935, the first IAU Moon map was published with 681 names. Since that time the IAU has been the sole authority that has presided over all changes and additions to lunar nomenclature. As a result of the joint endeavours of many generations of selenographers a unique pantheon has come into existence on the Moon, for it is here that names of deceased personalities can be added to commemorate their many and varied contributions to the development of science.

Global mapping of the Moon burgeoned during the 1960s, and the knowledge gained during that fruitful decade has led to an extension of the nomenclature to the far-side of the Moon. In 1970, at the XIV General Assembly of the IAU in Brighton, England, 513 additions were accepted, the majority being for formations on the far-side. A rare exception was also made: the IAU assigned to the lunar 'pantheon', and in so doing immortalized, the names of 12 living people: six American astronauts and six Soviet cosmonauts.

In 1973, at the XV General Assembly of the IAU in Sydney, lunar nomenclature was again the subject of extensive and controversial reform. Since 1935 the so-called subsidiary craters had been designated by using the capital letters of the Latin alphabet; for example, Mösting A was associated with the larger crater, Mösting. Also mountain peaks, domes, etc., were identified by lower-case Greek letters and rilles associated with nearby named formations by Roman numbers. These conventions were revised in 1973, when it was proposed that letter-designated craters should gradually be given the names of individuals. Also some small craters were to be bestowed with a male or female forename. Such were the requirements for detailed mapping of the Moon to a scale of 1:250000, as was used to produce the NASA Lunar Topographic Orthophotomaps. In such an atlas the scale is so large that manageable sheets or plates can depict only very small areas of the lunar surface, and for easy identification each sheet ought to contain at least one named formation. With this purpose in mind, up to 1988, 138 craters that had hitherto been designated by letters had been given names (e.g. Bowen instead of Manilius A). In future, hundreds, and as time goes by, thousands, of new names will appear...

In 1976, during the change-over period, the IAU agreed to compromise and allow charts containing the newly introduced substitute names also to include the original letter designations in brackets below the new names. For the sake of continuity with earlier selenographic literature it is convenient to retain the traditional capital letter designations of the subsidiary craters on modern maps as well. Up to 1988, a total of 6231 craters had been designated on the near-side of the Moon, of which 801 bear names and 5430 are identified by letters added to the name of a nearby prominent crater.

In addition to craters, the following types of formation are also usually designated by a Latin name:

Catena	crater chain	*Lacus*	lake
Dorsa	network of ridges	*Mare*	sea
Dorsum	mare ridge	*Mons*	mountain

Montes	mountain range or group of peaks	*Rima*	rille
		Rimae	network of rilles
Oceanus	ocean	*Rupes*	scarp
Palus	marsh	*Sinus*	bay
Promontorium	cape	*Vallis*	valley

Formations such as catena, rima, rimae, rupes and vallis usually derive their name from that of a nearby formation, exceptions having been made in the cases of the Rupes Altai and Rupes Recta, and the Vallis Bouvard and Vallis Schröteri. Some of the other types of formation have been assigned individual names that are unrelated to those of neighbouring features.

Lunar nomenclature has had a long and intricate history – full of mistakes, inaccuracies and amendments, and with many modifications – and a unique selection of names has been created. Soon we shall be referring to those people who, to all intents and purposes, have their names inscribed on the Moon. Recalling these names in historical succession constitutes a review of the processes of learning that have led to our present knowledge of the Earth and space.

NUMERICAL DATA ON THE MOON

Average distance of Moon from Earth	384 401 km = 60.27 × Earth's equatorial radii
Minimum distance of Moon at perigee	356 400 km
Maximum distance of Moon at apogee	406 700 km
Time taken for light to travel from Moon to Earth	1.3 s
Eccentricity of Moon's orbit around Earth	0.0549
Mean distance of centre of gravity of Earth–Moon system from centre of Earth	4670 km
Mean angular diameter of Moon in sky (geocentric)	31′ 05.2″
Angular diameter of Moon at perigee	33′ 28.8″
Angular diameter of Moon at apogee	29′ 23.2″
Stellar magnitude of Full Moon, *m*	–12.55
Inclination of Moon's orbit to ecliptic	5° 8′ 43.4″
Sidereal orbital period (i.e. with respect to stars)	27.321661d = 27d 7h 43m 11.5s
Synodic month (from New Moon to New Moon)	29.530588d = 29d 12h 44m 2.8s
Anomalistic month (from perigee to perigee)	27.554550d = 27d 13h 18m 33.1s
Rotation period of line of nodes (retrograde motion)	18.61 years
Rotation period of point of perigee (direct motion)	8.85 years
Average orbital velocity of Moon around Earth	3681 km/h = 1.023 km/s
Average angular velocity of Moon's motion in sky	33′/h
Mean diurnal motion of Moon with respect to stars	13.176358°
Mean interval between two successive meridian passages of Moon	24 h 50.47m
Optical libration in longitude	7° 54′
Optical libration in latitude	6° 50′
Total surface area of Moon theoretically observable from Earth	59%
Inclination of Moon's equator to plane of ecliptic	1° 32.5′
Inclination of Moon's equator to its orbital plane	6° 41′
Diameter of Moon	3476 km
Circumference of Moon at equator	10 920 km
Surface area of Moon	37.96×10^6 km^2 = 0.074 of surface area of Earth
Volume of Moon	21.99×10^9 km^3 = 2.03% of volume of Earth
Mass of Moon	7.352×10^{25} g = 1/81.3 of mass of Earth
Mean density of Moon	3.341 g.cm^{-3} = 0.606 of density of Earth
Moon's surface gravity	162.2 cm.s^{-2} = 16.5% of Earth's surface gravity
Moon's escape velocity	2.38 km.s^{-1} (11.2 km.s^{-1} on the Earth)
Illumination on surface of Earth by Full Moon	0.25 lux
Illumination on surface of Moon by Full Earth	16 lux
Surface temperature on night hemisphere of Moon	–170 to –185°C
Maximum temperature on sunlit side of Moon	+130°C
Constant temperature at depth of 1 m under surface	–35°C
Total surface area of lunar maria	16.9% of Moon's surface
Total surface area of lunar maria on near-side of Moon (one hemisphere)	31.2% of near-side of Moon
Total surface area of lunar maria on far-side of Moon (one hemisphere)	2.6% of far-side of Moon

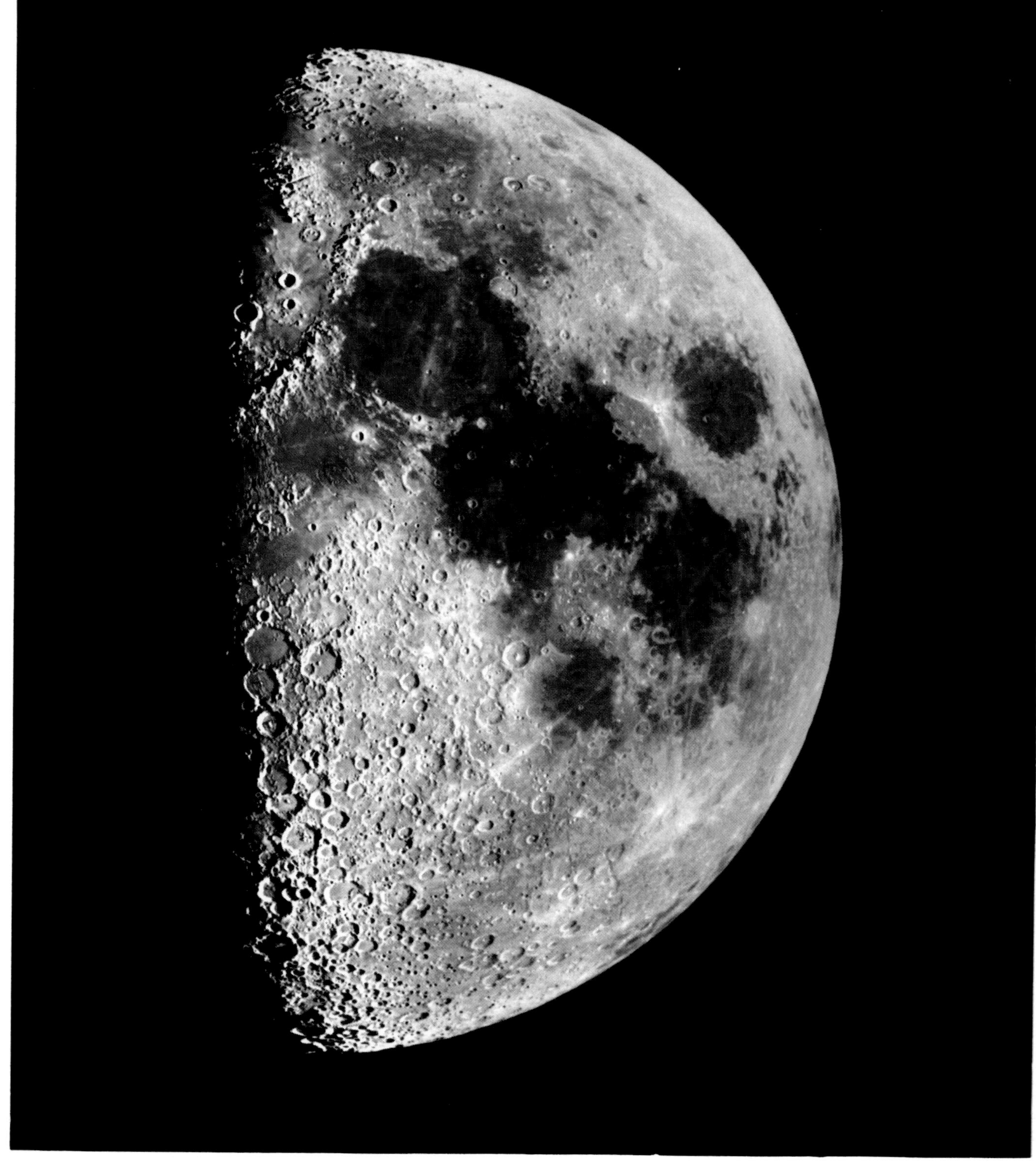

Becoming acquainted with the Moon

The Moon at First Quarter

The most suitable time to get acquainted with the Moon's surface features is at or close to First or Last Quarter, when the terminator regions are dramatically illuminated. Further away from this day-and-night boundary, towards the illuminated edge of the lunar disk, the shadows cast by mountains and crater walls gradually diminish until they disappear, giving an illusion of smoothness and absence of relief. A useful method of finding one's way around the Moon's disk is to learn the shapes, relative positions and names of the dark seas, bays and lakes. With a little practice a keen eye can identify the positions of large craters, groups or chains of craters, craters surrounded by bright rays, and long mountain ranges.

Our map shows the names of only the most prominent formations, which serve, collectively, as a guide to the lunar sur-

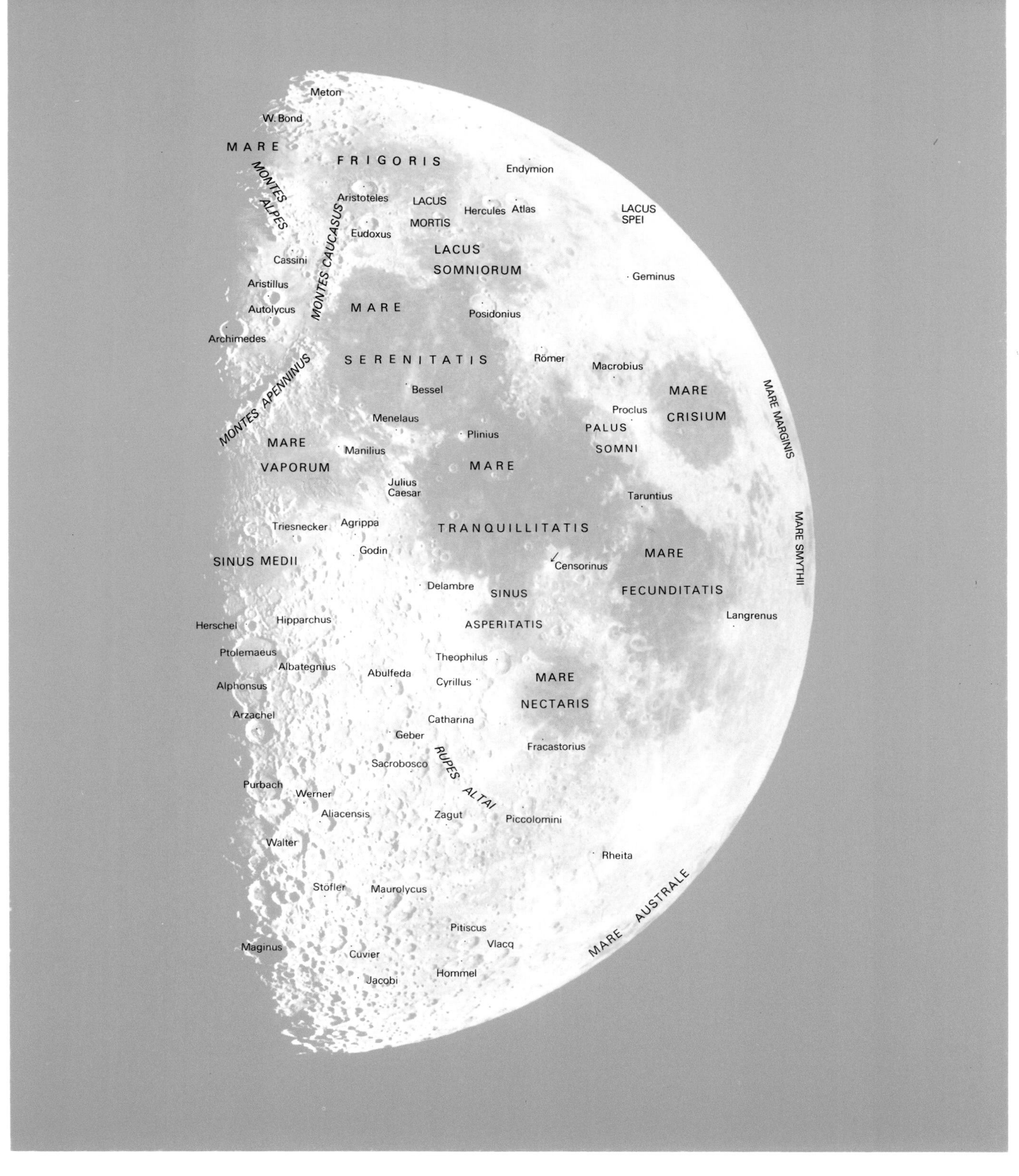

face. All of them can be seen easily with binoculars or field glasses. It must be borne in mind that the position, and therefore the appearance, of the lunar maria can be markedly different to that shown on the map, owing to the effects of libration (see p. 8). The variations in the appearance of the Mare Crisium, in particular, can be very striking because of the changing values of libration in longitude. From a short-list of craters we should try to remember the trio Theophilus, Cyrillus and Catharina, and the pairs Atlas and Hercules, and Aristoteles and Eudoxus. Other useful lunar landmarks include Proclus, a bright crater with a ray system just west of Mare Crisium, the walled plain Posidonius, and the extensive Altai scarp, which runs south from Catharina towards the sizeable crater Piccolomini.

The Moon at Last Quarter

At one time people generally believed (and some still do) that the weather on Earth is influenced by the Moon in accordance with an old adage which states that 'when the Moon is waxing the weather will be fine and when it is waning the weather will be cloudy, rainy, stormy and generally unpleasant'. This explains why the maria coming into view on the waxing crescent leading to First Quarter were given names associated with fine weather, e.g. Mare Tranquillitatis (Sea of Tranquillity) and Mare Serenitatis (Sea of Serenity), while on the other side of the lunar disk the maria visible on the waning crescent were labelled with 'rainy' names, e.g. Mare Imbrium (Sea of Rains), Mare Nubium (Sea of Clouds) and Oceanus Procellarum (Ocean of Storms).

On the western side of the visible disk there is a fine contrast between the dull surface of the large maria and the bright ray systems that originate from the relatively young craters Coper-

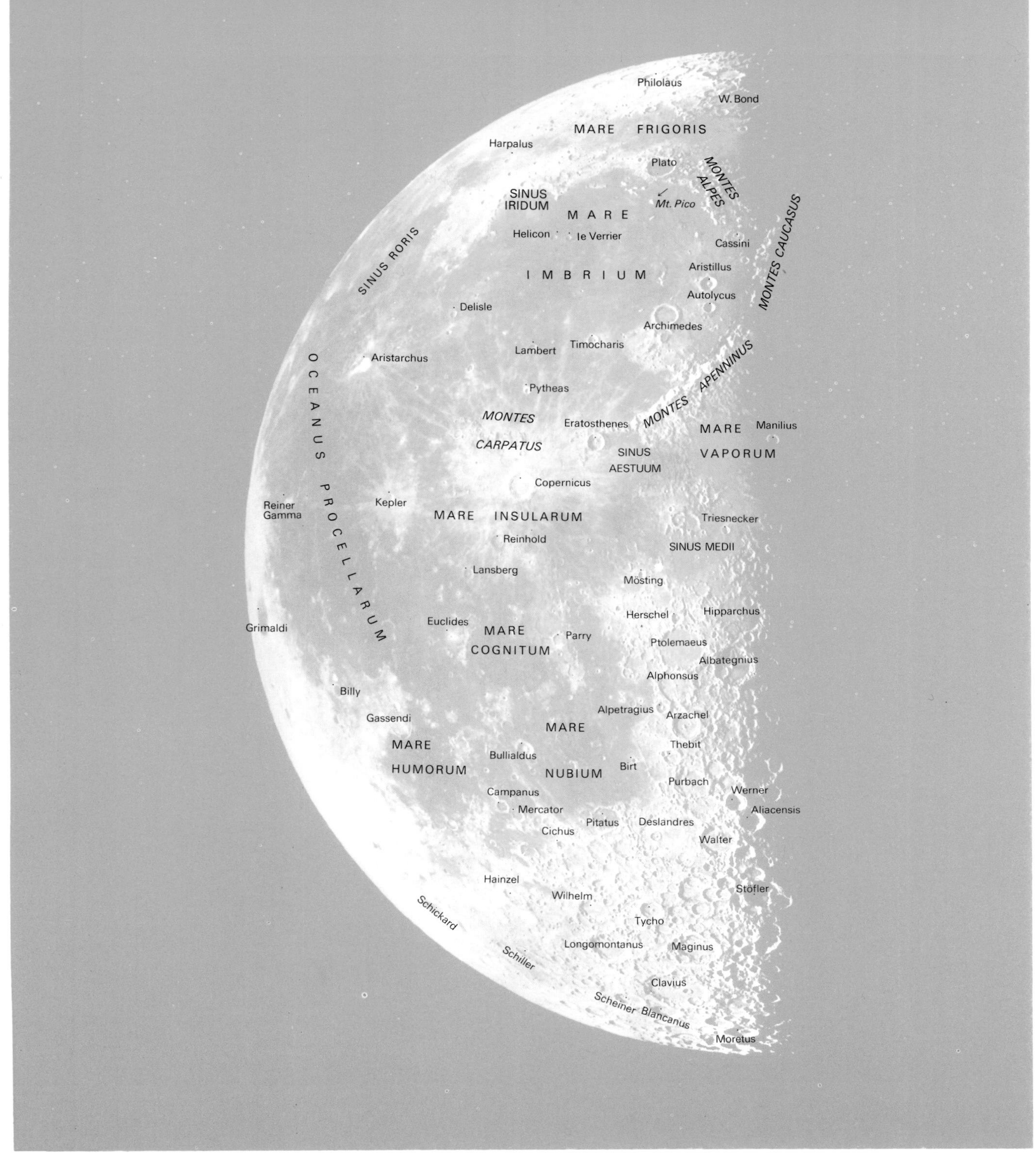

nicus, Kepler and Aristarchus. The crater Tycho dominates the southern hemisphere with its vast system of bright rays, the longest of which can be traced into the northern hemisphere and across Mare Serenitatis. South of Tycho is the enormous walled plain Clavius. Other points convenient for orientation purposes include the dark-floored craters Plato and Grimaldi, and large craters such as Archimedes, Eratosthenes, Bullialdus and the trio of craters Ptolemaeus, Alphonsus and Arzachel. Orientation is also assisted by the long mountain ranges that border the Mare Imbrium: the Carpathians (Montes Carpathus), the Apennines (Montes Apenninus), the Caucasus (Montes Caucasus) and the Alps (Montes Alpes). Often observed close to the terminator of the waning crescent is the very conspicuous Sinus Iridum (Bay of Rainbows), which adjoins the Mare Imbrium to the north.

The Full Moon

During the phase of Full Moon no shadows can be observed on the lunar disk from the Earth. Instead, the near-side is transformed into a complex pattern of areas of contrasting brightness or *albedo*. These include the already discussed dark maria, bright continents, ray craters, etc. At this phase it seems incongruous that some sizeable craters all but disappear, while others that are normally small and inconspicuous become bright and easily identifiable under the high-angle illumination. The latter can be used as measuring points to monitor the passage of the Earth's shadow across the Moon's disk during lunar eclipses. While such an event is progressing, the observer records the times when the craters enter, or emerge from, the Earth's shadow. Naturally, since no two eclipses are identical, the actual sequence of crater contacts is unique to each eclipse. Opposite is a list of 50 suitable features that are shown on the adjacent map, which also includes information about albedo differences; the identification numbers proceed from west to east.

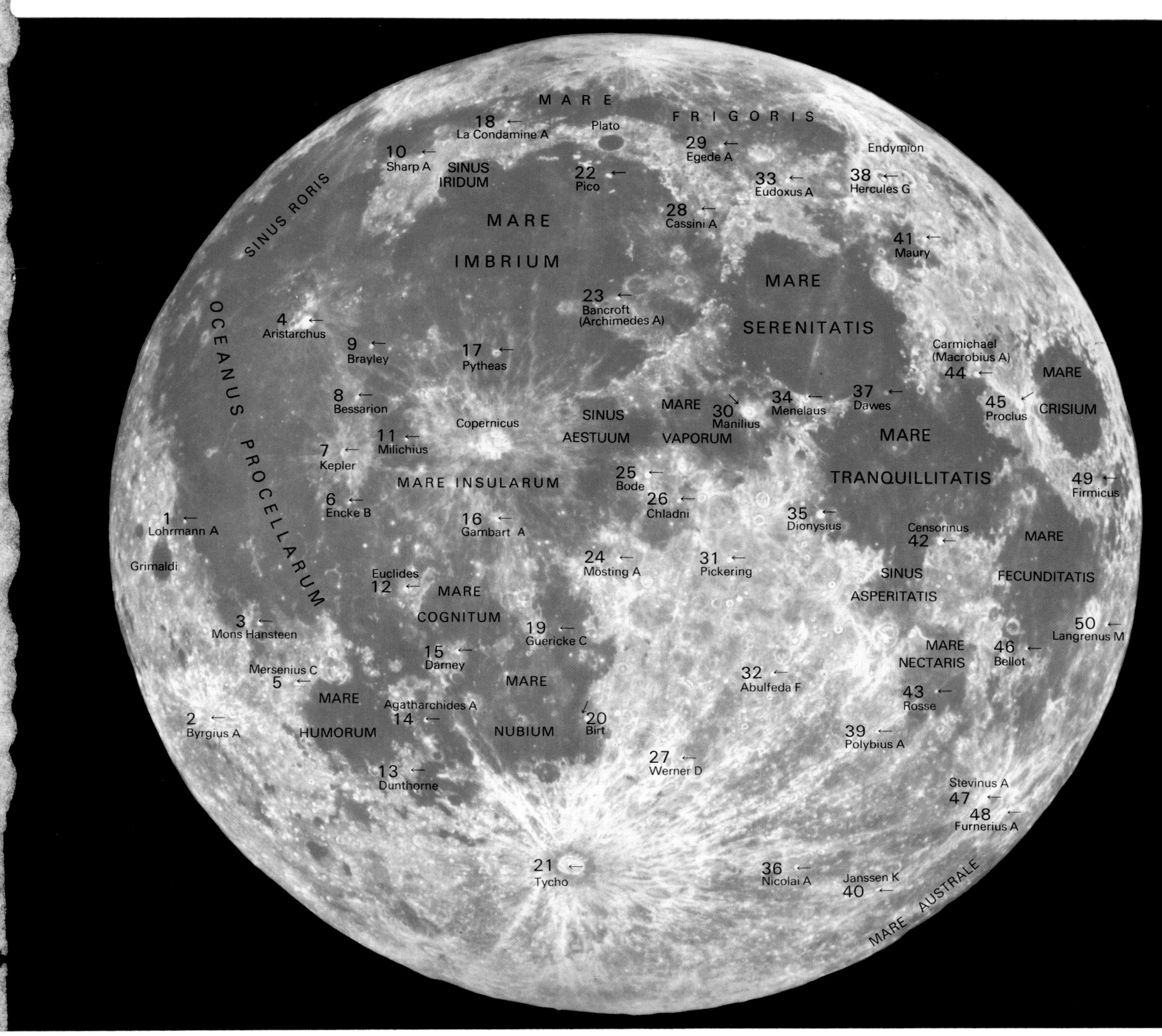

Lunar features useful for timing the progress of a lunar eclipse

1. Lohrmann A
2. Byrgius A
3. Hansteen, Mons[1]
4. Aristarchus
5. Mersenius C
6. Encke B
7. Kepler
8. Bessarion
9. Brayley
10. Sharp A
11. Milichius
12. Euclides
13. Dunthorne
14. Agatharchides A
15. Darney
16. Gambart A
17. Pytheas
18. la Condamine A
19. Guericke C
20. Birt
21. Tycho[2]
22. Pico, Mons[1]
23. Bancroft[3]
24. Mösting A
25. Bode
26. Chladni
27. Werner D
28. Cassini A
29. Egede A
30. Manilius[2]
31. Pickering
32. Abulfeda F
33. Eudoxus A
34. Menelaus
35. Dionysius
36. Nicolai A
37. Dawes
38. Hercules G
39. Polybius A
40. Janssen K
41. Maury
42. Censorinus
43. Rosse
44. Carmichael[4]
45. Proclus
46. Bellot
47. Stevinus A
48. Furnerius A
49. Firmicus[5]
50. Langrenus M

[1] mountain, [2] centre of a crater, [3] formerly Archimedes A, [4] formerly Macrobius A, [5] dark-floored crater

Atlas of the near-side of the Moon

Explanatory note

The main part of the present atlas consists of a detailed map of the near-side of the Moon, subdivided into 76 sections. This is a traditional map based on the orthographic projection, and it depicts the Moon as it appears from the Earth at zero libration. The outlines of craters and other features are therefore practically the same as they appear through a telescope, so the comparison between the map and reality presents no problems. The libration zones, however, cannot be shown completely here because the edge of the map is formed by the meridians at 90°E and 90°W; special maps of the libration zones are to be found on pp. 182–189.

A network of selenographic co-ordinates (see p. 16) is superimposed on all map sections. The parallels of latitude are drawn as segments of straight lines; the meridians of longitude as ellipses. North is always uppermost, and east is to the right. This is the same view as would be observed through a non-inverting telescope. Circular craters at the centre of the lunar disk are shown as circles, while those towards the edge are distorted into ellipses by the angle of view. This must be remembered when the dimensions of a foreshortened crater are measured, for its true diameter will correspond to the length of the major (longer) axis of the ellipse. In general, lengths measured along arcs whose centres coincide with the centre of the lunar disk are not distorted. To give meaning to this rule, the graphical scale used throughout the book and on all sections of the map corresponds to a lunar diameter of 1448 mm and a scale of 1 : 2 400 000.

The layout of the map is shown opposite and a key for rapid location of a particular area is presented on the front endpaper. Each section of the map is numbered and bears the name of a prominent crater within it. In addition, the numbers of adjacent sections are printed in red alongside, and at the bottom there is a miniature map of the near-side of the Moon with the position of the particular section marked upon it.

The Moon's surface is drawn as it appears under morning illumination, i.e. when sunlight is falling upon it from the east (from the right). To emphasize the undulating character of the relief, shadows of west-facing slopes are shown. Accordingly, we can distinguish between elevated and depressed formations, e.g. mountains and craters. Albedo differences are indicated by light or dark shading.

The map presents the official nomenclature of lunar formations as accepted by the IAU up to the end of 1988. For the sake of continuity with previous selenographic literature we also include the traditional (though unofficial) letter designations (see p. 18). So as to determine unambiguously which name corresponds to a given letter we have obeyed the rule introduced by Beer and Mädler. *The adjacent letter, designating a subsidiary small crater, is printed on the side nearest to the primary crater from which it derives its name.* The centres of these associated craters are marked by black dots and, as a rule, the letter is located on a line linking one of these points with the name of the 'parent' crater. In places where confusion could occur, an arrow pointing to the relevant name is added to the corresponding letter.

In some map sections the places where unmanned, soft-landing lunar probes and the manned *Apollo* lunar-landing missions touched down are marked.

Each of the left-hand pages, opposite the map sections, contains a description of the principal features of the adjacent mapped area. In addition, the reader's attention is drawn to features of special interest, as well as to lunar landing sites, etc. Also on these pages is an alphabetical list of the personalities after whom the craters, etc. were named, which contains surnames and forenames, dates of birth and death, nationalities, professions, and other biographical information. In order to make identification on the map section easier, the rounded selenographic co-ordinates of the features are given and, for quick reference, craters are classified as walled plains, ring mountains, etc. These descriptions of craters are accompanied by their diameter in kilometres and, where applicable, their maximum depth, measured from rim to floor in metres. For example, 12 km/2440 m means that a crater's diameter is 12 km and its depth is 2440 m. Among the data on small craters, the reader will find considerable diversity in the diameter-to-depth ratios. For amateur astronomers the smaller and, sometimes, the smallest craters can serve as convenient objects for testing the resolving power of a telescope.

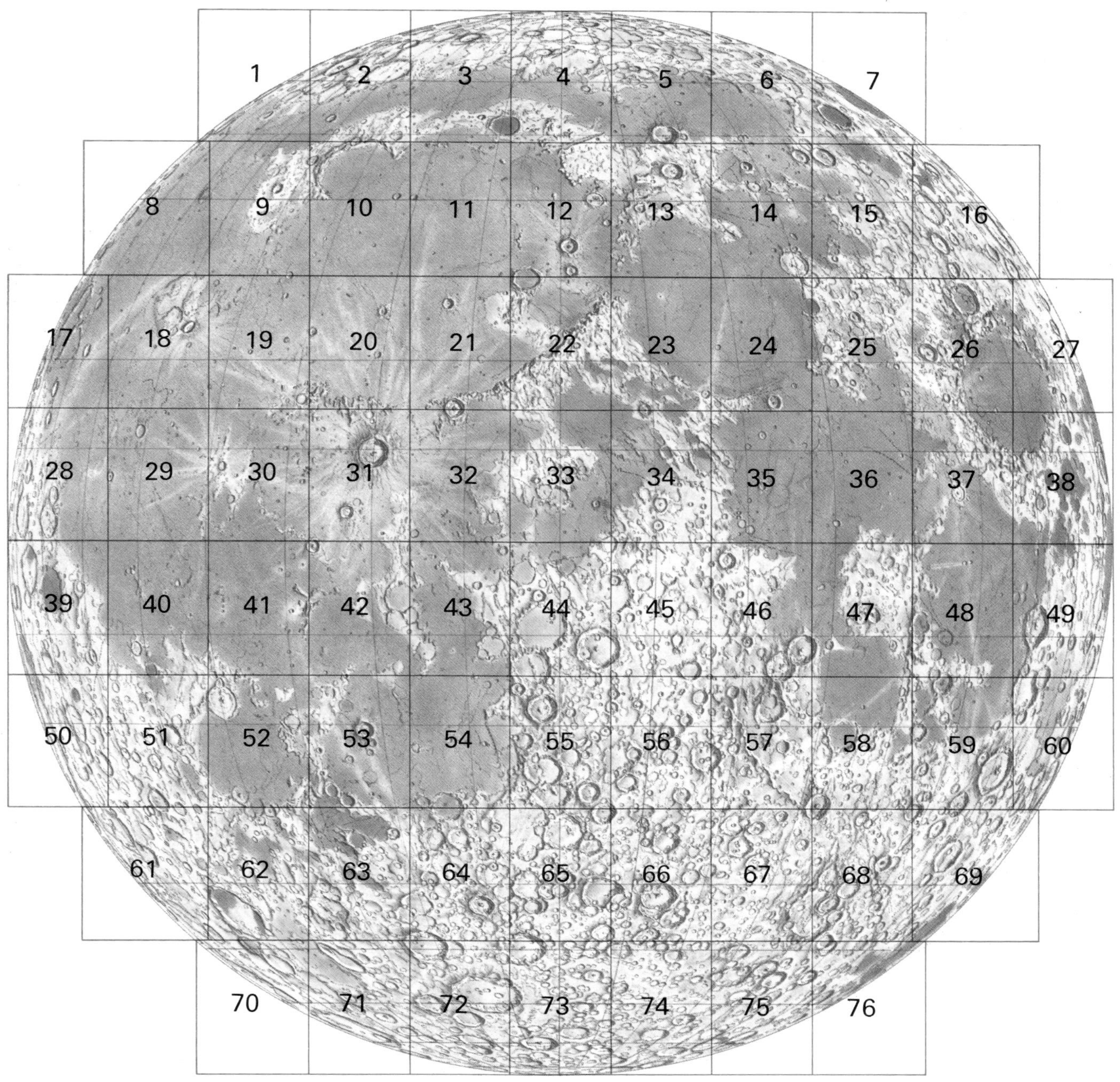

Layout of the map of the near-side of the Moon in 76 sections.

1. MARKOV

The map shows a section of the north-western portion of the near-side of the Moon including the Sinus Roris. The craters near the limb, in a libration zone of the Moon, are seen best shortly before Full Moon.

Cleostratus [60.4°N, 77.0°W] Cleostratus, *c.* 500 BC. Greek philosopher and astronomer. Improved Athenian calendar by introducing 8-year luni-solar cycle.
Crater (63 km).

Galvani [49.6°N, 84.6°W] Luigi Galvani, 1737–1798. Italian physicist and physician, specialist in comparative anatomy.
Crater (80 km).

Langley [51.1°N, 86.3°W] Samuel P. Langley, 1834–1906. American astronomer and physicist. Determined transparency of atmosphere for different wavelengths in the solar spectrum.
Crater (60 km).

Markov [53.4°N, 62.7°W] (1) Andrei A. Markov, 1856–1922. Russian mathematician, specialist in the theory of probability. (2) Alexander V. Markov, 1897–1968. Soviet astrophysicist. Photometry of the Moon.
Crater with a sharp rim (40 km).

Oenopides [57.0°N, 64.1°W] Œnopides of Chios, *c.* 500–430 BC. Greek astronomer and geometer. Discovery of the inclination of the ecliptic to the celestial equator is ascribed to him.
Walled plain (67 km).

Régnault [54.1°N, 88.0°W] Henri Victor Régnault, 1810–1878. French chemist and physicist. Development of steam engine.
Crater (47 km).

Repsold [51.4°N, 78.5°W] Johann G. Repsold, 1751–1830. German manufacturer of precision optical and mechanical apparatus and, in particular, astrometric devices.
Disintegrated crater (107 km).

Repsold, Rimae [51°N, 80°W]
Rilles, length 130 km.

Roris, Sinus (Bay of Dew). Riccioli's name for the mare area that links Mare Frigoris and Oceanus Procellarum.

Stokes [52.5°N, 88.1°W] Sir George G. Stokes, 1819–1903. British mathematician and physicist. Foundations of hydrodynamics, spectral analysis. Shape and gravitational field of the Earth.
Crater (51 km).

Volta [54.0°N, 84.9°W] Count Allessandro G. A. A. Volta, 1745–1827. Italian physicist. First electric battery in 1800.
Crater (113 km).

Xenophanes [57.6°N, 81.4°W] Xenophanes of Colophon, *c.* 570–478 BC. Greek philosopher, satirist and poet. Believed in flat Earth.
Crater (120 km).

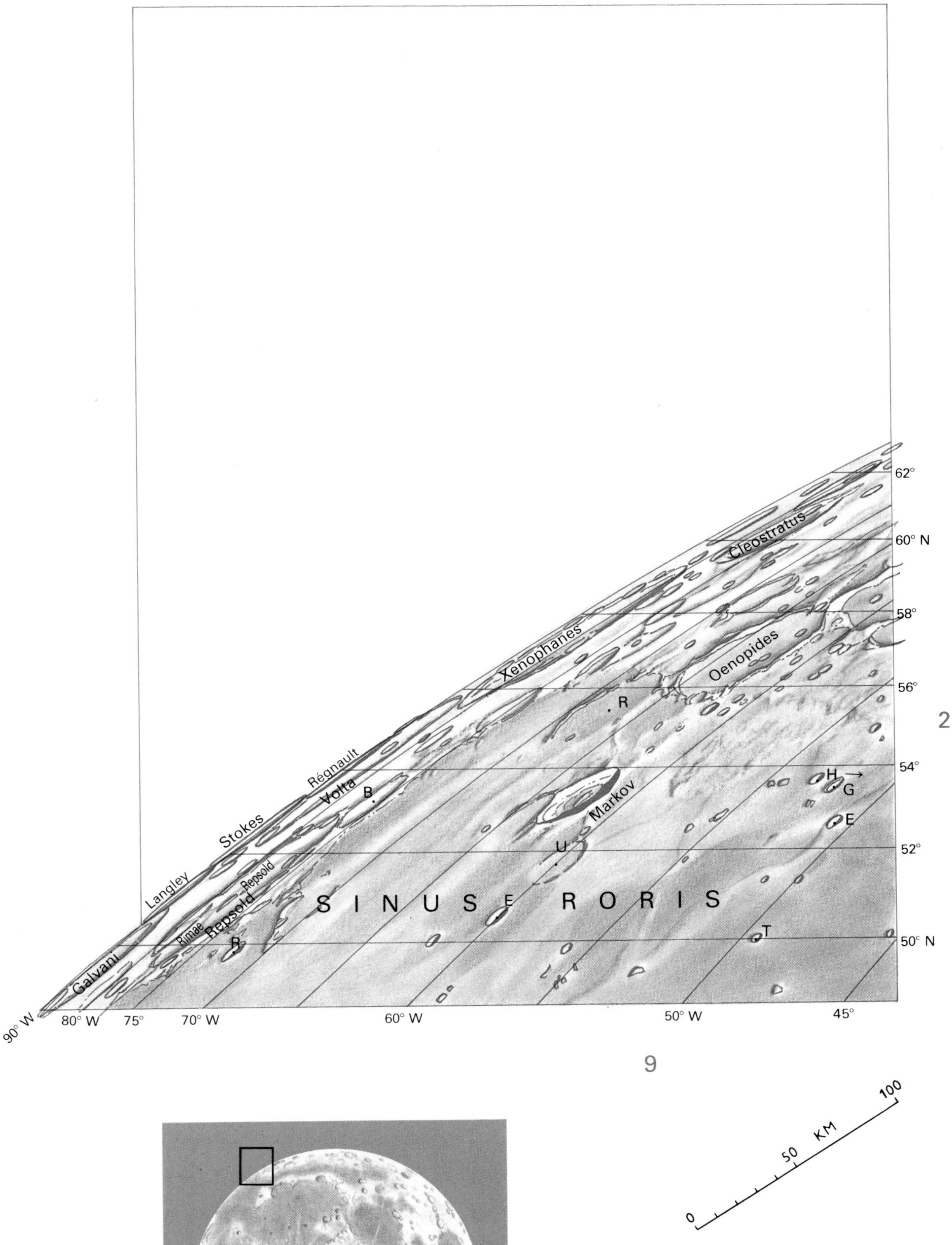

62°
Cleostratus
60° N
58°
Xenophanes
Oenopides
56°
R
2
Régnault
54°
Volta
B
H →
G
Markov
Stokes
E
52°
U
Langley
Repsold
Rimae
Repsold
SINUS E RORIS
Galvani
R
T
50° N
90° W 80° W 75° 70° W 60° W 50° W 45°

9

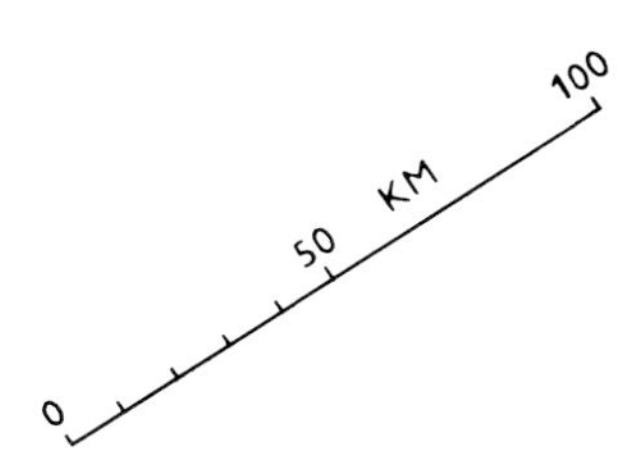

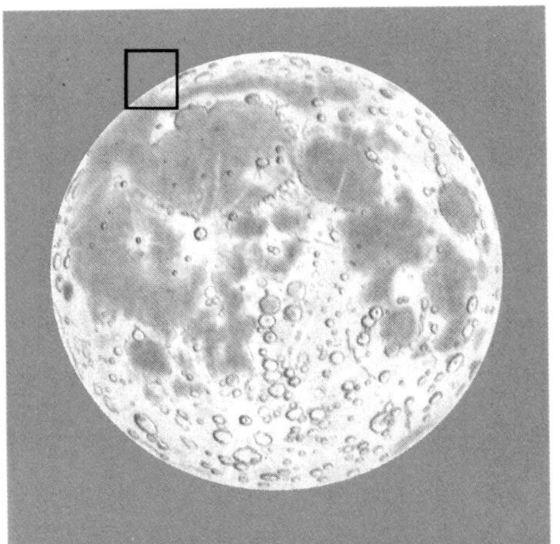

2. PYTHAGORAS

In this section of the Moon, close to the northern margin of the visible disk, Sinus Roris borders on Mare Frigoris. The crater Pythagoras dominates its surroundings with its terraced walls and central mountain.

Anaximander [66.9°N, 51.3°W] Anaximander, *c.* 610–546. BC. Greek philosopher of Miletus.
Crater (68 km).

Babbage [59.5°N, 56.8°W] Charles Babbage, 1792–1871. English mathematician and inventor of a calculating machine.
Walled plain (144 km).

Bianchini [48.7°N, 34.3°W] Francesco Bianchini, 1662–1729. Italian astronomer.
Crater (38 km).

Boole [63.7°N, 87.4°W] George Boole, 1815–1864. English mathematician.
Crater (63 km).

Bouguer [52.3°N, 35.8°W] Pierre Bouguer, 1698–1758. French hydrographer, geodesist and astronomer.
Crater (23 km).

Carpenter [69.4°N, 50.9°W] James Carpenter, 1840–1899. English astronomer.
Crater (60 km).

Cremona [67.5°N, 90.6°W] Luigi Cremona, 1830–1903. Italian mathematician.
Crater (85 km).

Desargues [70.2°N, 73.3°W] Gerard Desargues, 1593–1662. French mathematician and engineer.
Crater (85 km).

Foucault [50.4°N, 39.7°W] Léon Foucault, 1819–1868. French physician and physicist. First demonstration of the rotation of the Earth on its axis (Foucault pendulum).
Crater (23 km).

Harpalus [52.6°N, 43.4°W] Harpalus, *c.* 460 BC. Greek astronomer.
Ray crater (39 km).

Horrebow [58.7°N, 40.8°W] Peder Horrebow, 1679–1764. Danish mathematician and physicist.
Crater (24 km).

J. Herschel [62.1°N, 41.2°W] John Herschel, 1792–1871. English astronomer, son of William Herschel.
Disintegrated walled plain (156 km).

la Condamine [53.4°N, 28.2°W] Charles M. de La Condamine, 1701–1774. French physicist and astronomer.
Crater (37 km).

Maupertuis [49.6°N, 27.3°W] Pierre Louis de Maupertuis, 1698–1759. French mathematician and astronomer.
Disintegrated crater (46 km).

Pythagoras [63.5°N, 62.8°W] Pythagoras, *c.* 580–500 BC. Founder of a Greek school of philosophy and science. Transition from the flat to spherical Earth.
Very prominent crater (130 km).

Robinson [59.0°N, 45.9°W] John T. R. Robinson, 1792–1882. Irish astronomer and physicist.
Crater (24 km).

South [57.7°N, 50.8°W] James South, 1785–1867. English astronomer.
Disintegrated walled plain (108 km).

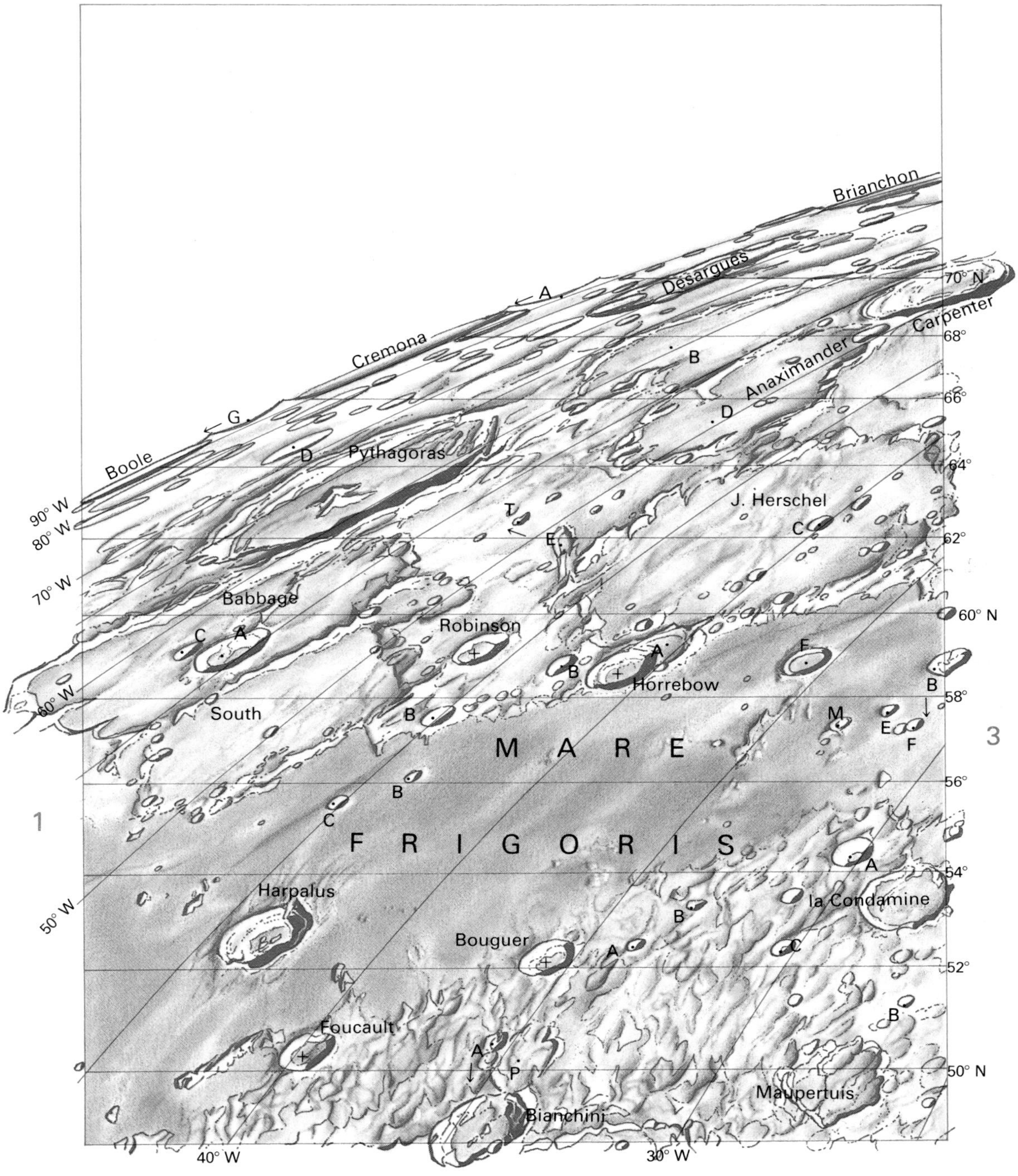

Briancho
Desargues
A
Cremona
70° N
Carpenter
68°
B
Anaximander
66°
D
G
Boole
D
Pythagoras
64°
90° W
80° W
J. Herschel
T
C
E
62°
70° W
Babbage
60° N
Robinson
C
A
A
F
B
60° W
Horrebow
58°
South
B
B
M
E
F
M A R E
B
56°
C
F R I G O R I S
A
54°
Harpalus
la Condamine
50° W
B
Bouguer
A
C
52°
B
Foucault
A
50° N
P
Maupertuis
Bianchini
40° W
30° W

1

3

10

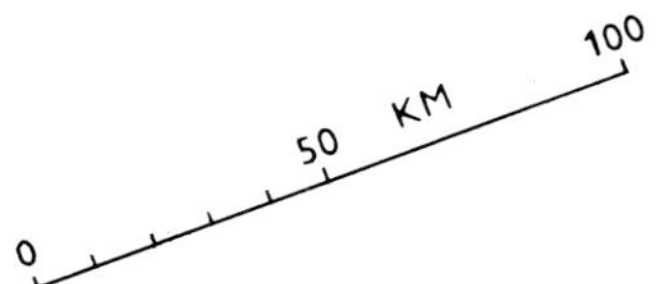

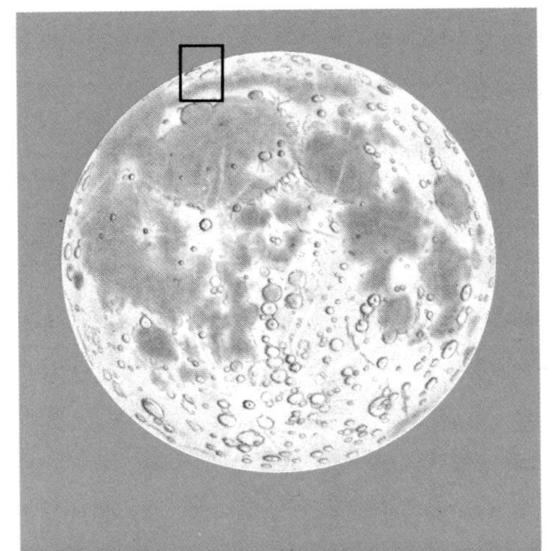

3. PLATO

Northern region of the Moon and the western part of Mare Frigoris. The lower part of the map is occupied by the walled plain Plato which has a very dark floor. Mare Frigoris and Mare Imbrium are separated by a narrow 'continental' strip.

Anaximenes [72.5°N, 44.5°W] Anaximenes, 585–528 BC. Greek philosopher of Miletus. Taught that the Earth was flat and that the Sun was hot because of the speed of its revolution around the Earth.
Crater (80 km).

Brianchon [74.8°N, 86.5°W] Charles J. Brianchon, 1783–1864. French mathematician.
Crater in a libration zone (145 km).

Fontenelle [63.4°N, 18.9°W] Bernard Le Bovier de Fontenelle, 1657–1757. French astronomer, popularizer of science, and one of the early members of the French Academy of Sciences.
Crater (38 km).

Frigoris, Mare (Sea of Cold). Riccioli's name for an elongated mare in the northern polar region.
The surface of M. Frigoris occupies an area of 436 000 sq km *(Lacus Mortis and the area west of the crater Hercules included – see map no. 14) and is comparable in size with the Black Sea on Earth.*

Maupertuis, Rimae [51°N, 22°W]
Rilles, length 100 km.

Mouchez [78.3°N, 26.6°W] Ernest A. B. Mouchez, 1821–1892. Officer of the French Navy, later Director of the Paris Observatory.
Remains of a crater (82 km).

Pascal [74.3°N, 70.1°W] Blaise Pascal, 1623–1662. French mathematician, physicist, and philosopher. Invented an adding machine.
Crater (106 km).

Philolaus [72.1°N, 32.4°W] Philolaus, end of the 5th century BC. Greek philosopher, adherent of Pythagorean astronomy. Taught that the Earth is moving. Believed that the centre of space is a 'central fire'.
Crater (71 km).

Plato [51.6°N, 9.3°W] Plato, *c.* 427–347 BC. Prominent Greek philosopher, pupil of Socrates. His astronomy is Pythagorean; conceived the Earth as a round body surrounded by planetary spheres and stars.
Walled plain (101 km).

Poncelet [75.8°N, 54.1°W] Jean V. Poncelet, 1788–1867. French mathematician.
Crater (69 km).

Sylvester [82.7°N, 79.6°W] James J. Sylvester, 1814–1897. British mathematician; number theory, analytical geometry.
Crater in a libration zone (58 km).

*The lunar crater **Plato.** Note that a part of the wall with triangular contours is separated from the western edge of the crater by a landslide of rock. There are four small craters between* 1.7 *and* 2.2 km *in diameter on the floor of Plato.*

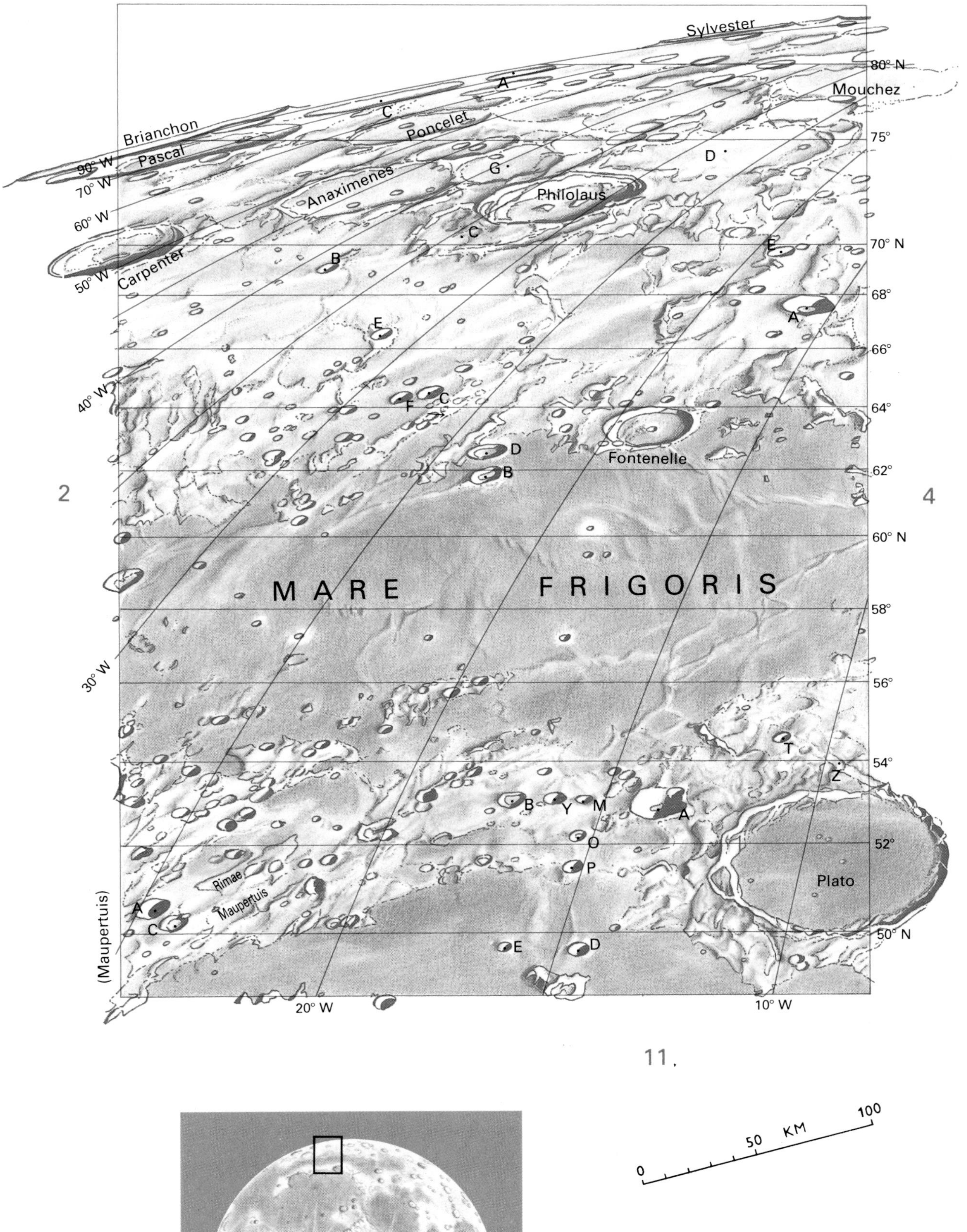

Sylvester
80° N
Mouchez
A
C
Brianchon
Poncelet
75°
90° W
Pascal
D
70° W
G
Anaximenes
Philolaus
60° W
C
70° N
E
B
Carpenter
50° W
68°
A
E
66°
40° W
F
C
64°
D
Fontenelle
B
62°

2

4

60° N

MARE FRIGORIS

58°
30° W
56°
T
54°
Z
B
Y
M
A
O
52°
P
Plato
Rimae
Maupertuis
A
C
(Maupertuis)
50° N
E
D
20° W
10° W

11.

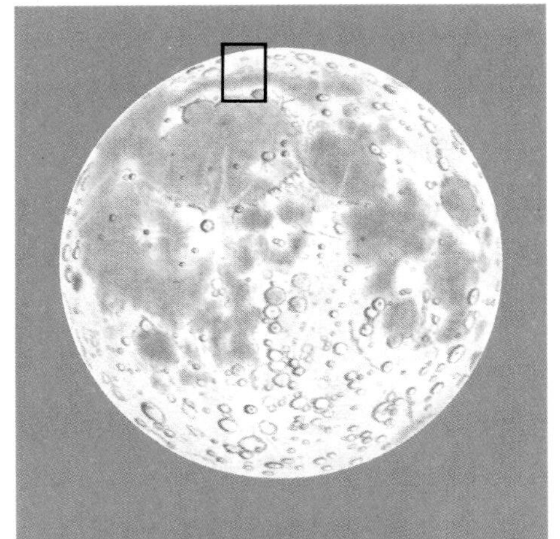

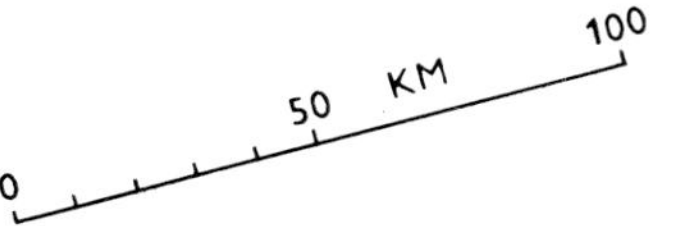

4. ARCHYTAS

Area surrounding North Pole, central part of Mare Frigoris, northern part of the Alps.

Alpes, Vallis [49°N, 3°E] (Alpine Valley)
Length 180 km, *cleft in floor.*

Anaxagoras [73.4°N, 10.1°W] Anaxagoras, 500–428 BC. Greek philosopher.
Ray crater (51 km).

Archytas [58.7°N, 5.0°E] Archytas, *c.* 428–347 BC. Greek philosopher, statesman and geometer.
Crater (32 km).

Archytas, Rima [53°N, 5°E]
Rille, length 90 km.

Barrow [71.3°N, 7.7°E] Isaac Barrow, 1630–1677. English mathematician, friend of Sir Isaac Newton.
Crater (93 km).

Birmingham [65.1°N, 10.5°W] John Birmingham, 1829–1884. Irish selenographer.
Remains of a crater (92 km).

Byrd [85.3°N, 9.8°E] Richard E. Byrd, 1888–1957. American polar explorer, pilot.
Walled plain (94 km).

Challis [79.5°N, 9.2°E] James Challis, 1803–1862. English astronomer.
Crater (56 km).

Epigenes [67.5°N, 4.6°W] Epigenes, third century BC. Greek astronomer.
Crater (55 km).

Gioja [83.3°N, 2.0°E] Flavio Gioja, *c.* 1302. Italian sea captain.
Crater (42 km).

Goldschmidt [73.0°N, 2.9°W] Hermann Goldschmidt, 1802–1866. German amateur astronomer.
Walled plain (120 km).

Hermite [86.4°N, 87.3°W] Charles Hermite, 1822–1901. French mathematician.
Crater (110 km).

Main [80.8°N, 10.1°E] Robert Main, 1808–1878. English astronomer.
Crater (46 km).

Meton [73.8°N, 19.2°E] Meton, *c.* 432 BC. Greek astronomer and mathematician.
Remains of a walled plain (122 km).

Peary [88.6°N, 33.0°E] Robert E. Peary, 1856–1920. American polar explorer.
Walled plain (74 km).

Plato, Rimae [51°N, 2°W]
Isolated rilles east of the crater Plato.

Protagoras [56.0°N, 7.3°E] Protagoras, *c.* 485–410 BC. Greek philosopher.
Crater (22 km).

Scoresby [77.7°N, 14.1°E] William Scoresby, 1789–1857. English navigator, oceanographer.
Crater (56 km).

Timaeus [62.8°N, 0.5°W] Timaeus, *c.* 400 BC. Pythagorean philosopher, friend of Plato.
Crater (33 km).

Trouvelot [49.3°N, 5.8°E] Étienne L. Trouvelot, 1827–1895. French astronomer
Crater (9 km).

W. Bond [65.3°N, 3.7°E] William C. Bond, 1789–1859. American astronomer.
Walled plain (158 km).

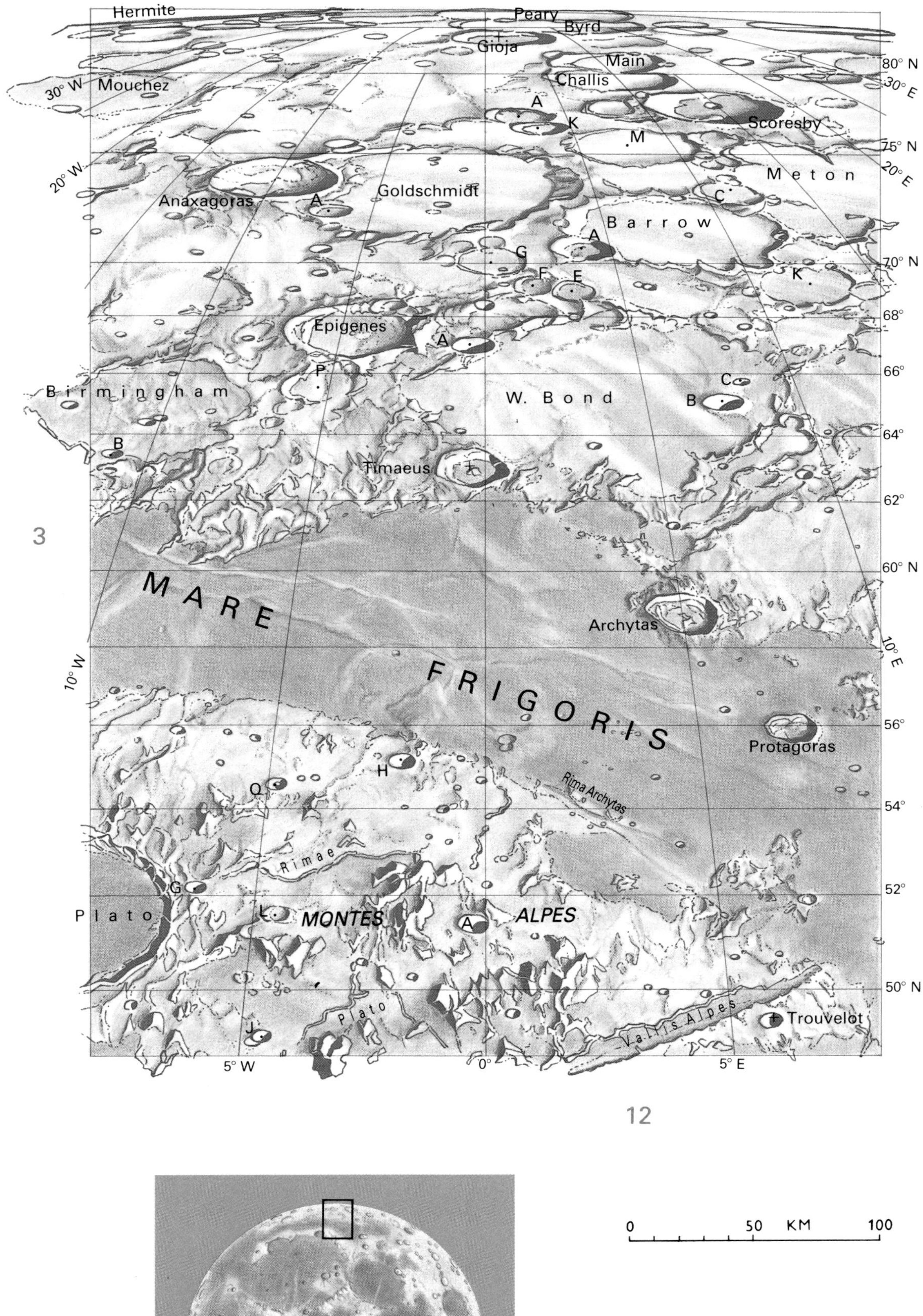

4

Hermite
Peary
Byrd
Gioja
Main
80° N
Challis
30° E
30° W
Mouchez
A
K
Scoresby
M
75° N
20° W
Meton
20° E
Anaxagoras
A
Goldschmidt
C
Barrow
A
G
70° N
F
E
K
68°
Epigenes
A
66°
P
C
Birmingham
W. Bond
B
64°
B
Timaeus
62°

3

5

60° N
MARE
Archytas
10° E
10° W
FRIGORIS
56°
Protagoras
H
Q
Rima Archytas
54°
Rimae
52°
G
Plato
L
MONTES
A
ALPES
50° N
Plato
Vallis Alpes
J
Trouvelot
5° W
0°
5° E

12

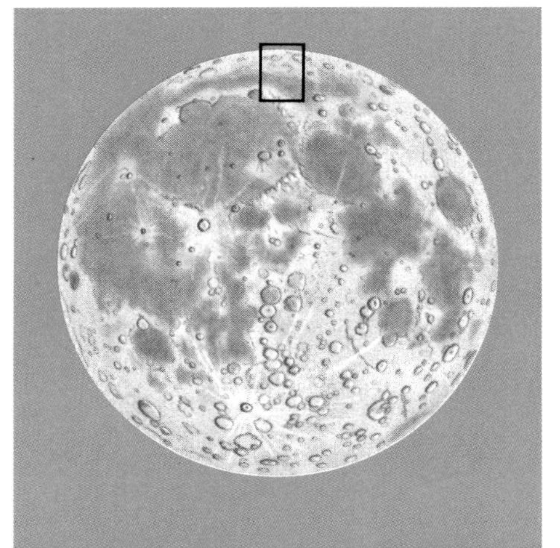

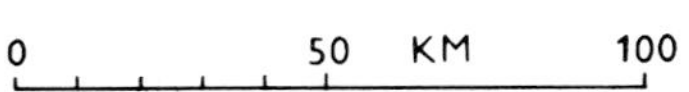

5. ARISTOTELES

The eastern part of Mare Frigoris and the area east of the north pole of the Moon. The most interesting formation in this section is the crater Aristoteles.

Aristoteles [50.2°N, 17.4°E] Aristoteles, *c.* 384–322 BC. Greek philosopher, whose teaching influenced Europe for several centuries.
Crater with terraced walls (87 km).

Arnold [66.8°N, 35.9°E] Christoph Arnold, 1650–1695. German amateur astronomer.
Crater (95 km).

Baillaud [74.6°N, 37.5°E] Benjamin Baillaud, 1848–1934. French astronomer.
Flooded crater (90 km).

C. Mayer [63.2°N, 17.3°E] Christian Mayer, 1719–1783. Austrian astronomer.
Prominent crater (38 km).

Democritus [62.3°N, 35.0°E] Democritus, *c.* 460–370 BC. Greek philosopher (atomic theory).
Prominent crater (39 km).

de Sitter [80.1°N, 39.6°E] Willem de Sitter, 1872–1934. Notable Dutch astronomer.
Crater (65 km).

Euctemon [76.4°N, 31.3°E] Euctemon, *c.* 432 BC. Astronomer from Athens, contemporary of Meton.
Crater (62 km).

Galle [55.9°N, 22.3°E] Johann G. Galle, 1812–1910. German astronomer. Discovered Neptune on the basis of Le Verrier's calculations on 23 September 1846.
Crater (21 km).

Kane [63.1°N, 26.1°E] Elisha K. Kane, 1820–1857. American traveller and explorer.
Flooded crater (55 km).

Mitchell [49.7°N, 20.2°E] Maria Mitchell, 1818–1889. American astronomer.
Crater (30 km).

Moigno [66.4°N, 28.9°E] François N. M. Moigno, 1804–1884. French mathematician and physicist.
Crater (37 km).

Nansen [81.3°N, 95.3°E] Fridtjof Nansen, 1861–1930. Norwegian polar explorer.
Crater in a libration zone (122 km).

Neison [68.3°N, 25.1°E] Edmund Neison, 1851–1940. English selenographer.
Crater (53 km).

Petermann [74.2°N, 66.3°E] August Petermann, 1822–1878. German geographer.
Crater (73 km).

Peters [68.1°N, 29.5°E] Christian A. F. Peters, 1806–1880. German astronomer.
Crater (15 km).

Sheepshanks [59.2°N, 16.9°E] Anne Sheepshanks, 1789–1876. Sister of an English astronomer, a benefactress to astronomy.
Crater (25 km).

Sheepshanks, Rima [58°N, 24°E]
Rille, length 200 km.

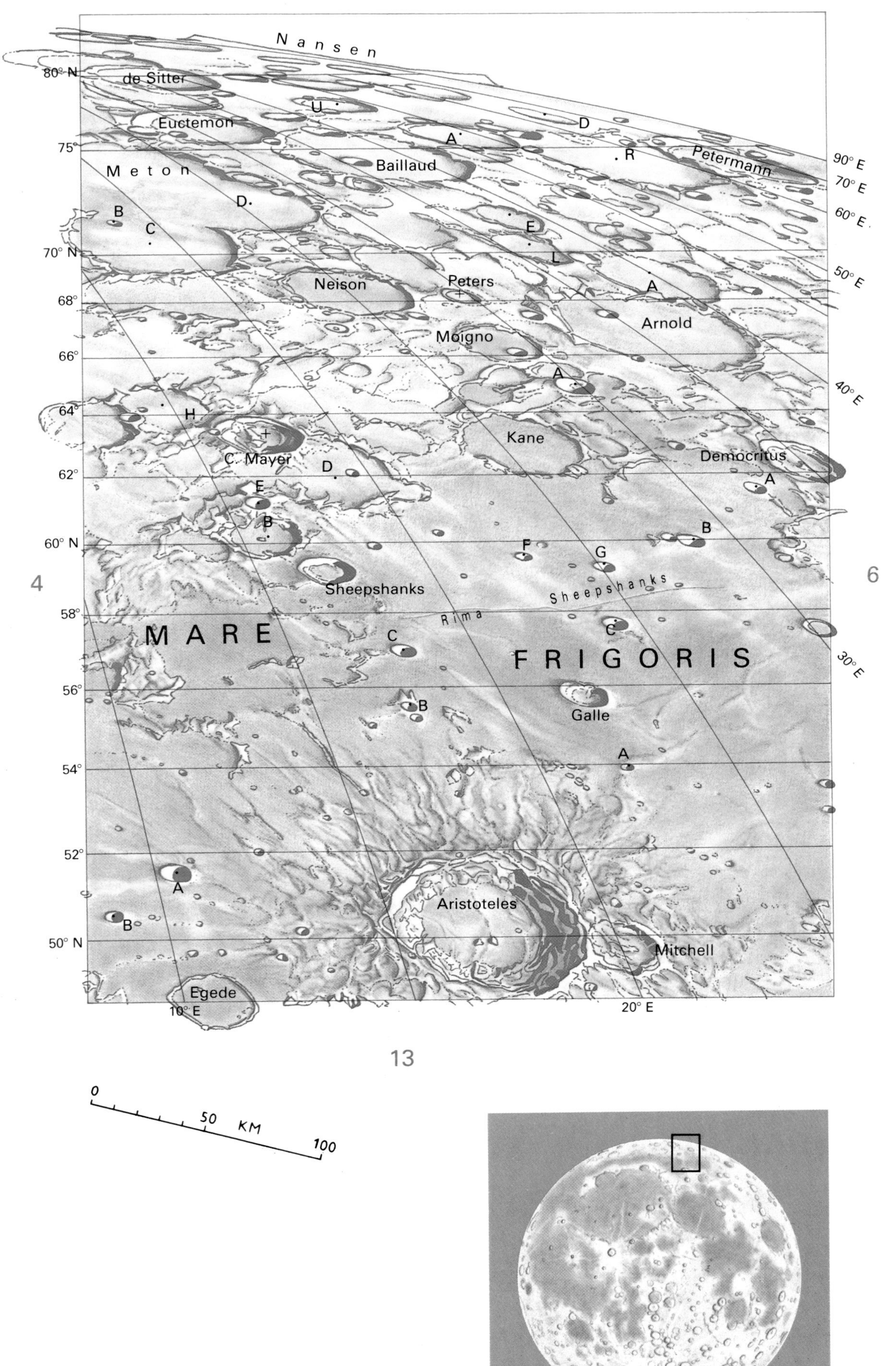
Nansen
80° N
de Sitter
U
Euctemon
D
A
R
75°
Baillaud
Petermann
90° E
70° E
Meton
60° E
B
D
C
E
70° N
L
Neison
Peters
A
50° E
68°
Arnold
Moigno
66°
A
40° E
64°
H
Kane
C. Mayer
Democritus
62°
D
A
E
B
B
60° N
F
G
4
6
Sheepshanks
Sheepshanks
58°
Rima
MARE
C
C
FRIGORIS
30° E
56°
B
Galle
A
54°
52°
A
Aristoteles
B
50° N
Mitchell
Egede
10° E
20° E
13
0
50
KM
100

6. STRABO

The map features a section of the north-eastern region of the Moon and the areas surrounding the eastern edge of Mare Frigoris. The pair of craters Strabo and Thales, and the prominent group of three craters, Strabo N, B and L, are useful for orientation.

Baily [49.7°N, 30.4°E] Francis Baily, 1774–1844. English businessman, who from 1825 devoted himself fully to astronomy. He was the first to describe the phenomenon 'Baily's beads', which he observed during a total eclipse of the Sun in 1836.
Crater with a disintegrated wall (27 km).

Cusanus [72.0°N, 70.8°E] Nikolaus Krebs Cusanus, 1401–1464. German by origin, mathematician and cardinal. Opposed the concept of a geocentric universe.
Crater with a flooded floor (63 km).

de la Rue [59.1°N, 53.0°E] Warren de la Rue, 1815–1889. Englishman, one of the pioneers of astrophotography.
Disintegrated walled plain (136 km).

Frigoris, Mare (Sea of Cold). See p. 32.

Gärtner [59.1°N, 34.6°E] Christian Gärtner, *c.* 1750–1813. German mineralogist and geologist.
Remains of a walled plain (102 km).

Gärtner, Rima
Rille inside a crater, length 30 km.

Hayn [64.7°N, 85.2°E] Friedrich Hayn, 1863–1928. German astronomer. Improved upon the existing theory of rotation of the Moon, mapped the limb areas of the Moon.
Crater (87 km).

Keldysh (Hercules A) [51.2°N, 43.6°E] Mstislav V. Keldysh, 1911–1978. Soviet mathematician, mechanic and engineer, prominent theoretician and organizer of Soviet astronautics.
Regular crater with a sharp rim (33 km).

Schwabe [65.1°N, 45.6°E] Heinrich Schwabe, 1789–1875. German astronomer, discovered the eleven-year cycle of solar activity.
Crater with a flooded floor (25 km).

Strabo [61.9°N, 54.3°E] Strabo, *c.* 64 BC – *c.* 24 AD. Greek geographer and historian. His *Geography* survived as the most significant work of its kind.
Prominent crater with a terraced wall and flooded floor (55 km).

Thales [61.8°N, 50.3°E] Thales of Miletus, about 624–*c.* 543 BC. Founder of Greek geometry, philosopher. Taught that water is the principle of everything.
Regular crater with a sharp rim (32 km).

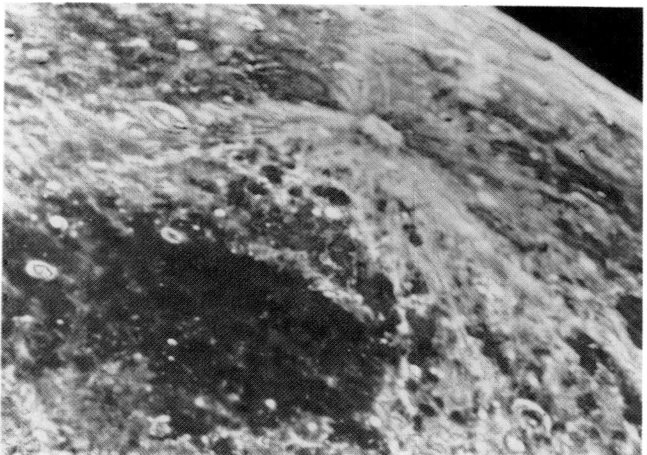

*Under high illumination the bright ray crater **Thales** stands out close to the eastern edge of Mare Frigoris.*

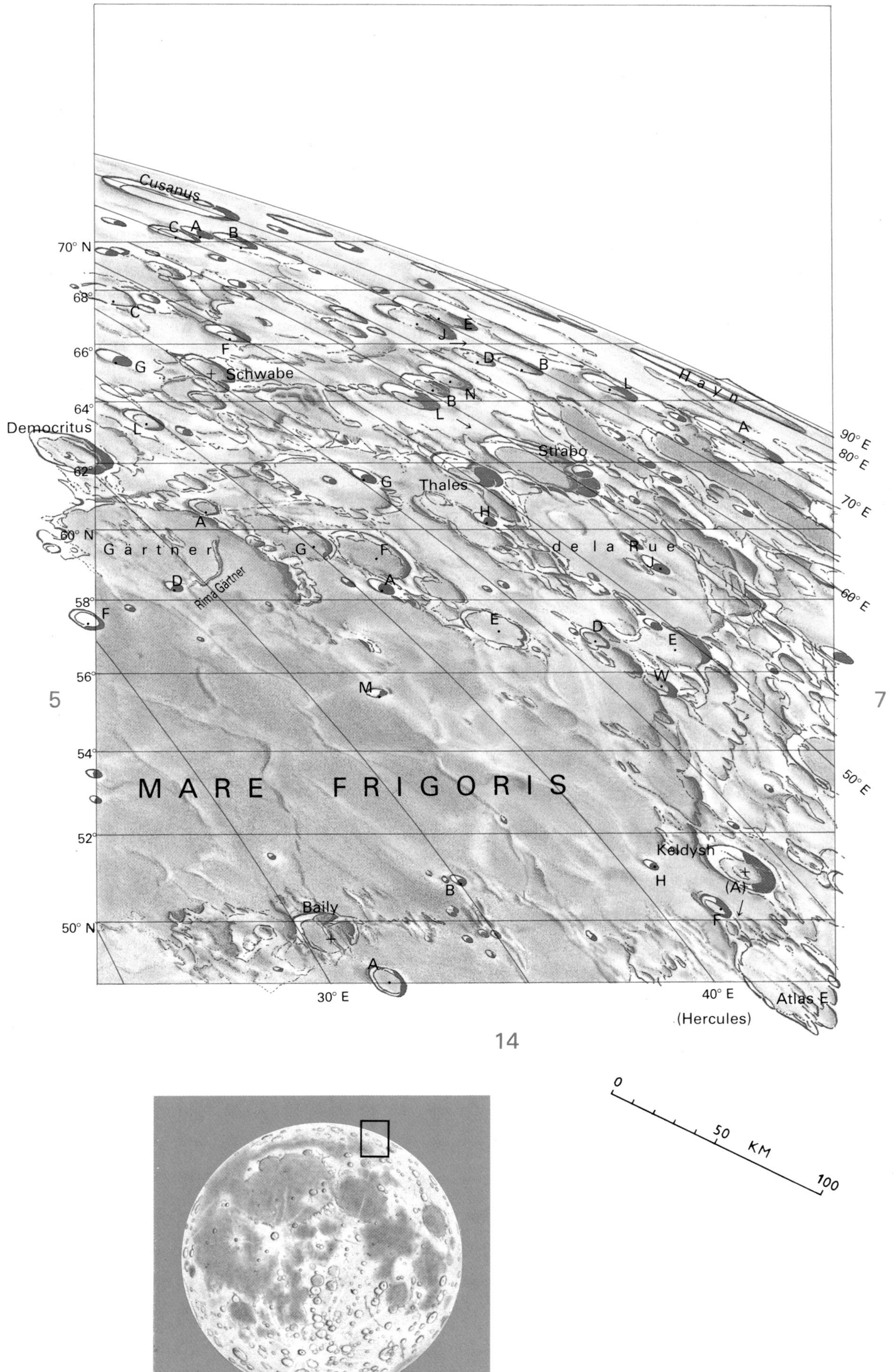

Cusanus
70° N
68°
66°
Schwabe
64°
Democritus
62°
Strabo
Thales
Hayn
90° E
80° E
70° E
60° N
Gärtner
Rima Gärtner
de la Rue
58°
60° E
56°
5
7
54°
MARE FRIGORIS
50° E
52°
Keldysh
(A)
Baily
50° N
30° E
40° E
(Hercules)
Atlas E
14
0
50
KM
100

7. ENDYMION

At the north-eastern edge of the Moon there are two very prominent dark spots: the floor of the flooded crater Endymion and the Mare Humboldtianum. Under oblique illumination a mountain range stands out, forming a nearly continuous wall around the M. Humboldtianum: this wall is partly intersected by a wall of the crater Belkovich, which lies in the libration zone.

Belkovich [61.5°N, 90.0°E] Igor V. Belkovich, 1904–1949. Soviet astronomer, specialist in selenodesy. Observation and calculation of the shape and rotational elements of the Moon.

Walled plain with central peaks and two larger craters on the circumference of the wall (198 km).

Endymion [53.6°N, 56.5°E] Endymion. A young shepherd who, according to a Greek legend, went to sleep on Mount Latmos; his beauty so aroused the cold heart of Selene, Goddess of the Moon, that she came down the Earth and kissed Endymion, who slept on forever.

Very prominent crater with sizeable wall and flooded dark floor (125 km).

Humboldtianum, Mare (Humboldt's Sea) [57°N, 80°E]. Alexander von Humboldt, 1769–1859. German naturalist and explorer. In South America in 1799 he observed the Leonid meteor shower. Exploratory expeditions to the rivers Orinoco and Amazon, the Andes, Mexico and Siberia. Mädler gave Humboldt's name to this formation because he recognized a symbolic parallel between Humboldt's explorations of unknown terrestrial continents and the way that this lunar mare seems to form a link between the known and unknown hemispheres of the Moon.

The eastern edge of Mare Humboldtianum extends to longitude 90°E *and hence its visibility is much affected by libration. Mare Humboldtianum is a dark flooded centre of a lunar basin with a concentric outer wall of diameter approximately* 640 km; *this wall runs from the crater Strabo (see map 6), east of Endymion and continues south-eastward; around the crater Mercurius E it turns eastward and finally passes over to the lunar far-side. The diameter of Mare Humboldtianum is approximately* 160 km *and it occupies an area of* 22 000 sq km.

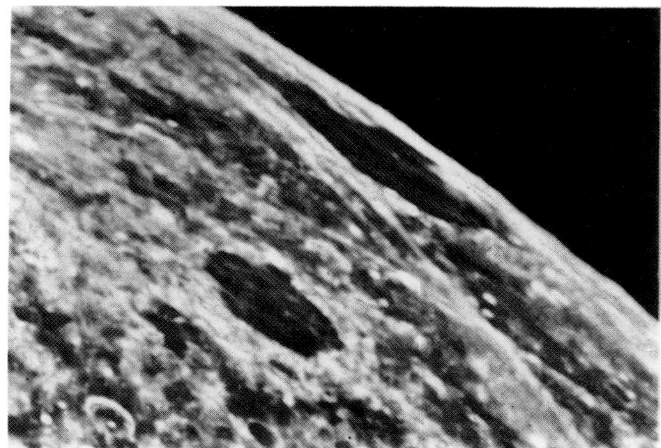

*At favourable libration the small **Mare Humboldtianum** is observable on the northeastern limb of the Moon. Farther from the limb the dark floor of the crater **Endymion** is visible. The photograph shows the are under high illumination.*

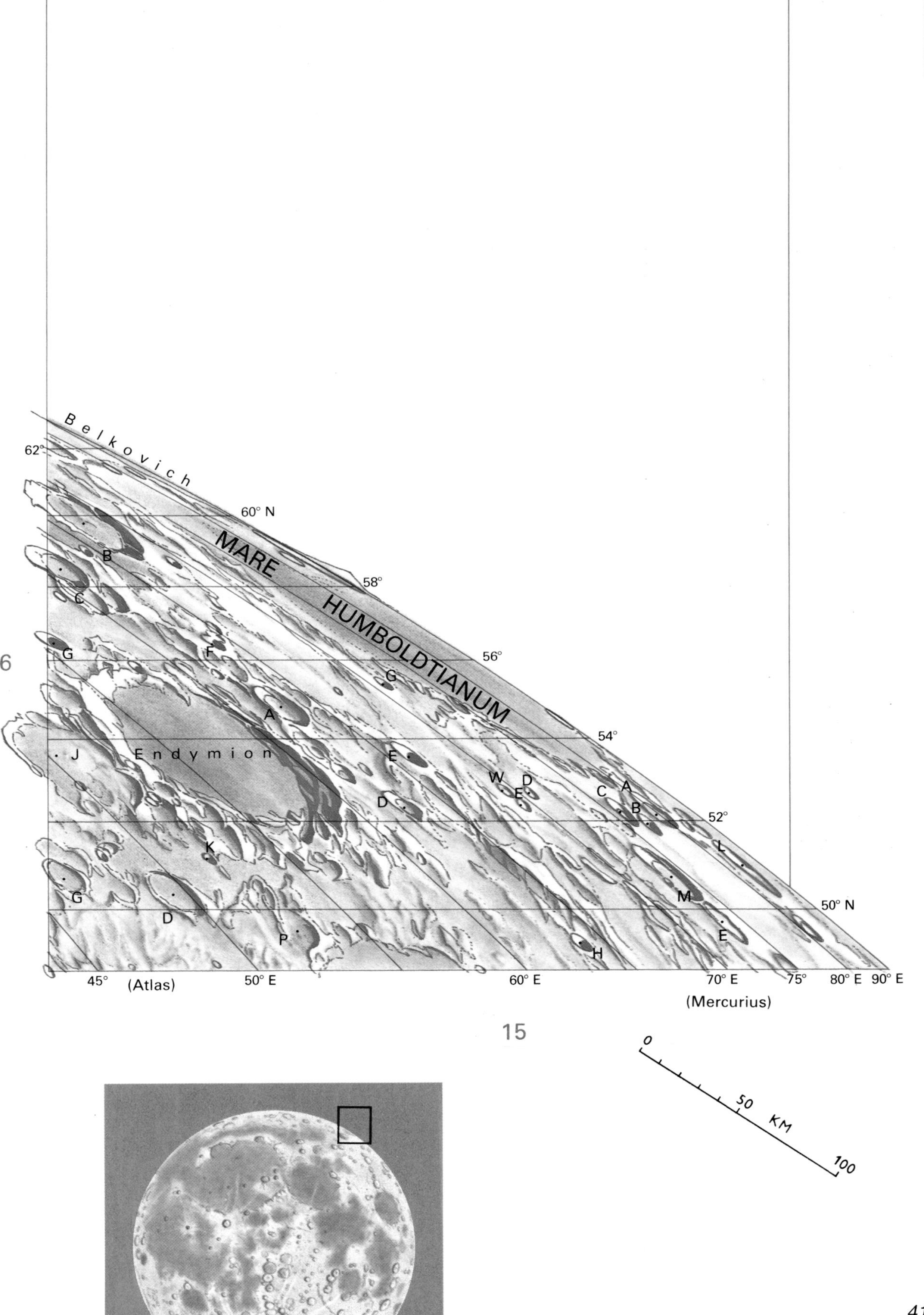

Belkovich
62°
60° N
58°
56°
54°
52°
50° N
MARE HUMBOLDTIANUM
Endymion
B
C
G
F
G
A
J
E
W
D
E
C
A
B
D
K
L
G
D
M
P
E
H
6
45°
(Atlas)
50° E
60° E
70° E
(Mercurius)
75°
80° E
90° E
15
0
50 KM
100

8. RÜMKER

This map is dominated by the dark surface of Oceanus Procellarum; the right-hand edge of the map features an unusual elevated formation, Rümker. Oceanus Procellarum is separated from the north-western limb of the Moon by a seemingly narrow 'continental' strip.

Aston [32.9°N, 87.7°W] Francis W. Aston, 1877–1945. British chemist and physicist, Nobel Laureate for chemistry in 1922. Discovered 212 isotopes.
Crater (43 km).

Bunsen [41.4°N, 85.3°W] Robert W. Bunsen, 1811–1899. German chemist. Pioneer in the application of spectral analysis in chemistry.
Disintegrated crater (52 km).

Dechen [46.1°N, 68.2°W] Ernst H. Karl von Dechen, 1800–1889. German mineralogist and geologist.
Circular crater (12 km).

Gerard [44.5°N, 80.0°W] Alexander Gerard, 1792–1839. English explorer, known for his expeditions to the Himalayas and Tibet.
Remains of a crater (90 km).

Harding [43.5°N, 71.7°W] Karl Ludwig Harding, 1765–1834. German astronomer. Discovered asteroid Juno in 1804.
Crater with a sharp rim (23 km).

Humason (Lichtenberg G) [30.7°N, 56.6°W] Milton L. Humason, 1891–1972. American astronomer.
Small crater (4 km).

Lavoisier [38.2°N, 81.2°W] Antoine Laurent Lavoisier, 1743–1794. French chemist, one of the founders of modern chemistry.
Crater (70 km).

Lichtenberg [31.8°N, 67.7°W] Georg Christoph Lichtenberg, 1742–1799. German physicist. Worked in the field of static electricity. Known also as a satirist.
Crater (20 km).

Naumann [35.4°N, 62.0°W] Karl Friedrich Naumann, 1797–1873. German geologist.
Small crater (9.6 km).

Nielsen (Wollaston C) [31.8°N, 51.8°W] (1) Axel V. Nielsen, 1902–1970, Danish astronomer. (2) Harald H. Nielsen, 1903–1973, American physicist.
Small crater (10 km).

Procellarum Oceanus (Ocean of Storms). See p. 84.

Rümker, Mons [41°N, 58°W] Karl Ludwig Christian Rümker, 1788–1862. German astronomer, director of naval school in Hamburg.
Unique complex of lunar domes; the diameter of the formation is c. 70 km.

Scilla, Dorsum (Scilla's Ridge) [32°N, 60°W] Agostino Scilla, 1639–1700. Italian geologist.
Mare ridge, length about 120 km.

Ulugh Beigh [32.7°N, 81.9°W] Muhammad Taragaj Ulug-bek, 1394–1449. Uzbek astronomer and mathematician, grandson of the conqueror Timur. Founded astronomical school, built an observatory with 40 m quadrant near Samarkand.
Disintegrated, flooded crater (54 km).

Whiston, Dorsa (Whiston's Ridges) [30°N, 57°W]. William Whiston, 1667–1752. British mathematician.
System of mare ridges, length about 120 km.

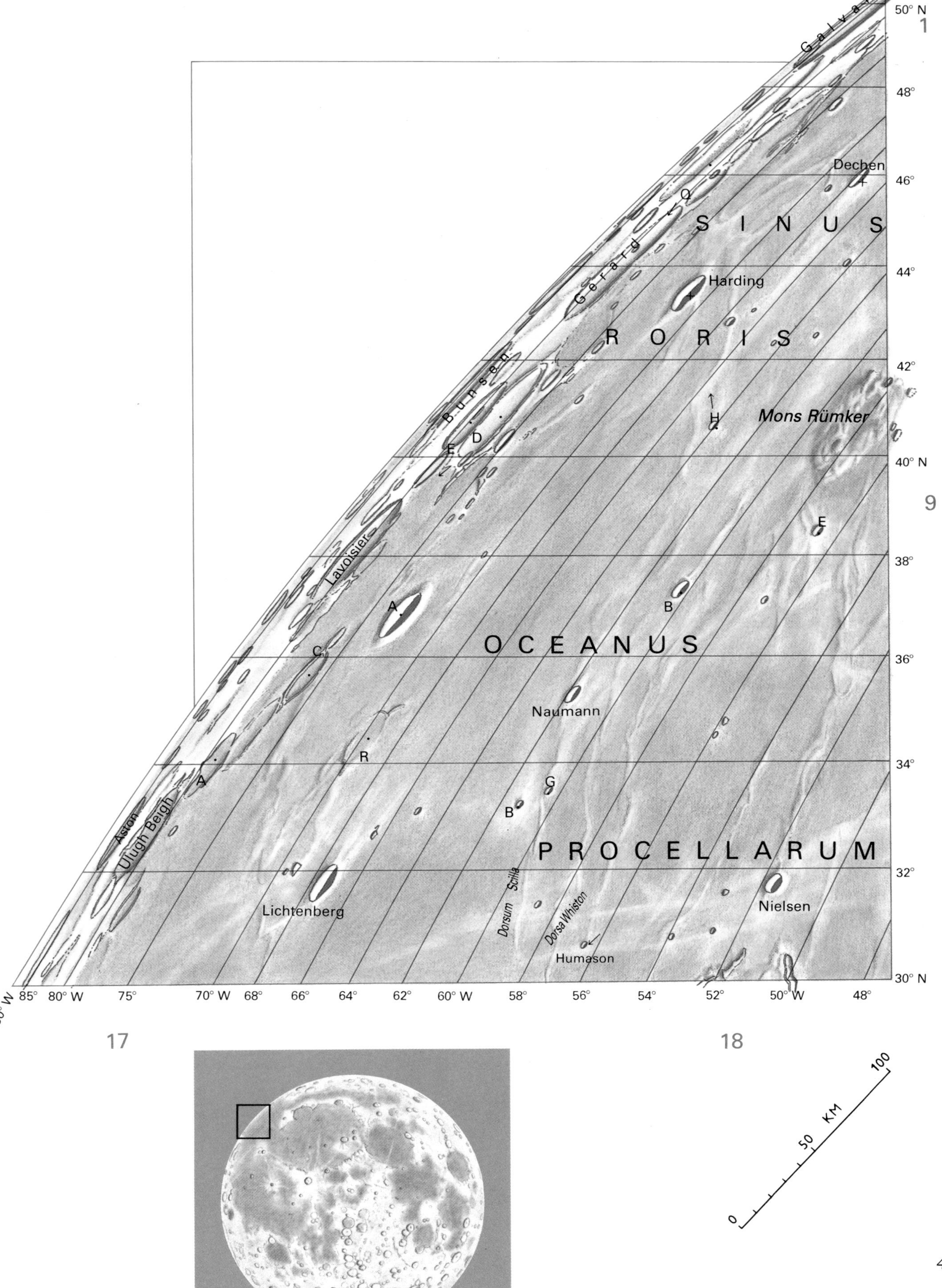
1
9
17
18
Galvani
Gerard
Bunsen
Lavoisier
Aston
Ulugh Beigh
Dechen
SINUS
RORIS
Harding
Mons Rümker
OCEANUS
PROCELLARUM
Naumann
Lichtenberg
Nielsen
Humason
Dorsum Scilla
Dorsa Whiston
Q
H
D
E
E
A
B
C
R
A
G
B
50° N
48°
46°
44°
42°
40° N
38°
36°
34°
32°
30° N
90° W
85°
80° W
75°
70° W
68°
66°
64°
62°
60° W
58°
56°
54°
52°
50° W
48°
0
50
KM
100

9. MAIRAN

The 'continental' area adjoining the Sinus Iridum stretches here into Oceanus Procellarum from the north-east (see map 10). A local curiosity is a group of peaks to the north of the crater Gruithuisen. The formation Gruithuisen Gamma is shaped like an upturned bath tub; in fact, it is a tall dome-like mountain massif with a circular base about 20 km in diameter, with a 900 m peak craterlet.

Bucher, Dorsum (Bucher's Ridge) [31°N, 39°W] W. H. Bucher, 1889–1965. Swiss geophysicist.
Mare ridge, length about 90 km.

Delisle [29.9°N, 34.6°W] Joseph N. Delisle, 1688–1768. French astronomer. At the invitation of the Russian Empress Catherine I he was put in charge of the new observatory at St Petersburg (1726–47). Suggested a method for determining the distance of the Sun from observations of the transits of Mercury and Venus.
Crater (25 km/2550 m).

Delisle, Rima [31°N, 33°W].
Rille, length 50 km.

Gruithuisen [32.9°N, 39.7°W] Franz von Gruithuisen, 1774–1852. German physician and astronomer. Dedicated but eccentric observer. Wrote a book about his 'discovery' of buildings and other evidence of 'Moon-dwellers'.
Crater (16 km/1860 m).

Gruithuisen Delta, Mons [36.0°N, 39.5°W].
Mountain massif, base diameter 20 km.

Gruithuisen Gamma, Mons [36.6°N, 40.5°W].
Dome-like mountain massif, base diameter 20 km.

Imbrium, Mare (Sea of Rains). See p. 48.

Louville [44.0°N, 46.0°W] Jacques E. d'Allonville, Chevalier de Louville, 1671–1732. French mathematician and astronomer. Discovered a method for calculating the precise circumstances of solar eclipses.
Eroded crater (36 km).

Mairan [41.6°N, 43.4°W] Jean J. Dortous de Mairan, 1678–1771. French astronomer. The study of the aurora borealis.
Crater with a sharp rim (40 km).

Mairan T [41.7°N, 48.2°W]
Dome with summit crater (3 km).

Mairan, Rima [38°N, 47°W]
Rille, length 100 km.

Procellarum, Oceanus (Ocean of Storms). See p. 84.

Roris, Sinus (Bay of Dew). Riccioli's name for the northern promontory of Oceanus Procellarum. A narrow cleft of *Rima Sharp* (length 210 km) runs out into the bay, which can be traced in *Lunar Orbiter* photographs to between the craters Mairan T and G.

Wollaston [30.6°N, 46.9°W] William Hyde Wollaston, 1766–1828. English scientist – medicine, chemistry, mineralogy and astronomy. Discovered palladium and rhodium, devised a goniometer and, in 1802, observed the Fraunhofer lines, which he mistook for the boundaries between different colours in the solar spectrum.
Crater with a sharp rim (10.2 km).

1

8

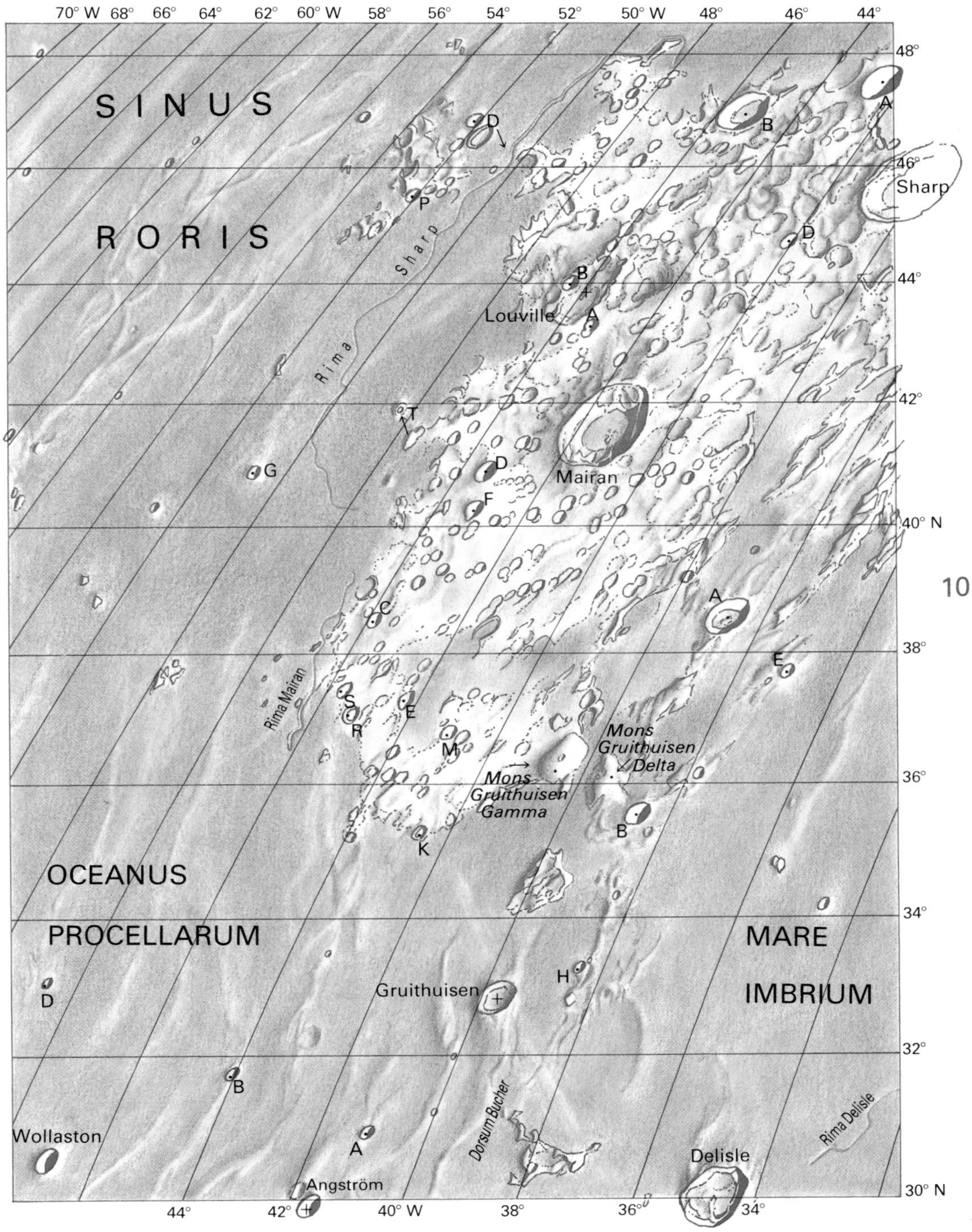

70° W
68°
66°
64°
62°
60° W
58°
56°
54°
52°
50° W
48°
46°
44°
48°
46°
44°
42°
40° N
38°
36°
34°
32°
30° N
44°
42°
40° W
38°
36°
34°
SINUS
RORIS
OCEANUS
PROCELLARUM
MARE
IMBRIUM
Sharp
Rima Sharp
Louville
Mairan
Rima Mairan
Mons Gruithuisen Delta
Mons Gruithuisen Gamma
Gruithuisen
Dorsum Bucher
Rima Delisle
Delisle
Wollaston
Angström
A
B
C
D
E
F
G
H
K
M
P
R
S
T

10

19

0
50
KM
100

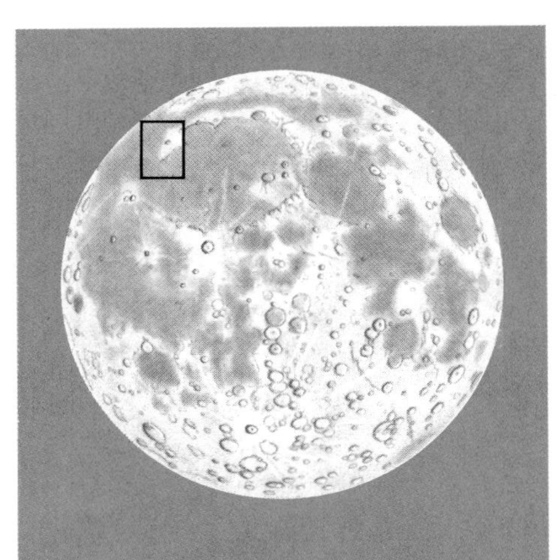

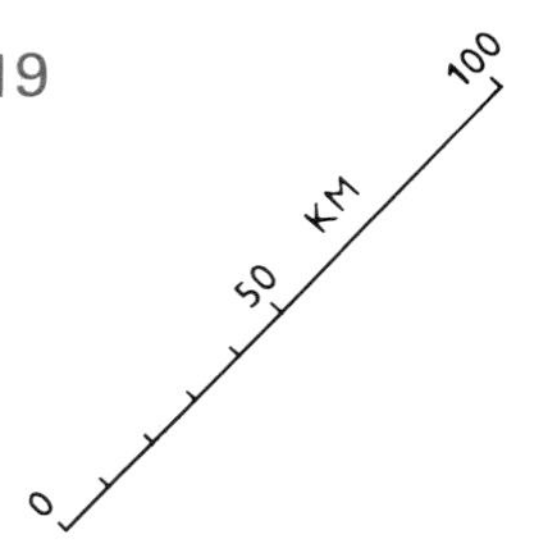

10. SINUS IRIDUM

The north-western part of the Mare Imbrium with the beautiful Sinus Iridum, and the Jura mountain range which forms its edge. Sinus Iridum is crossed by mare ridges. *Luna 17,* which landed to the south of Cape Heraclides, transported an automatic mobile laboratory, *Lunokhod 1,* to the Moon.

C. Herschel [34.5°N, 31.2°W] Caroline Herschel, 1750–1848. Sister of William Herschel. As a devoted assistant to her brother, she worked with him for 50 years. She also discovered eight comets.
Crater (13.4 km/1850 m).

Carlini [33.7°N, 24.1°W] Francesco Carlini, 1783–1862. Italian astronomer. Worked in the field of celestial mechanics, improved the theory of the motion of the Moon.
Crater with a sharp rim (11.4 km/2200 m).

Heim, Dorsum (Heim's Ridge) [31°N, 29°W] Albert Heim, 1849–1937. Swiss geophysicist.
Mare ridge, length about 130 km.

Heis [32.4°N, 31.9°W] Eduard Heis, 1806–1877. German astronomer, observer of variable stars.
Crater (14 km/1910 m); **Heis A** (6.1 km/650 m).

Helicon [40.4°N, 23.1°W] Helicon, fourth century BC. Greek mathematician and astronomer.
Crater (25 km/1910 m); *nearby is* **Helicon E** (2.4 km/470 m)

Heraclides, Promontorium (Cape Heraclides) [41°N, 34°W] Heraclides Ponticus, *c.* 390–310 BC. Pupil of Plato. Maintained that the Earth rotates on an axis.

Imbrium, Mare (Sea of Rains). See p. 48.

Iridum, Sinus (Bay of Rainbows) [45°N, 32°W]. Named by Riccioli.
Crater formation, diameter 260 km.

Jura, Montes (Jura Mountains) [47°N, 37°W]. Named by Debes. The mountain range borders Sinus Iridum as the wall of a flooded crater.

Laplace, Promontorium (Cape Laplace) [46°N, 26°W] Pierre Simon Laplace, 1749–1827. Outstanding French mathematician, disciple of Newton. Worked in the field of celestial mechanics. 'Nebula Hypothesis' of the origin of the solar system.

McDonald (Carlini B) [30.4°N, 20.9°W] (1) William J. McDonald, 1844–1926. American benefactor. (2) Thomas L. MacDonald, died in 1973. Scottish selenographer.
Small crater (8 km/1470 m).

Sharp [45.7°N, 40.2°W] Abraham Sharp, 1651–1742. English astronomer, assistant to Flamsteed at Greenwich Observatory.
Crater (40 km).

Zirkel, Dorsum (Zirkel's Ridge) [29°N, 24°W] Ferdinand Zirkel, 1838–1912. German geologist and mineralogist.
Mare ridge, length about 210 km (continued on map 20).

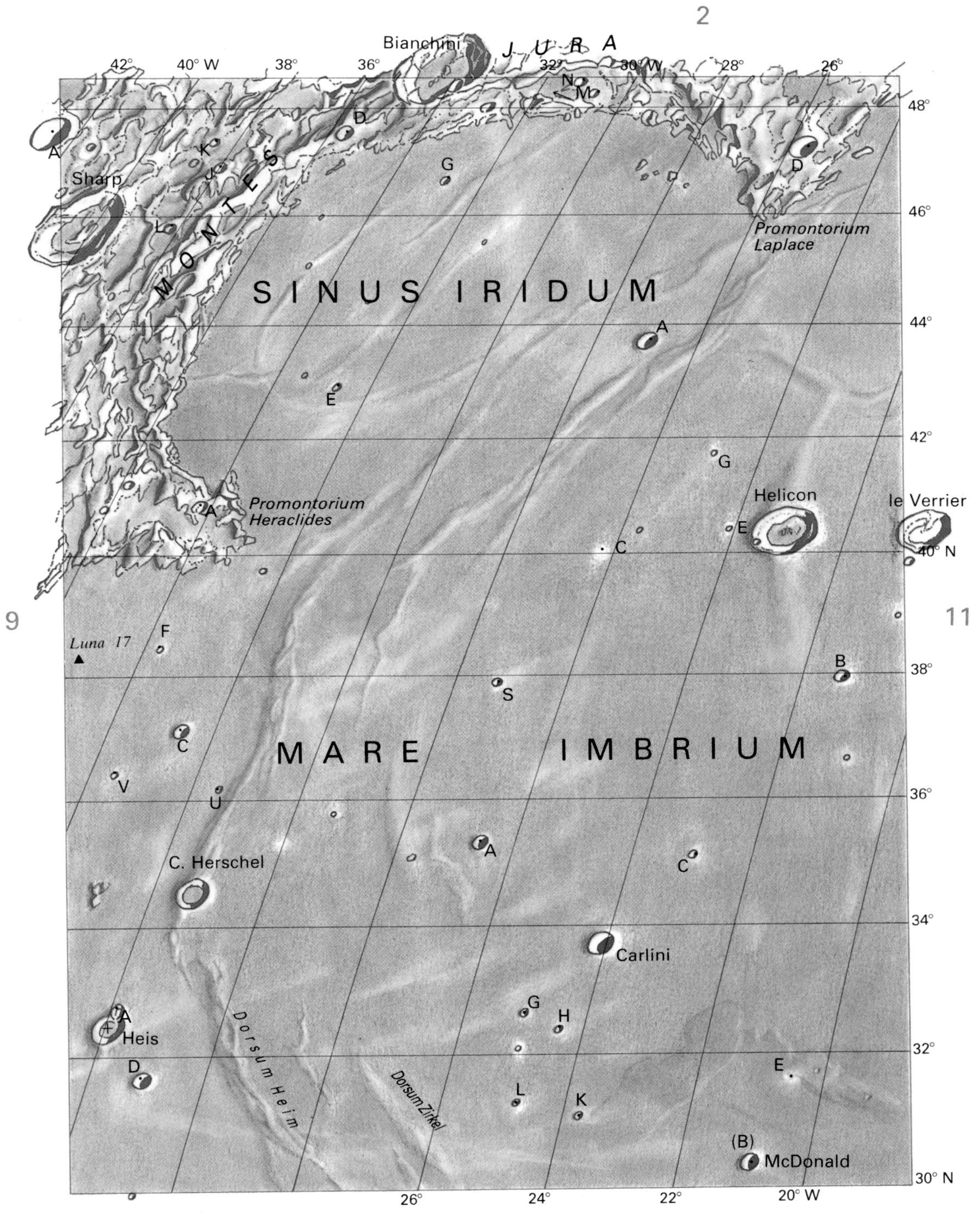

2

J U R A

Bianchini

42° 40° W 38° 36° 32° 30° W 28° 26°

N M

D

A

K

Sharp

J

D

G

M O N T E S

L

48°

46°

Promontorium Laplace

SINUS IRIDUM

44°

A

E

42°

G

Helicon

le Verrier

Promontorium Heraclides

A

E

C

40° N

9

11

Luna 17

F

B

38°

S

C

MARE IMBRIUM

V

U

36°

A

C. Herschel

C

34°

Carlini

G

H

A

Heis

32°

D

Dorsum Heim

Dorsum Zirkel

L

K

E

(B)

McDonald

30° N

26° 24° 22° 20° W

20

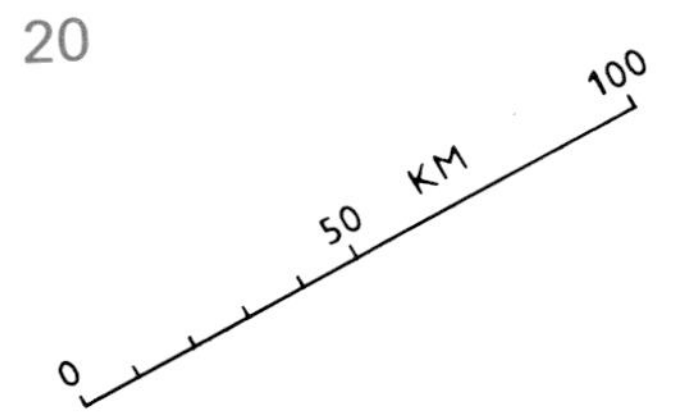

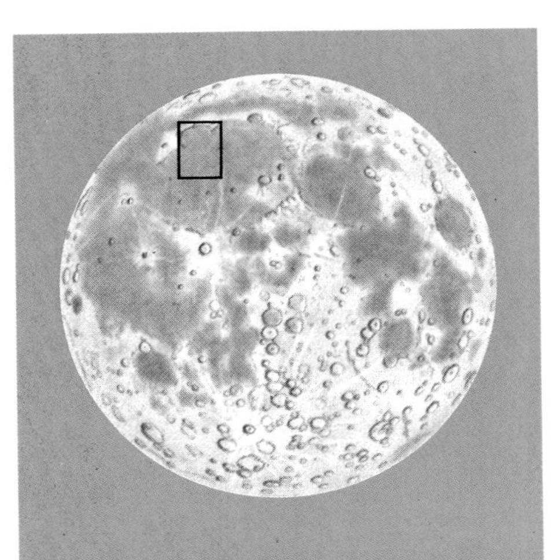

11. LE VERRIER

The central and northern areas of Mare Imbrium have no large craters, but a telescope pointed at the area shown along the northern edge of the map at lunar sunrise or sunset will reveal the long, pointed shadows of isolated peaks. In the past such shadows cast by a low Sun helped to convey the false impression that the lunar mountains are very high, steep and jagged.

Grabau, Dorsum (Grabau's Ridge) [30°N, 14°W] Amadeus W. Grabau, 1870–1946. American geophysicist.
Wrinkle ridge, length about 120 km.

Imbrium, Mare (Sea of Rains). Named by Riccioli. See p. 18.
With a surface area of 830 000 sq km *Mare Imbrium is the second largest mare after Oceanus Procellarum. At the same time it is the largest lunar basin. M. Imbrium is surrounded by a circular mountain range which opens in the west as it joins Oceanus Procellarum. The perimeter of M. Imbrium is formed by the Jura, Alps, Caucasus, Apennine and Carpathian mountain ranges. Tectonic clefts which originated during the formation of the basin, can be traced far to the south and south-east. The diameter of the basin is* 1250 km. *Remains of the inner wall of the basin are formed by Montes Recti, Montes Teneriffe, Pico and Montes Spitzbergen. One of the first mascons was discovered in the centre of this basin.*

Landsteiner (Timocharis F) [31.3°N, 14.8°W] Karl Landsteiner, 1868–1943. Austrian-born American pathologist, Nobel Laureate.
Small crater (6 km/1350 m).

le Verrier [40.3°N, 20.6°W] Urbain Jean le Verrier, 1811–1877. French mathematician and astronomer, calculated (independently of Adams) the orbit and position of Neptune.
Crater (20 km/2100 m).

Pico, Mons [46°N, 9°W] Mountain named by Schröter, who evidently had in mind 'Pico von Teneriffe'; he compared the height of this mountain with the height of other lunar mountain ranges.
2400 m *high, its base measures* 15 × 25 km.

Recti, Montes [48°N, 20°W] Straight Range. Named by Birt because of its shape.
Length about 90 km, *height up to* 1800 m.

Teneriffe, Montes (Teneriffe Mountains) [48°N, 13°W]. Name reminiscent of the mountains of Teneriffe, where Piazzi Smyth first tested telescopic observational conditions high above sea level.
Length about 110 km, *height up to* 2400 m.

Small craters: **le Verrier B** (5.1 km/980 m)
le Verrier D (9.1 km/1830 m)
le Verrier W (3.3 km/620 m).

3

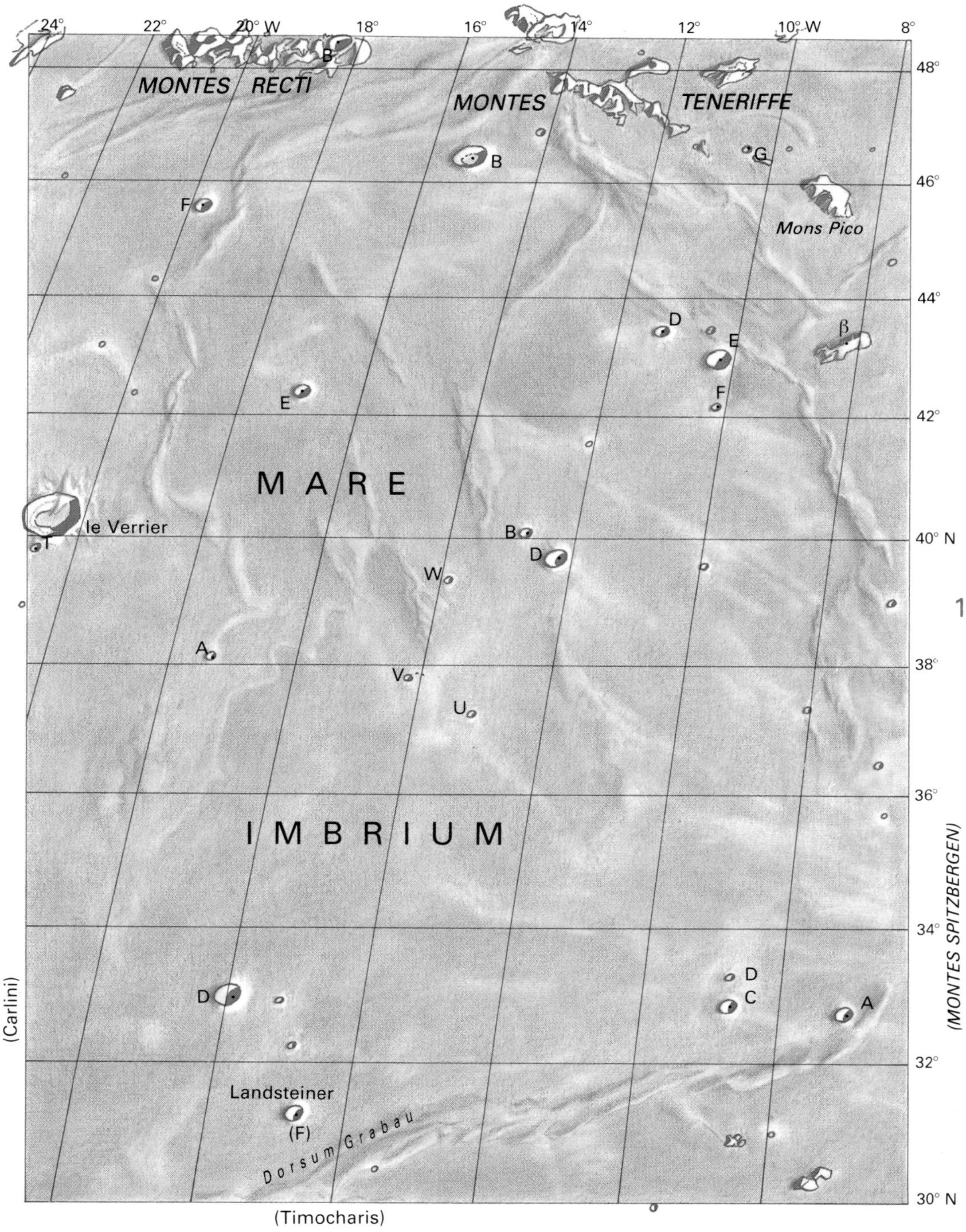

24° 22° 20° W 18° 16° 14° 12° 10° W 8°

MONTES RECTI

B

MONTES TENERIFFE

48°

B

G

46°

F

Mons Pico

44°

D

E

β

E

F

42°

MARE

le Verrier

T

B

D

40° N

W

10

12

A

38°

V

U

36°

IMBRIUM

(MONTES SPITZBERGEN)

34°

D

D

C

A

(Carlini)

32°

Landsteiner

(F)

Dorsum Grabau

30° N

(Timocharis)

21

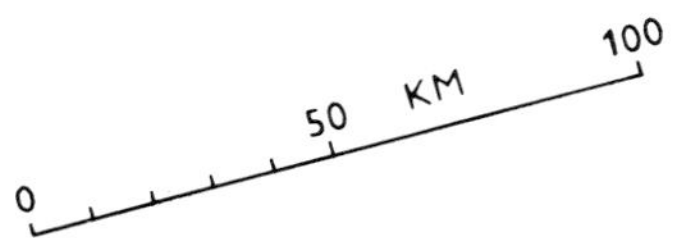

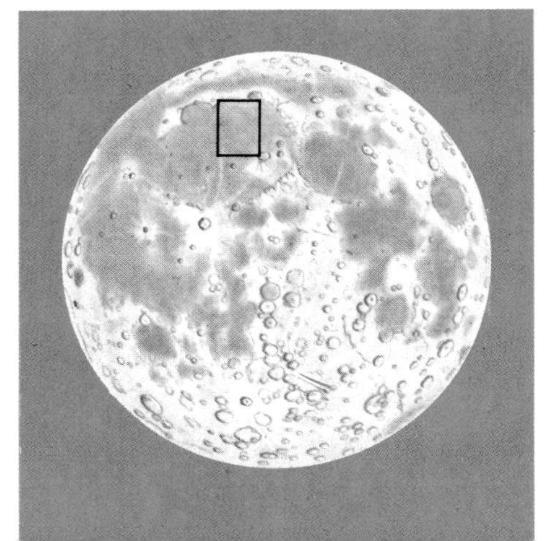

12. ARISTILLUS

The eastern edge of Mare Imbrium is one of the most interesting parts of the lunar surface. The lunar Alps with their well-known valley, the solitary mountain Piton and the group of three craters Archimedes, Autolycus and Aristillus are suitable objects for telescopic observation. *Luna 2* crash-landed close to the crater Autolycus.

Alpes, Montes (The Alps). Mountain range (length about 250 km) named by Hevelius. The heights of its peaks range from 1800 m to 2400 m. The following features are named:

Mons Blanc (Mt Blanc) [45°N, 1°E]
Mountain *3600 m high, base diameter* 25 km.

Promontorium Agassiz [42°N, 2°E] (Cape Agassiz) Louis J. R. Agassiz, 1807–1873. Swiss naturalist.

Promontorium Deville [43°N, 1°E]. (Cape Deville) Sainte-Claire Charles Deville, 1814–1876. French geologist.

Aristillus [33.9°N, 1.2°E] Aristillus, *c.* 280 BC. One of the earliest astronomers of the Greek school of Alexandria.
Ray crater (55 km/3650 m), *with a group of three peaks on the floor* (900 m).

Autolycus [30.7°N, 1.5°E] Autolycus, *c.* 330 BC. Greek astronomer and mathematician.
Crater (39 km/3430 m).

Cassini [40.2°N, 4.6°E] (1) Giovanni-Domenico Cassini, 1625–1712. Italian-born French astronomer, discovered four of Saturn's satellites and the so-called 'Cassini Division' in Saturn's rings. (2) Jacques J. Cassini, 1677–1756, son of Giovanni-Domenico Cassini, whom he succeeded as Director of the Paris Observatory.
Flooded crater (57 km/1240 m); **Cassini A** (17 km/2830 m)

Kirch [39.2°N, 5.6°W] Gottfried Kirch, 1639–1710. German astronomer, discovered a large comet in 1680.
Crater (11.7 km/1830 m).

Lunicus, Sinus (Bay of *Luna or 'Lunik'*). [32°N, 1°W] Site of the first touchdown of a space probe on the Moon (*Luna 2,* 1959). The name was allocated by the IAU in 1970.

Piazzi Smyth [41.9°N, 3.2°W] Charles Piazzi Smyth, 1819–1900. British, Astronomer Royal for Scotland. Author of the mysticism of numbers concerning Khufu's pyramid (Great Pyramid).
Crater (12.8 km/2530 m).

Piton, Mons [41°N, 1°W]. Mountain named after a peak in the Teneriffe massif.
Isolated mountain 2250 m *high, base diameter* 25 km.

Spitzbergen, Montes [35°N, 5°W] Spitzbergen. Mountain range named by M. Blagg because of its similarity in shape to the terrestrial Spitzbergen.
Chain of mountains (length 60 km*) reaching a height of* 1500 m.

Theaetetus [37.0°N, 6.0°E] Theaetetus, 415 –369 BC. Athenian philosopher, friend of Plato (on the Moon the crater is near the walled plain Plato).
Crater (25 km/2830 m).

Theaetetus, Rimae [33°N, 6°E]
Group of rilles, length about 50 km.

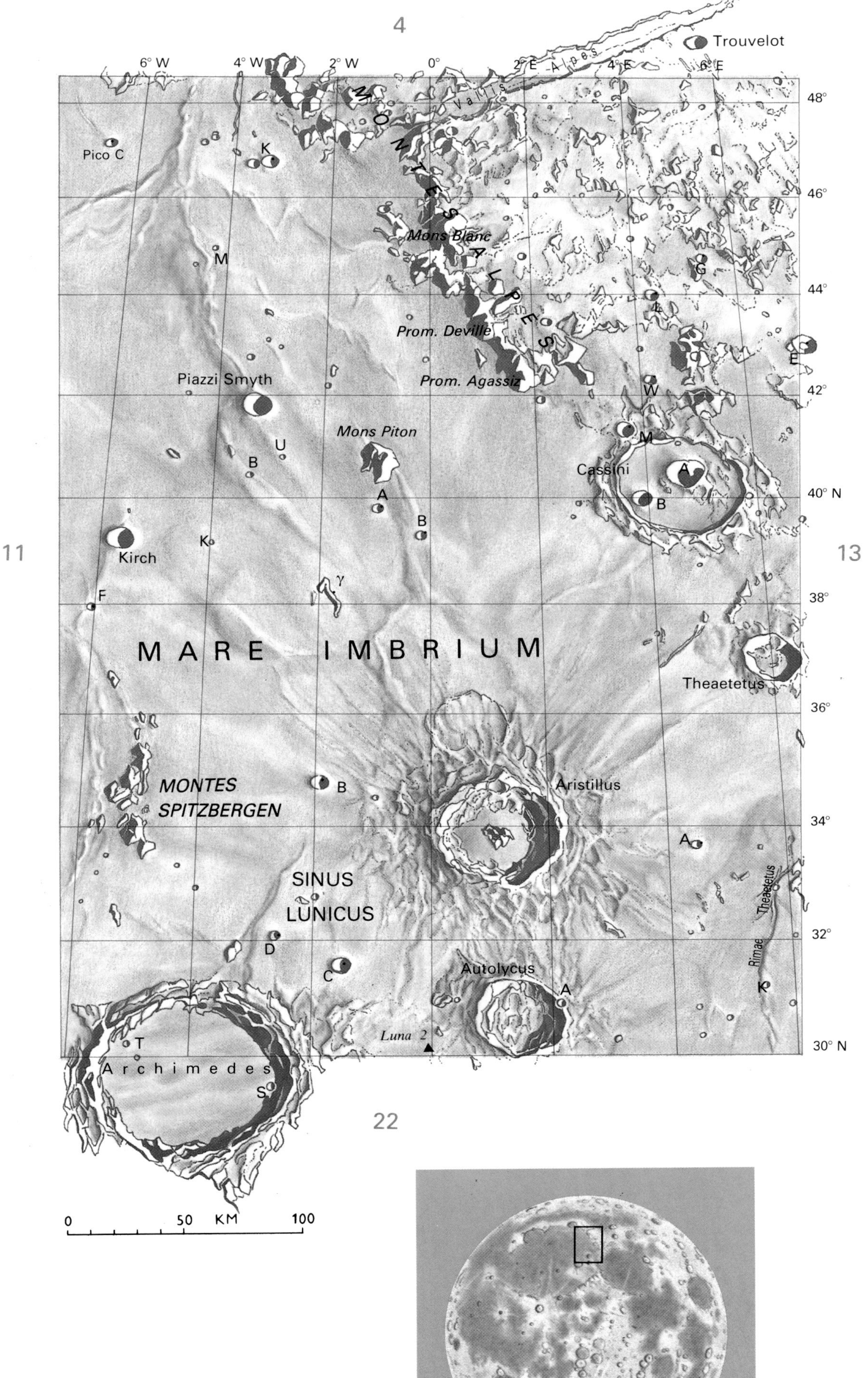
4
Trouvelot
6° W
4° W
2° W
0°
2° E
4° E
6° E
Vallis Alpes
48°
46°
44°
42°
40° N
38°
36°
34°
32°
30° N
Pico C
K
MONTES ALPES
Mons Blanc
M
G
Prom. Deville
E
Prom. Agassiz
Piazzi Smyth
W
Mons Piton
M
U
B
Cassini
A
A
B
B
11
K
Kirch
13
F
γ
MARE IMBRIUM
Theaetetus
Aristillus
MONTES SPITZBERGEN
B
A
SINUS LUNICUS
Rimae Theaetetus
D
Autolycus
C
A
K
Luna 2
T
Archimedes
S
22
0
50
KM
100

13. EUDOXUS

The left-hand part of the map is occupied by the Caucasus Mountains, which lie on the boundary of Mare Imbrium and Mare Serenitatis. The north-western part of the Mare Serenitatis with its bright rays from crater Aristillus (map 12) is situated at the base of the map. The large crater Eudoxus forms a striking companion to Aristoteles (map 5).

Alexander [40.3°N, 13.5°E] Alexander the Great of Macedon, 356–323 BC. Statesman and commander; his expeditions broadened Greek knowledge of the Earth. The city of Alexandria in Egypt, named after him, became a centre of science.
Greatly eroded walled plain (82 km).

Calippus [38.9°N, 10.7°E] Calippus, *c.* 330 BC. Greek astronomer, pupil of Eudoxus. Improved the system of homocentric spheres.
Crater (33 km/2690 m).

Calippus, Rima [37°N, 13°E]
Rille, length 40 km.

Caucasus, Montes [39°N, 9°E] Caucasus Mountains.
This mountain range, named by Mädler, is a direct continuation of the lunar Apennines, from which the Caucasus are separated by a 50 km *wide 'strait' between the M. Imbrium and the M. Serenitatis. The length of the Caucasus is about* 520 km *and the mountain range extends from the 'strait' (in the bottom left-hand corner of map 13) to the crater Eudoxus. The peaks of the Caucasus reach a height of* 6000 m *above the level of the adjacent 'seas'; from that height an observer's horizon would be* 110 km *away and he could view parts of the both maria.*

Egede [48.7°N, 10.6°E] Hans Egede, 1686–1758. Danish missionary, who worked for 15 years in Greenland.
Flooded crater with a low wall (37 km).

Eudoxus [44.3°N, 16.3°E] Eudoxus, *c.* 400–347 BC. Famous Greek astronomer, pupil of Plato, outstanding geometer. Devised a system of concentric spheres, rotating about the Earth, to explain the motions of celestial bodies.
Prominent crater with terraced walls (67 km).

Lamèch [42.7°N, 13.1°E] Felix Chemla Lamèch, 1894–1962. French astronomer and selenographer.
Crater (13 km/1460 m).

Serenitatis, Mare (Sea of Serenity).
Circular plain which has a surface area of 303 000 sq km, *the sixth largest among lunar 'seas'. It is smaller, however, than the terrestrial Caspian Sea, which covers an area of* 370 000 sq km. *The eastern part of M. Serenitatis is occupied by an extensive series of mare ridges (see map 24).*

Small craters: **Cassini C** (13.7 km/2420 m)
Eudoxus D (9.6 km/1300 m)
Linné F (5.0 km/1050 m)
Linné H (3.2 km/730 m)

5

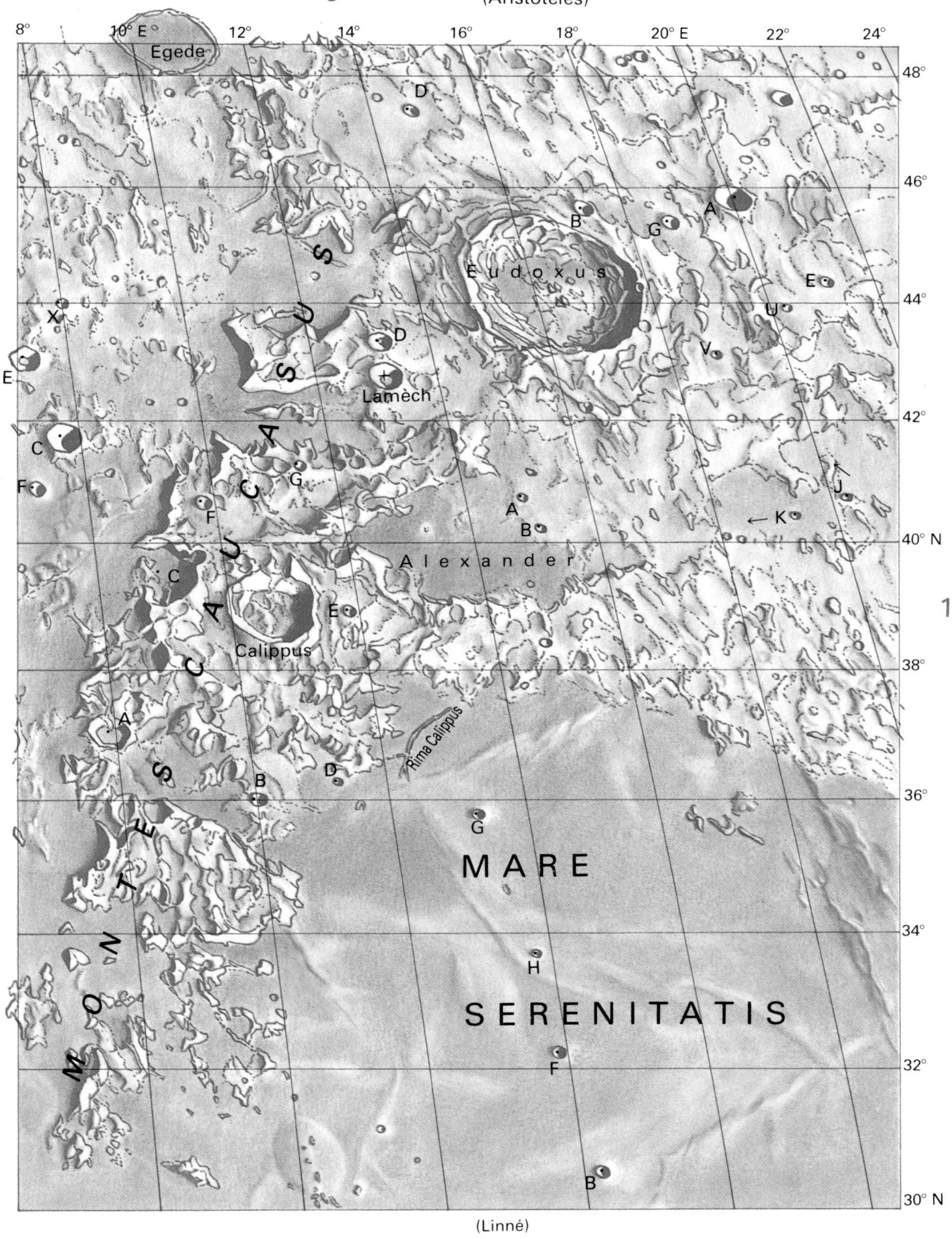

(Aristoteles)

8° 10° E 12° 14° 16° 18° 20° E 22° 24°

48° 46° 44° 42° 40° N 38° 36° 34° 32° 30° N

(Cassini)

12

14

Egede

Eudoxus

Lamèch

Alexander

Calippus

Rima Calippus

MONTES CAUCASUS

MARE SERENITATIS

(Linné)

23

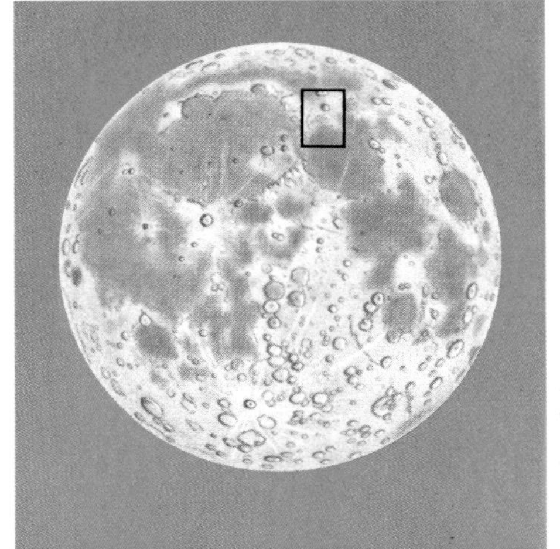

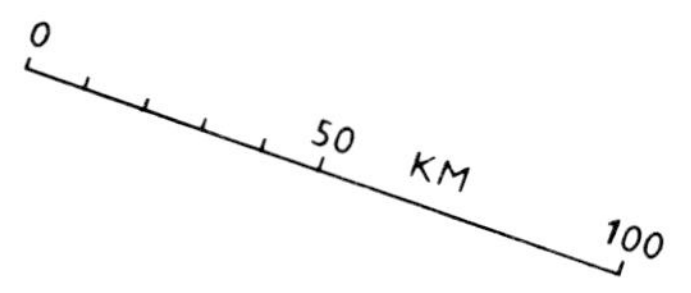

14. HERCULES

A very interesting landscape with a great variety of lunar formations. The clefts in the crater Bürg, and on the floor of the walled plain Posidonius, as well as the rilles Rimae Daniell, all demand attention. Dark areas include the lakes Lacus Mortis, Lacus Somniorum and the north-eastern edge of Mare Serenitatis.

Bürg [45.0°N, 28.2°E] Johann Tobias Bürg, 1766–1834. Austrian astronomer. Theory of the motion of the Moon.
Prominent crater in Lacus Mortis (40 km).

Bürg, Rimae [45°N, 26°E].
Rilles to the west of the crater Bürg, length up to 100 km.

Chacornac [29.8°N, 31.7°E] Jean Chacornac, 1823–1873. French astronomer, discovered six asteroids.
Crater with a disintegrated wall (51 km/1450 m).

Chacornac, Rimae
System of rilles inside Chacornac and to the south of the crater, length up to 120 km (see also map 25).

Daniell [35.3°N, 31.1°E] John Frederick Daniell, 1790–1845. English physicist and meteorologist, inventor of the hygrometer.
Oval crater (30 × 23 km/2070 m).

Daniell, Rimae [37°N, 26°E]
System of rilles, length up to 200 km.

Grove [40.3°N, 32.9°E] Sir William Robert Grove, 1811–1896. English lawyer, who carried out valuable research in physics.
Crater (28 km).

Hercules [46.7°N, 39.1°E] Hercules, hero of Greek mythology endowed with superhuman strength. Riccioli presumed that he was an astronomer who lived about 1560 BC.
Prominent crater with dark areas on its floor (69 km).

Luther (33.2°N, 24.1°E] Robert Luther, 1822–1900. German astronomer, discovered 24 minor planets.
Crater (9.5 km/1900 m).

Mason [42.6°N, 30.5°E] Charles Mason, 1730–1787. English astronomer, assistant at Greenwich Observatory.
Flooded, partly disintegrated crater (33 × 43 km).

Mortis, Lacus (Lake of Death) [45°N, 27°E] Named by Riccioli.
A formation 150 km in diameter, resembling a flooded crater, surface area 21 000 sq km. *On the floor are the rilles Rimae Bürg, mare ridges and faults.*

Plana [42.2°N, 28.2°E] Giovanni A. A. Plana, 1781–1864. Italian astronomer and mathematician.
Crater with a central peak (44 km).

Posidonius [31.8°N, 29.9°E] Posidonius, 135–51 BC. Greek philosopher, geographer and astronomer.
Prominent walled plain (95 km/2300 m).

Posidonius, Rimae
System of rilles inside the crater.

Serenitatis, Mare (Sea of Serenity). See p. 52.

Somniorum, Lacus (Lake of Dreams). Named by Riccioli.
Irregular contours, indefinite borders, surface area about 70 000 sq km.

Williams [42.0°N, 37.2°E] Arthur Stanley Williams, 1861–1938. English lawyer and a diligent observer of the planets, especially Jupiter.
Disintegrated crater (36 km).

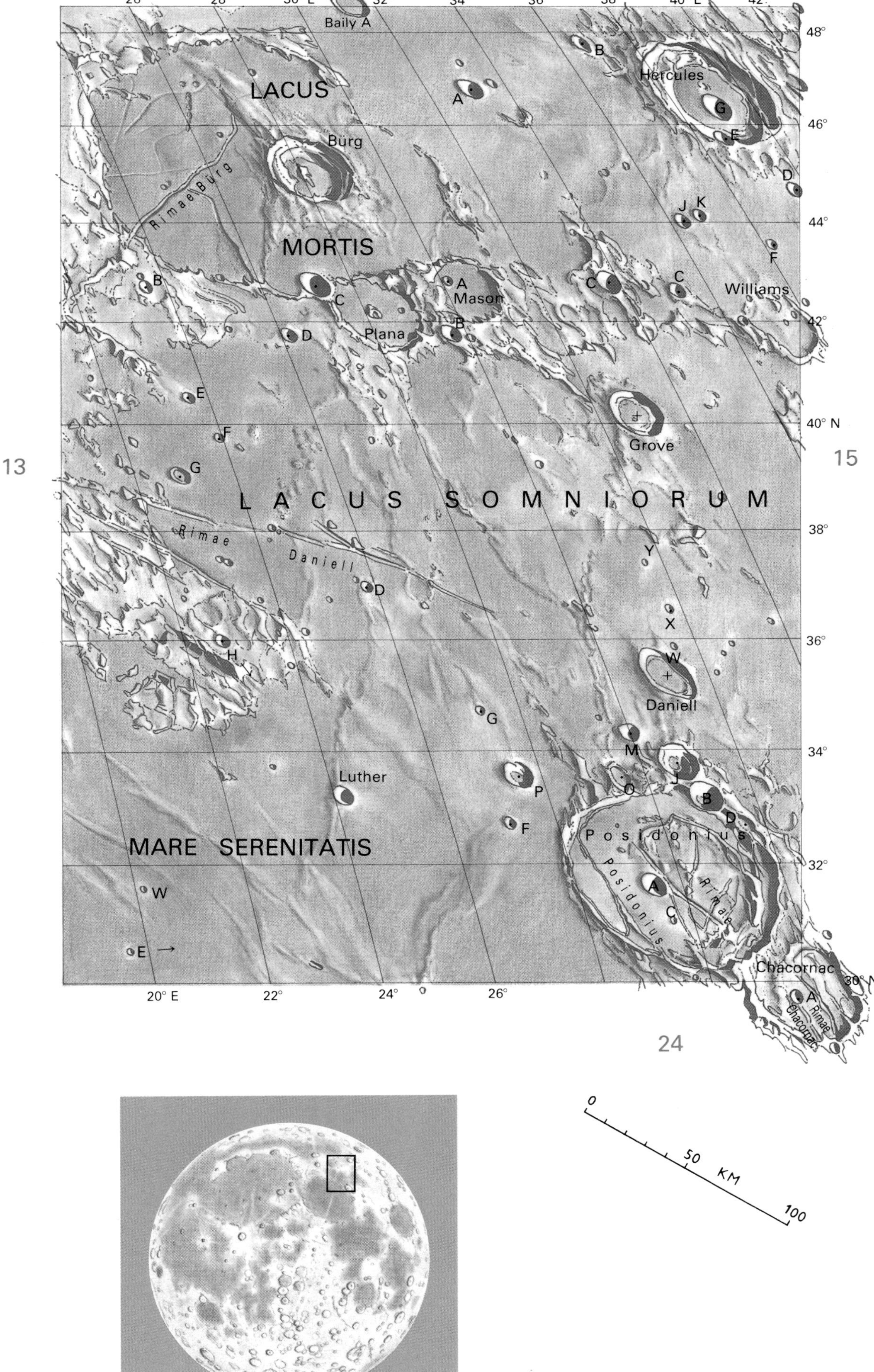
6
13
15
24
26°
28°
30° E
32°
34°
36°
38°
40° E
42°
48°
46°
44°
42°
40° N
38°
36°
34°
32°
30° N
20° E
22°
24°
26°
Baily A
LACUS
MORTIS
LACUS SOMNIORUM
MARE SERENITATIS
Bürg
Rimae Bürg
Hercules
Williams
Mason
Plana
Grove
Rimae Daniell
Daniell
Luther
Posidonius
Rimae Posidonius
Chacornac
Rimae Chacornac
0
50
KM
100

15. ATLAS

A landscape close to the north-eastern limb of the Moon. The eastern promontory of Lacus Somniorum has a wide cleft Rima G. Bond. Hills along the base of the map link up with the Taurus mountain range.

Atlas [46.7°N, 44.4°E] Atlas. According to Greek mythology, one of the Titans, standing at the western end of the Earth and bearing the sky on his shoulder. According to Riccioli, Atlas was a Moroccan king, interested in astronomy, who lived about 1580 BC.
Crater (87 km).

Atlas, Rimae
System of clefts on the floor of Atlas.

Berzelius [36.6°N, 50.9°E] Jöns J. Berzelius, 1779–1848. Swedish chemist, author of modern chemical nomenclature.
Crater (51 km).

Carrington [44.0°N, 62.1°E] Richard C. Carrington, 1826–1875. English astronomer, determined the rotation period of the Sun.
Crater (30 km).

Cepheus [40.8°N, 45.8°E] Cepheus. Mythological king of Ethiopia, whose name was also given to a constellation.
Crater (40 km).

Chevallier [44.9°N, 51.2°E] Temple Chevallier, 1794–1873. French by origin, Director of Durham Observatory in England.
Disintegrated, flooded crater (52 km).

Franklin [38.8°N, 47.7°E] Benjamin Franklin, 1706–1790. American statesman, diplomat and physicist. Invented the lightning conductor.
Prominent crater (56 km).

G. Bond [32.4°N, 36.2°E] George P. Bond, 1826–1865. American astronomer. Postulated that the rings of Saturn must be fluid.
Crater (20 km/2780 m).

G. Bond, Rima [33°N, 35°E]
Cleft, length 150 km.

Hall [33.7°N, 37.0°E] Asaph Hall, 1829–1907. American astronomer; discovered the moons of Mars.
Disintegrated, flooded crater (39 km/1140 m).

Hooke [41.2°N, 54.9°E] Robert Hooke, 1635–1703. Outstanding English physicist, experimenter and inventor.
Flooded crater (37 km).

Kirchhoff [30.3°N, 38.8°E] Gustav R. Kirchhoff, 1824–1887. German physicist, discovered the basic principles of spectroscopic analysis.
Crater (25 km/2590 m).

Maury [37.1°N, 39.6°E] (1) Matthew F. Maury, 1806–1873. American oceanographer. (2) Antonia C. Maury, 1866–1952. American astronomer, pioneer in the classification of stellar spectra.
Crater (17.6 km/3270 m).

Mercurius [46.6°N, 66.2°E] Mercury – legendary messenger of the Gods.
Crater (68 km).

Oersted [43.1°N, 47.2°E] Hans C. Oersted, 1777–1851. Danish physicist and philosopher.
Flooded crater (42 km).

Shuckburgh [42.6°N, 52.8°E] Sir George Shuckburgh, 1751–1804. British astronomer.
Crater (39 km).

Temporis, Lacus (Lake of Time) [46°N, 57°E]
Diameter about 250 km.

7

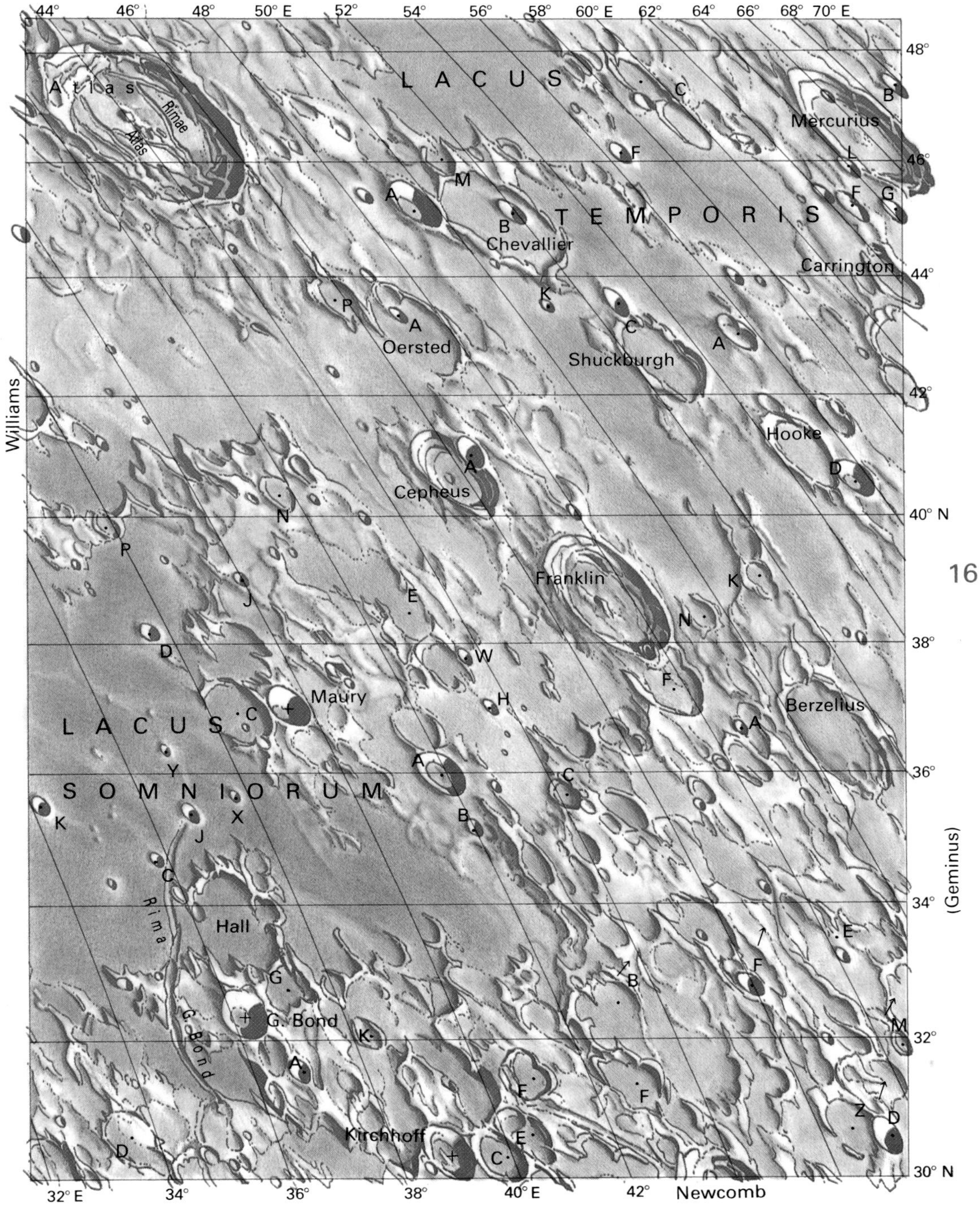

44° 46° 48° 50° E 52° 54° 56° 58° 60° E 62° 64° 66° 68° 70° E

48°
46°
44°
42°
40° N
38°
36°
34°
32°
30° N

32° E 34° 36° 38° 40° E 42°

Atlas
Rimae Atlas
LACUS TEMPORIS
Mercurius
Chevallier
Carrington
Oersted
Shuckburgh
Williams
Hooke
Cepheus
Franklin
Berzelius
Maury
LACUS SOMNIORUM
(Geminus)
Rima Hall
Hall
G. Bond
Rima G. Bond
Kirchhoff
Newcomb

14

16

25

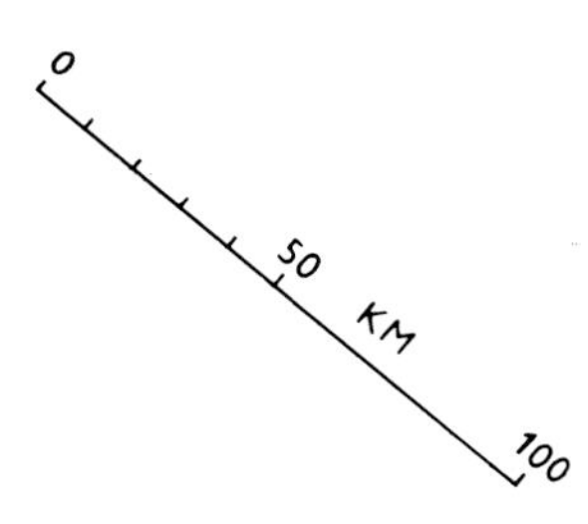

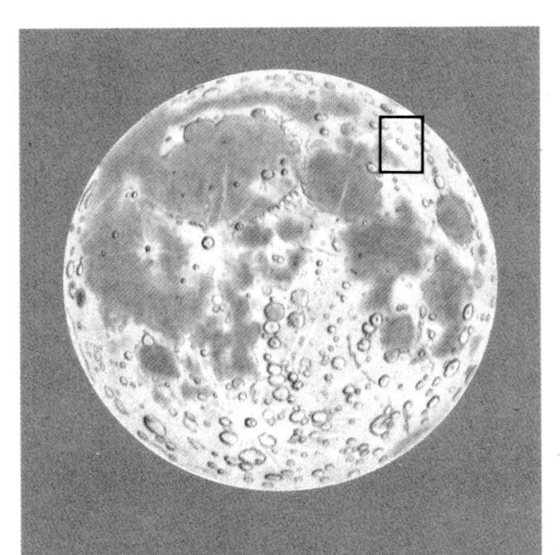

16. GAUSS

The north-eastern limb of the Moon, containing the large walled plain Gauss and prominent craters Geminus, Berosus and Hahn. To the east of the crater Schumacher is a dark spot, Lacus Spei.

Beals [37.3°N, 86.5°E] Carlyle F. Beals, 1899–1979. Canadian astronomer.
Crater (48 km).

Bernoulli [35.0°N, 60.7°E] (1) Jacques Bernoulli, 1654–1705. (2) Jean Bernoulli, 1667–1748. Two brothers, Swiss mathematicians, Dutch by origin.
Crater (47 km).

Berosus [33.5°N, 69.9°E] Berosus of Chaldea, third century BC. Babylonian priest, historian and astronomer; noted the synchronous rotation of the Moon.
Flooded crater (74 km).

Boss [45.8°N, 89.2°E] Lewis Boss, 1846–1912. American astronomer, compiler of positional star catalogues.
Crater (47 km).

Burckhardt [31.1°N, 56.5°E] Johann K. Burckhardt, 1773–1825. German astronomer, worked in the time service.
Crater (57 km).

Gauss [35.9°N, 79.1°E] Karl Friedrich Gauss, 1777–1855. Famous German mathematician, physicist, geodesist and theoretical astronomer; Director of the Observatory in Göttingen. Invented the magnetometer.
Walled plain (177 km).

Geminus [34.5°N, 56.7°E] Geminus, *c.* 70 BC. Greek astronomer.
Prominent crater (86 km).

Hahn [31.3°N, 73.6°E] Friedrich, Graf von Hahn, 1741–1805. German amateur astronomer, diligent observer.
Crater (84 km).

Messala [39.2°N, 59.9°E] Ma-sa-Allah (or Mashalla), died *c.* 815 AD. Jewish astronomer and astrologer, author of textbooks which were still used in Europe during the Middle Ages.
Walled plain (124 km).

Riemann [39.5°N, 87.2°E] Georg Bernhard Riemann, 1826–1866. German mathematician, developed 'Riemannian geometry' and the fundamental mathematics (calculus) for the geometry used in modern physics.
Remains of a walled plain (110 km).

Schumacher [42.4°N, 60.7°E] Heinrich Christian Schumacher, 1780–1850. Danish-born German astronomer; founder of the specialist periodical *Astronomische Nachrichten.*
Flooded, eroded crater (61 km).

Spei, Lacus (Lake of Hope) [43°N, 65°E]
Diameter 80 km.

Zeno [45.2°N, 72.9°E] Zeno, *c.* 335–263 BC. Greek philosopher and astronomer, who correctly explained the reasons for solar and lunar eclipses.
Crater (65 km).

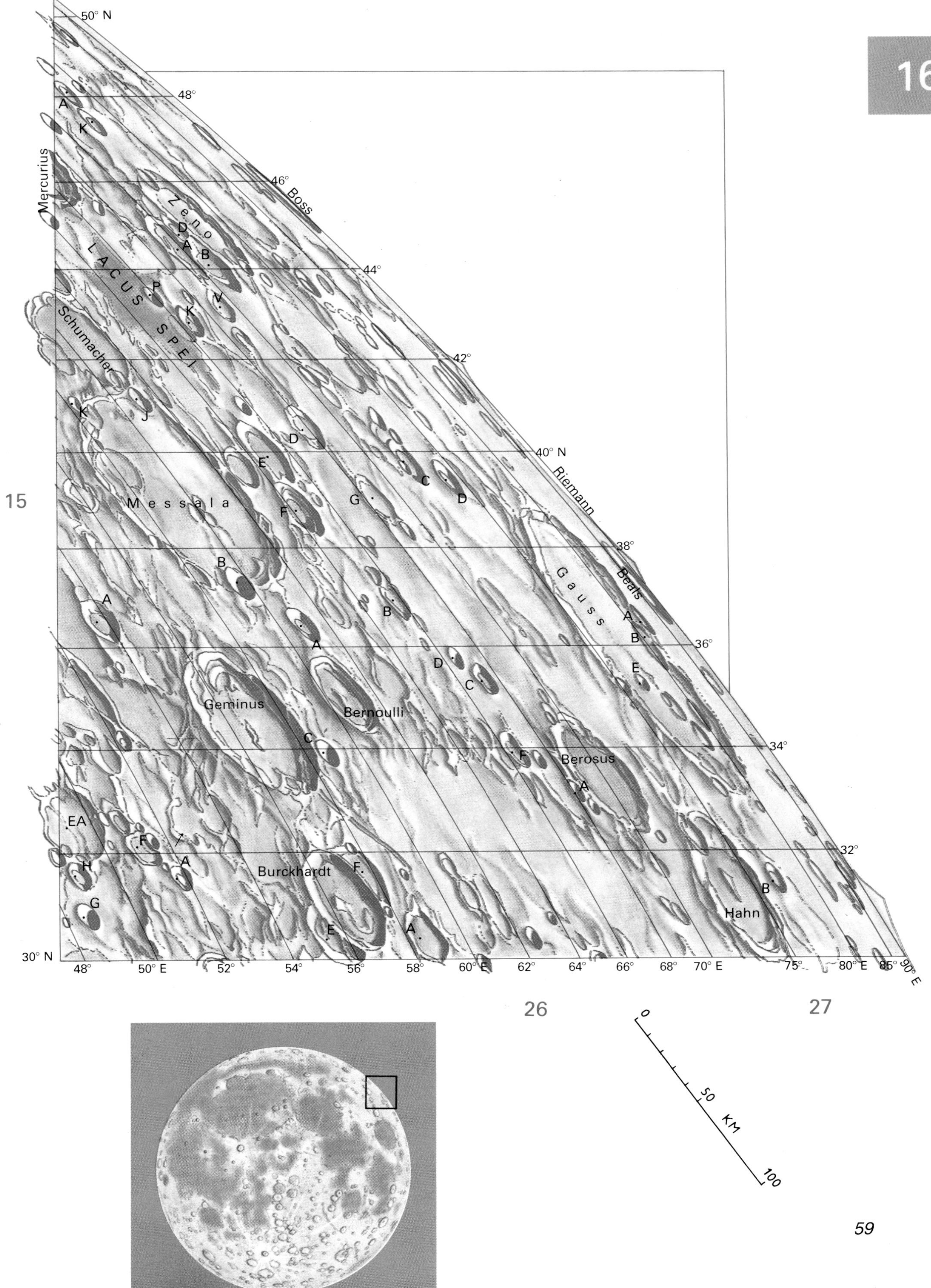
16
15
26
27
50° N
48°
46°
44°
42°
40° N
38°
36°
34°
32°
30° N
48°
50° E
52°
54°
56°
58°
60° E
62°
64°
66°
68°
70° E
75°
80° E
85°
90° E
Mercurius
Boss
Zeno
LACUS SPEI
Schumacher
Messala
Riemann
Gauss
Beals
Geminus
Bernoulli
Berosus
Burckhardt
Hahn
0
50
KM
100

17. STRUVE

Western part of Oceanus Procellarum and the western limb of the Moon. Bright rays radiate from the crater Olbers. The area contains the soft-landing site of the Soviet automatic probe *Luna 13*.

Balboa [19.1°N, 83.2°W] Vasco N. de Balboa, *c.* 1475–1517. Spanish explorer and conquistador, the first European to sight and reach the Pacific Ocean.
Flooded crater (70 km).

Bartels [24.5°N, 89.8°W] Julius Bartels, 1899–1964. German geophysicist.
Crater (55 km).

Briggs [26.5°N, 69.1°W] Henry Briggs, 1556–1630. English mathematician, developed the use of Napierian logarithms.
Crater (37 km).

Dalton [17.1°N, 84.3°W] John Dalton, 1766–1844. English chemist and physicist.
Crater (61 km).

Eddington [21.5°N, 71.8°W] Sir Arthur S. Eddington, 1882–1944. Prominent English astrophysicist and mathematician; internal structure of stars, relativity.
Remains of a flooded walled plain (125 km).

Einstein [16.6°N, 88.5°W] Albert Einstein, 1879–1955. Of German origin. One of the world's greatest theoretical physicists; author of the general and special theories of relativity.
Walled plain (170 km) *with a central crater* (45 km), *in a libration zone partly beyond* 90°W.

Krafft [16.6°N, 72.6°W] Wolfgang Ludwig Krafft, 1743–1814. Astronomer and physicist of German origin, worked all his life in St Petersburg.
Crater with flooded floor (51 km).

Krafft, Catena [15°N, 72°W]
Crater chain, length 60 km.

Russell [26.5°N, 75.4°W] (1) John Russell, 1745–1806. British painter, amateur astronomer and selenographer. (2) Henry Norris Russell, 1877–1957. American astronomer, co-author of Hertzsprung-Russell diagram.
Remains of walled plain (103 km).

Seleucus [21.0°N, 66.6°W] Seleucus, *c.* 150 BC. Babylonian astronomer, defender of the heliocentric theory.
Prominent crater (43 km).

Struve [23.0°N, 76.6°W] (1) Friedrich G. Wilhelm von Struve, 1793–1864. German-Russian astronomer. Director of Pulkovo Observatory. Observed double stars and measured the parallax of stars. (2) Otto von Struve, 1819–1905. Son of the former and Director of Pulkovo Observatory. (3) Otto Struve, 1897–1963. Russian-born American astrophysicist, grandson of Friedrich (1).
Remains of a flooded walled plain (170 km).

Voskresenskiy [28.0°N, 88.1°W] Leonid A. Voskresenskiy, 1913–1965. Eminent Soviet specialist in the field of rocket technology.
Flooded crater (50 km).

8

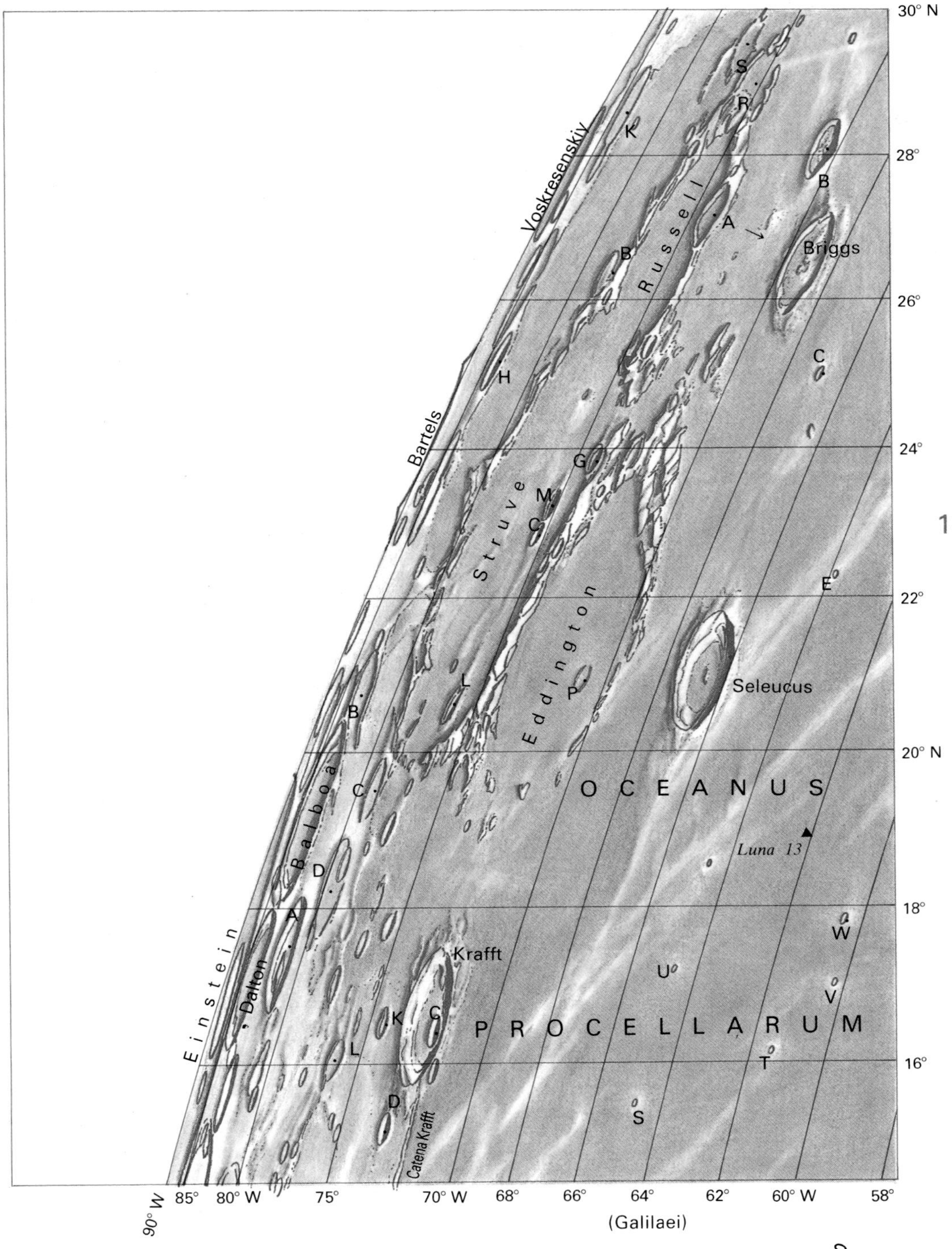

30° N
28°
26°
24°
22°
20° N
18°
16°

Voskresenskiy
K
S
R
Russell
A
B
Briggs
C
H
Bartels
G
M
C
Struve
Eddington
E
L
P
Seleucus
B
C
Balboa
D
OCEANUS
Luna 13
A
Einstein
Dalton
Krafft
W
U
V
K
C
L
PROCELLARUM
T
D
S
Catena Krafft

90° W
85°
80° W
75°
70° W
68°
66°
64°
62°
60° W
58°
(Galilaei)

18

28

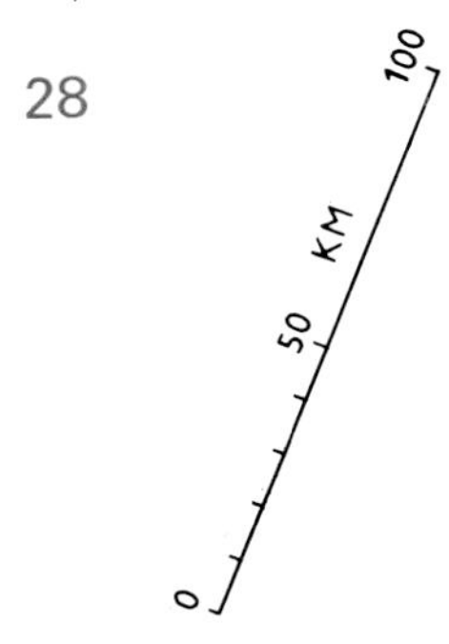

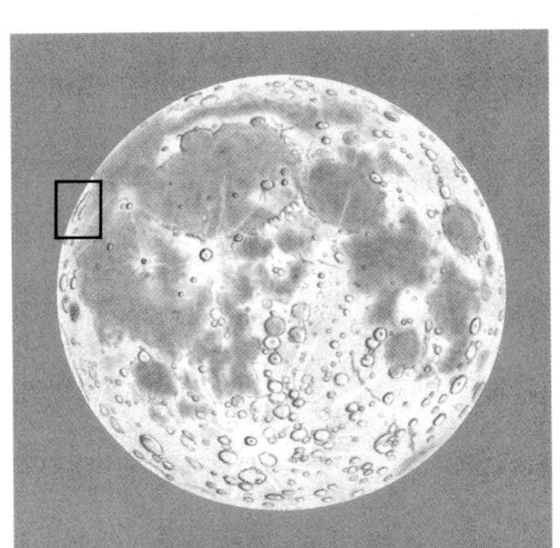

18. ARISTARCHUS

In this part of Oceanus Procellarum there are exceptionally interesting formations; these include Schröter's Valley, which is visible through even quite a small telescope, and the long sinuous rilles Rima Marius and Rimae Aristarchus. Herodotus Omega is a lunar dome and numerous domes can be found in the vicinity of the crater Marius (see also map 29). This area is dominated by the crater Aristarchus.

Aristarchus [23.7°N, 47.4°W] Aristarchus, *c.* 310–230 BC. Greek astronomer from Samos; the first to teach that the Earth revolves around the Sun and rotates on its axis.
Extraordinarily bright crater, visible also on the night side of the Moon in Earthshine; centre of bright ray system; the crater is believed to have originated about 450 million years ago (40 km/3000 m).

Aristarchus, Rimae [28°N, 47°W].
System of rilles, length 120 km.

Agricola, Montes [29°N, 54°W] Georgius Agricola, 1494–1555. German physician and naturalist.
Elongated mountain range, length 160 km.

Burnet, Dorsa [27°N, 57°W] Thomas Burnet, 1635–1715. English naturalist.
System of wrinkle ridges, length 200 km.

Freud [25.8°N, 52.3°W] Sigmund Freud, 1856–1939. Austrian physician and psychoanalyst.
Crater (3 km).

Golgi (Schiaparelli D) [27.8°N, 60.0°W] Camillo Golgi, 1843–1926. Italian physician, Nobel Laureate.
Small crater (5 km).

Herodotus [23.2°N, 49.7°W] Herodotus, *c.* 485–425 BC. Greek historian from Halicarnassus (Asia Minor), called the 'Father of History'.
Flooded crater (35 km).

Herodotus, Mons [27°N, 53°W]
Mountain, base diameter 5 km.

Marius, Rima [17°N, 49°W].
Typical sinuous rille, named after the nearby crater Marius (map 29). The rille starts about 25 km *north-west of the crater Marius C, where it is about* 2 km *wide, winds north and by the crater Marius B then turns west and narrows down to* 1 km. *The rille ends about* 40 km *west of the crater Marius P, where its width is only* 500 m. *Total length of the rille is about* 250 km.

Niggli, Dorsum (Niggli's Ridge) [29°N, 52°W]. Paul Niggli, 1888–1953. Swiss naturalist.
Mare ridge, length 50 km.

Raman (Herodotus D) [27.0°N, 55.1°W] Chandrasekhara V. Raman, 1888–1970. Indian physicist.
Crater (11 km).

Schiaparelli [23.4°N, 58.8°W] Giovanni V. Schiaparelli, 1835–1910. Italian astronomer. Discovered relationship between meteor showers and comets. In 1877 discovered the *canali* (so-called 'canals') on Mars; developed terminology used in Mars charts.
Crater (24 km).

Schröteri, Vallis [26°N, 51°W].
The largest sinuous valley on the Moon, named after the German selenographer Johannes Schröter (crater, see p. 91). The valley starts 25 km *north of the crater Herodotus, and resembles a dry river bed with numerous meanders. Starting at a crater* 6 km *in diameter, the valley widens to* 10 km *to form a shape that observers have called the 'Cobra's head'. From this it gradually narrows to* 500 m *and, still narrowing, terminates at a* 1000 m *high precipice on the edge of a tetragonal 'continent'. The total length of this flat-floored valley is* 160 km *and its maximum depth is about* 1000 m. *Another sinuous rille on the floor of the valley cannot be seen from the Earth.*

Toscanelli (Aristarchus C) [27.9°N, 47.5°W] Paolo Dal Pozza Toscanelli, 1397–1482. Italian physician and cartographer.
Crater (7 km).

Toscanelli, Rupes [27°N, 47°W]
Fault, length 70 km.

Väisälä (Aristarchus A) [25.9°N, 47.8°W] Yrjo Väisälä, 1891–1971. Finnish astronomer.
Crater (8 km).

Zinner (Schiaparelli B) [26.6°N, 58.8°W] Ernst Zinner, 1886–1970. German astronomer.
Crater (4 km).

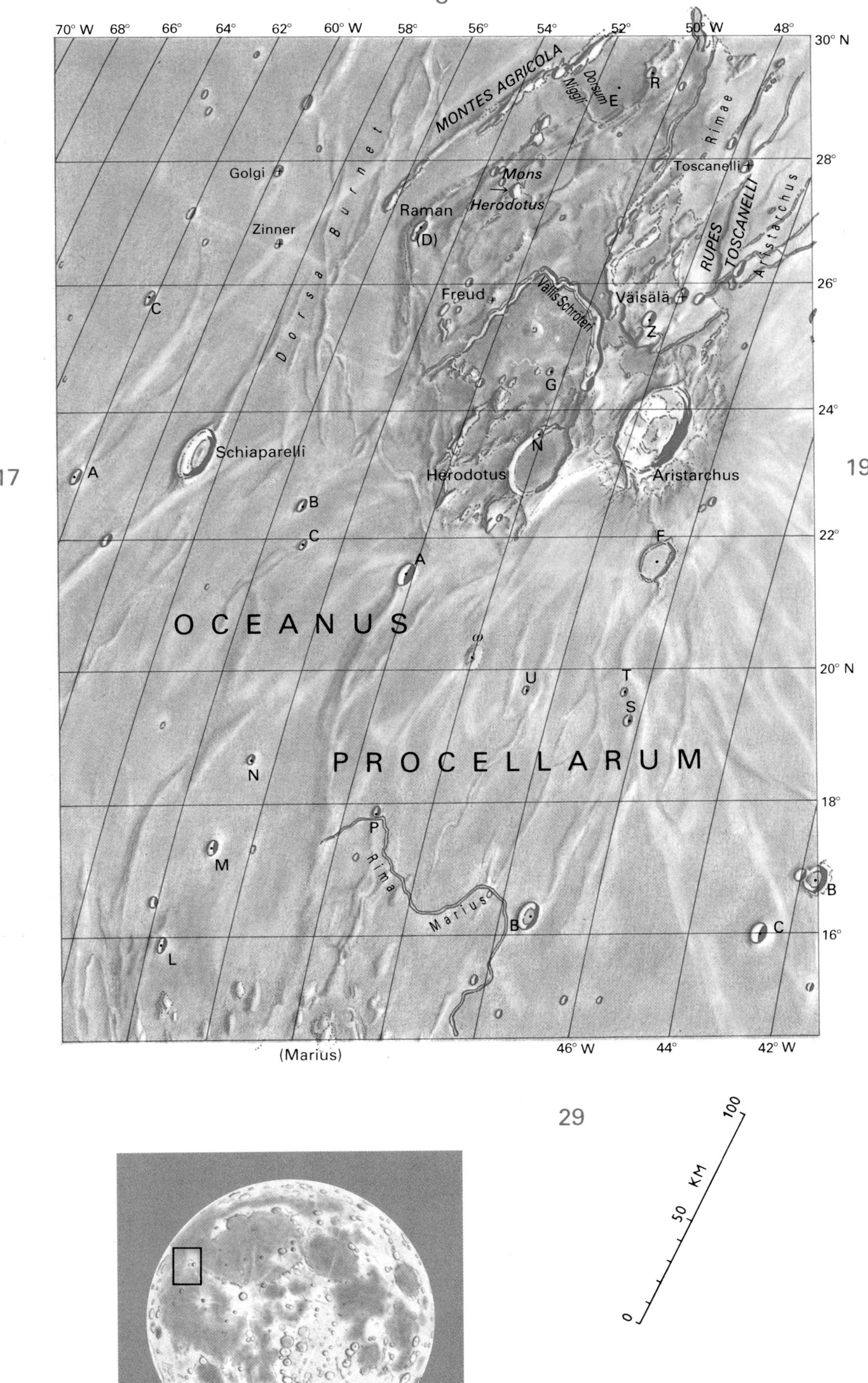
8
17
19
29
70° W
68°
66°
64°
62°
60° W
58°
56°
54°
52°
50° W
48°
30° N
28°
26°
24°
22°
20° N
18°
16°
46° W
44°
42° W
MONTES AGRICOLA
Dorsum
Niggli
E
R
Rimae
Toscanelli
Golgi
Mons
Herodotus
Raman
(D)
Zinner
Dorsa Burnet
RUPES
TOSCANELLI
Aristarchus
Freud
Vallis Schröteri
Väisälä
C
Z
G
Schiaparelli
A
N
Herodotus
Aristarchus
B
C
F
A
OCEANUS
ω
U
T
S
N
PROCELLARUM
P
Rima
Marius
M
B
B
C
L
(Marius)
0
50
KM
100

19. BRAYLEY

South-western part of Mare Imbrium with numerous wrinkle ridges and a network of bright streaks radiating mainly from the craters Copernicus and Aristarchus. An interesting object for larger telescopes is the system of sinuous rilles near the crater Prinz.

Ångström [29.9°N, 41.6°W] Anders J. Ångström, 1814–1874. Swedish physicist.
Crater (9.8 km/2030 m).

Arduino, Dorsum [26°N, 36°W] Giovanni Arduino, 1713–1795. Italian naturalist.
Wrinkle ridge, length 110 km.

Argand, Dorsa [28°N, 40°W] Emile Argand, 1879–1940. Swiss naturalist.
System of wrinkle ridges, length 150 km.

Aristarchus, Rimae. See map 18.

Artsimovich (Diophantus A) [27.6°N, 36.6°W] Lev A. Artsimovich, 1909–1973. Soviet physicist.
Crater (9 km/860 m).

Bessarion [14.9°N, 37.3°W] Johannes Bessarion, 1389–1472. Greek scholar.
Crater (10.2 km/2000 m).

Brayley [20.9°N, 36.9°W] Edward W. Brayley, 1801–1870. English professor of physical geography and meteorology at the Royal Institution, London.
Crater (14.5 km/2840 m).

Brayley, Rima [23°N, 36°W]
Narrow rille, not observable with small telescopes, length 240 km.

Delisle. See map 9.

Delisle, Mons [29°N, 36°W]
Mountain, base diameter 30 km.

Diophantus [27.6°N, 34.3°W] Diophantus, *c.* fourth century AD. Greek mathematician from Alexandria; established principle for solving mathematical equations.
Crater (18.5 km/2970 m).

Diophantus, Rima [29°N, 33°W]
Narrow rille, length 140 km.

Fedorov [28.2°N, 37.0°W] A. P. Fiodorov, 1872–1920. Russian specialist in rocketry.
Crater (7 km).

Harbinger, Montes (Harbinger Mountains) [27°N, 41°W]
Group of isolated mountains at the edge of Mare Imbrium, area of about 90 sq km.

Imbrium, Mare (Sea of Rains) See p. 48.

Ivan (Prinz B) [26.9°N, 43.3°W] Russian male name.
Crater (4 km).

Jehan (Euler K) [20.7°N, 31.9°W] Turkish female name.
Crater (4.8 km/730 m).

Krieger [29.0°N, 45.6°W] Johann N. Krieger, 1865–1902. German selenographer; drawings of the details of lunar surface features on enlarged photographs.
Flooded crater (22 km).

Louise [28.5°N, 34.2°W] German female forename.
Crater (1.5 km).

Natasha (Euler P) [20.0°N, 31.3°W] Russian female name.
Crater (12 km/290 m).

Prinz [25.5°N, 44.1°W] Wilhelm Prinz, 1857–1910. German selenographer. Comparative studies of lunar and terrestrial surfaces.
Remains of a flooded crater (47 km/1010 m).

Prinz, Rimae [27°N, 43°W].
System of sinuous rilles, observable with larger telescopes (length of rilles up to 80 km).

Procellarum, Oceanus (Ocean of Storms). See p. 84.

Rocco (Krieger D) [28.9°N, 45.0°W] Italian male name.
Crater (4.4 km/880 m).

Ruth [28.7°N, 45.1°W] Jewish female forename.
Crater (3 km).

T. Mayer [15.6°N, 29.1°W] Tobias Mayer, 1723–1762. German selenographer, author of an accurate map of the Moon.
Crater (33 km/2920 m).

Van Biesbroeck (Krieger B) [28.7°N, 45.6°W] George A. Van Biesbroeck, 1880–1974. Belgian-born American astronomer.
Crater (10 km).

Vera (Prinz A) [26.3°N, 43.7°W] Latin female forename.
Crater (4.9 km/180 m): *the longest rille of Rimae Prinz originates at this crater.*

Vinogradov, Mons (Euler, Mons) [22.4°N, 43.4°W] Alexander P. Vinogradov, 1895–1975. Soviet geochemist.
Group of mountains, base diameter 25 km.

9

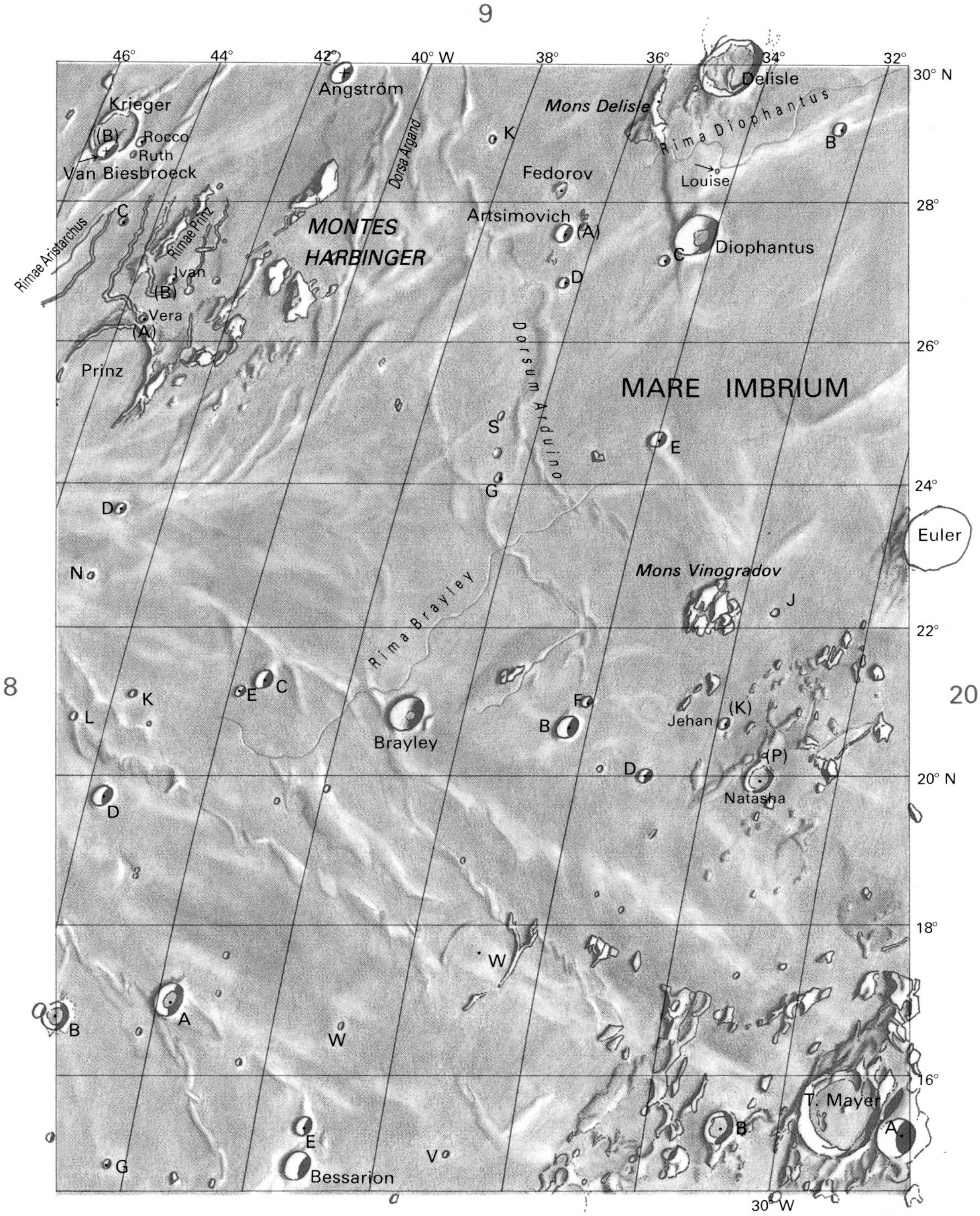

46° 44° 42° 40° W 38° 36° 34° 32°

30° N

Angström
Krieger
(B)
Rocco
Ruth
Van Biesbroeck
Mons Delisle
Delisle
Rima Diophantus
B
K
Dorsa Argand
Fedorov
Louise
28°
C
Rimae Aristarchus
Rimae Prinz
MONTES HARBINGER
Artsimovich
(A)
C
Diophantus
Ivan
(B)
D
Vera
(A)
Dorsum Arduino
26°
Prinz
MARE IMBRIUM
S
E
G
24°
D
Euler
N
Mons Vinogradov
J
Rima Brayley
22°
18
K
E
C
F
20
L
Brayley
B
Jehan
(K)
(P)
D
D
Natasha
20° N
18°
W
B
A
W
16°
T. Mayer
B
A
E
V
G
Bessarion
30° W

30

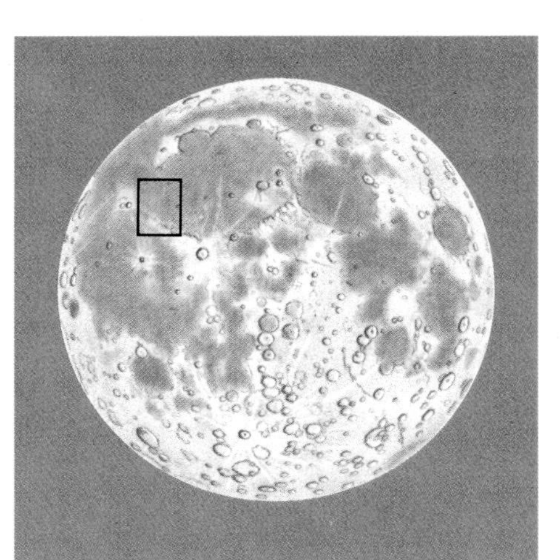

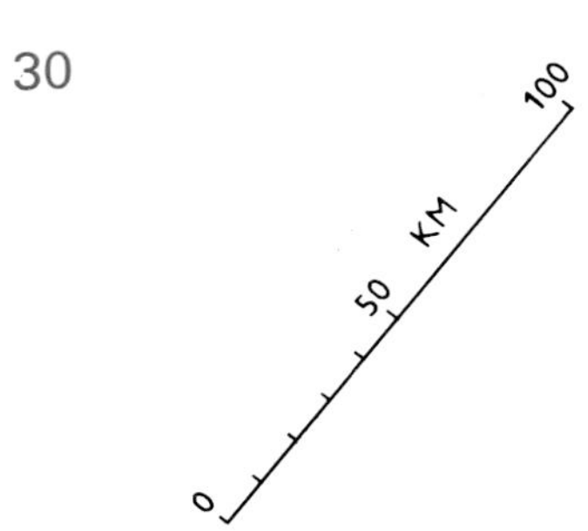

20. PYTHEAS

The southern part of Mare Imbrium, bordered by the Carpathian Mountains (see also map 31). A complex pattern of bright rays in this area originate from the crater Copernicus (see map 31). Interesting objects for observation include a crater chain north of the crater Stadius M at the bottom right of the map, and the 'ghost' crater Lambert R, visible only when close to the terminator.

Artemis [25.0°N, 25.4°W] Greek goddess, sister of Apollo.
Crater (2 km).

Carpatus, Montes (Carpathian Mountains) [15°N, 25°W] Mädler's name for a lunar mountain range which follows the southern border of Mare Imbrium.
Length about 400 km.

Caventou (La Hire D) [29.8°N, 29.4°W] Joseph B. Caventou, 1795–1877. French chemist and pharmacologist.
Crater (3 km/400 m).

Draper [17.6°N, 21.7°W] Henry Draper, 1837–1882. American astronomer, one of the pioneers of astrophotography and spectroscopy, photographed the Moon and spectra of the stars; the first to photograph the great nebula in Orion.
Crater (8.8 km/1740 m).

Euler [23.3°N, 29.2°W] Leonhard Euler, 1707–1783. Swiss mathematician; worked in the fields of pure and applied mathematics and celestial mechanics.
Crater (28 km/2240 m).

Imbrium, Mare (Sea of Rains). See p. 48.

La Hire, Mons [28°N, 25°W] Philippe de La Hire, 1640–1718. French mathematician, surveyor and astronomer.
Isolated mountain massif (10 × 20 km).

Lambert [25.8°N, 21.0°W] Johann H. Lambert, 1728–1777. German mathematician and astronomer.
Prominent crater with terraced walls (30 km/2690 m).

Pytheas [20.5°N, 20.6°W] Pytheas of Massalia, *c.* 350 BC. Greek navigator who sailed far to the north of Britain. First Greek to associate the tides with the Moon.
Crater with a sharp rim and a hilly floor (20 km/2530 m).

Stille, Dorsa [27°N, 19°W] Hans Stille, 1876–1966. German naturalist.
System of wrinkle ridges, length 80 km.

Verne [24.9°N, 25.3°W] Latin male forename.
Crater (2 km).

Zirkel, Dorsum [29°N, 24°W] Ferdinand Zirkel, 1838–1912. German geologist and mineralogist.
Wrinkle ridge, length 210 km.

Small craters: **Draper C** (7.8 km/1610 m)
Pytheas A (6.0 km/1180 m)
Pytheas D (5.2 km/370 m)
Pytheas G (3.4 km/490 m)

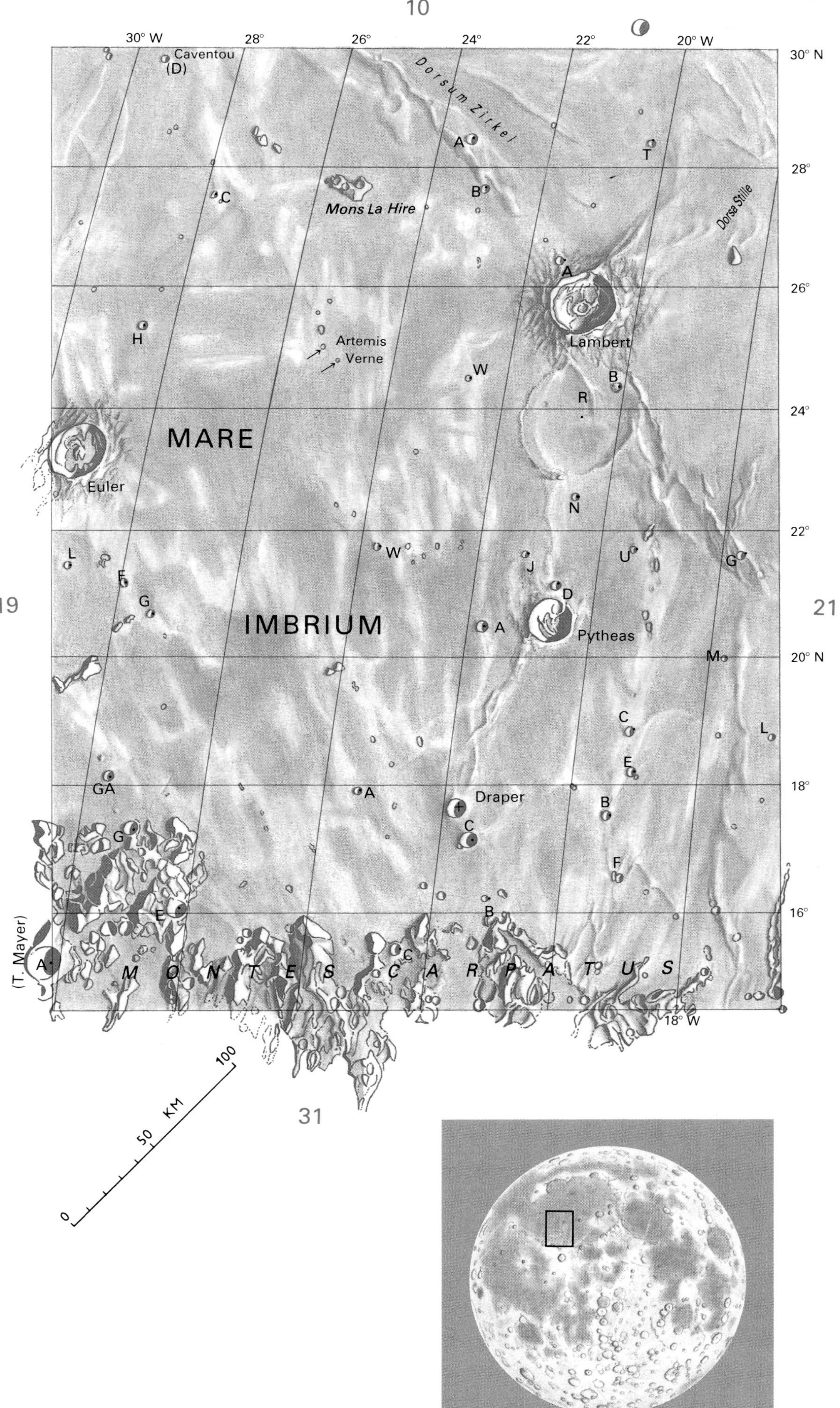
10
30° W
28°
26°
24°
22°
20° W
30° N
28°
26°
24°
22°
20° N
18°
16°
18° W
Caventou
(D)
Dorsum Zirkel
A
T
B
C
Mons La Hire
Dorsa Stille
A
Lambert
H
Artemis
Verne
W
B
R
MARE
Euler
N
L
W
J
U
G
F
G
D
19
IMBRIUM
A
Pytheas
21
M
C
L
E
GA
A
Draper
B
C
G
F
E
B
(T. Mayer)
A
C
M O N T E S C A R P A T U S
100
50 KM
0
31

21. TIMOCHARIS

The south-eastern part of Mare Imbrium and south-western promontory of the Apennines, terminated by the crater Eratosthenes. Bright rays from the crater Copernicus are visible on the surface of Mare Imbrium under high illumination. A system of wrinkle ridges and a small dome can be seen under oblique illumination about 15 km south-south-east of the crater Beer.

Apenninus, Montes (Apennines). See map 22.

Bancroft (Archimedes A) [28.0°N, 6.4°W] W. D. Bancroft, 1867–1953. American chemist.
Crater (13.1 km/2490 m).

Beer [27.1°N, 9.1°W] Wilhelm Beer, 1797–1850. German selenographer, collaborator of Mädler, with whom he published a map of the Moon and a monograph *Der Mond* (1837).
Circular crater with a sharp rim (10.2 km/1650 m).

Eratosthenes [14.5°N, 11.3°W] Eratosthenes, *c.* 275–195 BC. Greek mathematician, geographer and astronomer; the first to determine the circumference of the Earth.
Prominent crater with large terraced walls and central peaks (58 km/3570 m).

Feuillée [27.4°N, 9.4°W] Louis Feuillée, 1660–1732. French naturalist, Director of Marseilles Observatory.
Circular crater with a sharp rim (9.5 km/1810 km).

Heinrich (Timocharis A) [24.8°N, 15.3°W] Vladimír Heinrich, 1884–1965. Czechoslovak astronomer.
Crater (7.4 km/1420 m).

Higazy, Dorsum [28°N, 17°W] Riad Higazy, 1919–1967. Egyptian naturalist.
Wrinkle ridge, length 60 km.

Imbrium, Mare (Sea of Rains). See p. 48.

Macmillan (Archimedes F) [24.2°N, 7.8°W] William Duncan Macmillan, 1871–1948. American mathematician and astronomer.
Crater (7.5 km/360 m).

Pupin (Timocharis K) [23.8°N, 11.0°W] Mihajlo Pupin, 1858–1935. Yugoslav physicist, worked in the USA.
Crater (2 km/400 m).

Sampson [29.7°N, 16.5°W] Ralph Allen Sampson, 1866–1939. British astronomer; Astronomer Royal for Scotland.
Crater (1.5 km).

Timocharis [26.7°N, 13.1°W] Timocharis, *c.* 280 BC. Greek astronomer of the Alexandrian school.
Prominent crater with a sharp rim and terraced walls (34 km/3110 m).

Timocharis, Catena [29°N, 13°W]
Narrow crater chain, length about 50 km.

Wallace [20.3°N, 8.7°W] Alfred R. Wallace, 1823–1913. English naturalist and explorer.
Remains of flooded crater (26 km).

Wolff, Mons [17°N, 7°W] Christian von Wolff, 1679–1754. German philosopher and mathematician.
Mountain massif in the south-west promontory of the Apennines, diameter 35 km.

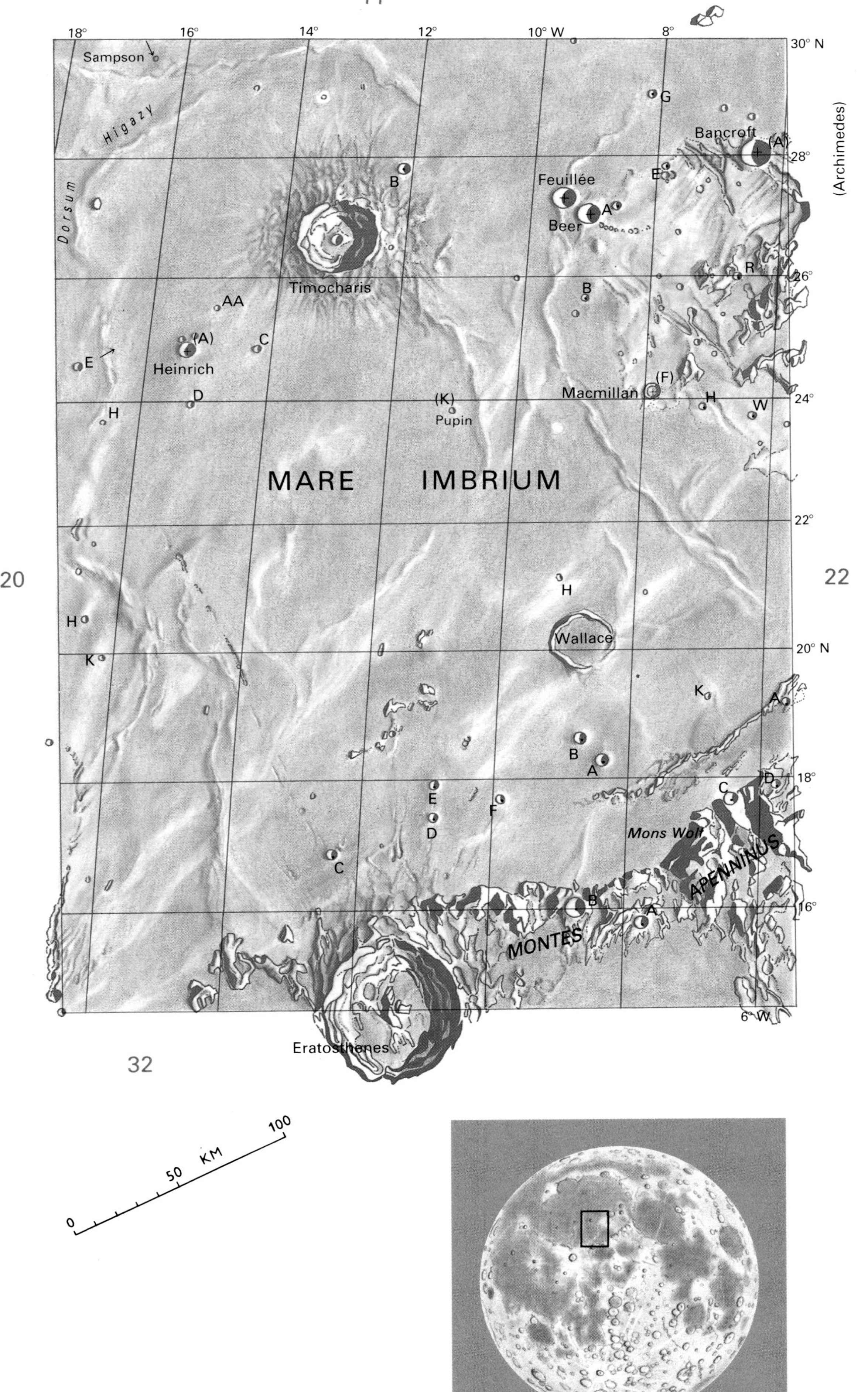
11
18°
16°
14°
12°
10° W
8°
30° N
Sampson
Higazy
Dorsum
G
Bancroft
(A)
28°
(Archimedes)
B
Feuillée
E
A
Beer
Timocharis
R
26°
B
AA
(A)
C
E
Heinrich
(F)
Macmillan
D
(K)
Pupin
H
H
W
24°
MARE IMBRIUM
22°
20
22
H
H
Wallace
20° N
K
K
A
B
A
18°
C
D
E
F
D
C
Mons Wolf
APENNINUS
MONTES
B
A
16°
6° W
Eratosthenes
32
0
50
KM
100

22. CONON

The largest lunar mountain range, the Apennines, and the prominent crater Archimedes dominate this part of the Moon close to the prime meridian. It was at the foot of the Apennines, close to Rima Hadley, that the *Apollo 15* expedition landed.

Ampère, Mons [19°N, 4°W] André M. Ampère, 1775–1836. French physicist. The unit of electric current bears his name.
Mountain massif in the central part of Apennines, length 30 km.

Apenninus, Montes (Apennines) [20°N, 3°W] Name given by Hevelius to the impressive mountain range on the south-east border of Mare Imbrium.
The Apennines form part of the wall of Mare Imbrium into which they descend relatively steeply (about 30°). The slopes of the Apennines towards Mare Vaporum are gradual. The height of some mountain peaks exceeds 5000 m, *the length of the range is* 600 km.

Aratus [23.6°N, 4.5°E] Aratus, *c.* 315–245 BC. Popular Greek poet. Author of the oldest description of 48 ancient constellations.
Crater (10.6 km/1860 m).

Archimedes [29.7°N, 4.0°W] Archimedes, *c.* 287–212 BC. Greek mathematician and physicist of Syracuse. He discovered the principle of hydrostatic equilibrium.
Very prominent flooded crater with terraced walls (83 km/2150 m).

Archimedes, Montes [26°N, 5°W]
Mountain range south of Archimedes, extending over an area about 140 km *in diameter.*

Archimedes, Rimae [27°N, 4°W]
System of rilles south-east of Archimedes, length approximately 150 km.

Bancroft (Archimedes A). See map 21.

Běla [24.7°N, 2.3°E] Slavic female forename.
Elongated crater (11 × 2 km) *at the beginning of Rima Hadley.*

Bradley, Mons [22°N, 1°E] James Bradley, 1692–1762. English astronomer, discovered the aberration of the light of stars and the effect known as nutation.
Mountain massif, close to the crater Conon, length 30 km.

Bradley, Rima [23°N, 2°W]
Prominent straight rille, length 130 km.

Conon [21.6°N, 2.0°E] Conon, *c.* 260 BC. Greek mathematician and astronomer, friend of Archimedes.
Prominent crater with a sharp rim (22 km/2320 m).

Conon, Rima [18°N, 2°E]
Sinuous rille in the Sinus Fidei, length 45 km.

Felicitatis, Lacus (Lake of Happiness) [19°N, 5°E]
Diameter 90 km.

Fidei, Sinus (Bay of Faith) [18°N, 2°E].
Length about 70 km.

Fresnel, Promontorium (Cape Fresnel) [29°N, 5°E] Augustin J. Fresnel, 1788–1827. French physicist, distinguished in optics (Fresnel's lens').
Northern promontory of the Apennines.

Fresnel, Rimae [28°N, 4°E]
System of rilles, length about 90 km.

Galen (Aratus A) [21.9°N, 5.0°E] Galenos from Pergamum, *c.* 129–200 AD. Greek physician.
Crater (10 km).

Hadley, Mons [27°N, 5°E] John Hadley, 1682–1743. English pioneer of the reflecting telescope and the reflecting quadrant.
Mountain massif in the northern part of the Apennines, length 25 km.

Hadley Delta, Mons [26°N, 4°E]
Mountain massif, at the landing site of Apollo 15.

Hadley, Rima [25°N, 3°E]
Sinuous rille, length 80 km.

Huxley (Wallace B) [20.2°N, 4.5°W] Thomas H. Huxley, 1825–1895. British biologist.
Crater (4 km/840 m).

Huygens, Mons [20°N, 3°W] Christiaan Huygens, 1629–1695. Dutch astronomer and optician, recognised the true identity of Saturn's rings.
Mountain massif in the central part of the Apennines, height 5400 m, *length* 40 km.

Marco Polo [15.4°N, 2.0°W] Marco Polo, 1254–1324. Famous Venetian traveller to the Far East.
Remains of an elongated crater (28 × 21 km).

Putredinis, Palus (Marsh of Decay) [27°N, 0°] Named by Riccioli.
Diameter 180 km.

Santos-Dumont (Hadley B) [27.7°N, 4.8°E] Alberto Santos Dumont, 1873–1932. Brazilian aeronautical expert.
Crater (9 km).

Spurr (Archimedes K) [27.9°N, 1.2°W] Josiah E. Spurr, 1870–1950. American geologist.
Remains of a flooded crater (13 km).

Yangel' (Manilius F) [17.0°N, 4.7°E] Mikhail K. Yangel', 1911–1971. Soviet specialist in rocket propulsion.
Crater (9 km).

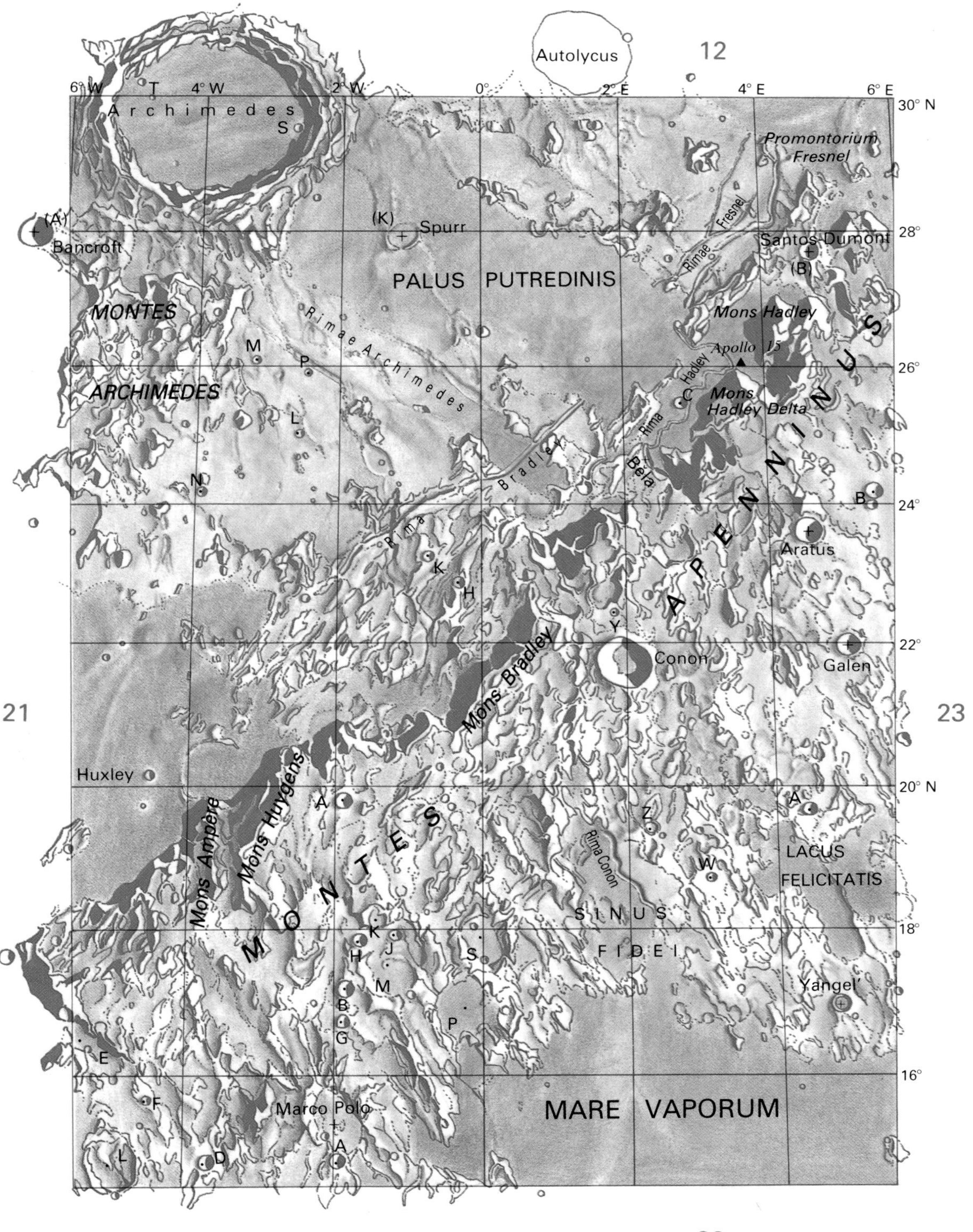
12
Autolycus
6° W
T
4° W
2° W
0°
2° E
4° E
6° E
30° N
Archimedes
S
Promontorium Fresnel
(A)
Bancroft
(K)
Spurr
28°
Rimae Fresnel
Santos-Dumont
(B)
PALUS PUTREDINIS
MONTES
ARCHIMEDES
Rimae Archimedes
Mons Hadley
Apollo 15
M
P
26°
Hadley
C
Mons Hadley Delta
L
Rima Bradley
Rima Bela
N
B
24°
APENNINUS
Aratus
K
H
Y
22°
Conon
Galen
Mons Bradley
21
23
Huxley
20° N
A
Mons Ampère
Mons Huygens
MONTES
Z
Rima Conon
W
LACUS FELICITATIS
SINUS
FIDEI
18°
K
H
J
S
M
B
Yangel'
P
G
E
16°
F
Marco Polo
MARE VAPORUM
A
L
D
33
0
50
KM
100

23. LINNÉ

The western part of Mare Serenitatis offers the observer crater pits, long wrinkle ridges, and also the 'mysterious' crater Linné, about which much has been written. There is a series of clefts west of the crater Sulpicius Gallus, which parallel the Haemus mountain range.

Banting (Linné E) [26.6°N, 16.4°E] Sir Frederick G. Banting, 1891–1941. Canadian physician.
Crater (5 km/1100 m).

Bobillier (Bessel E) [19.6°N, 15.5°E] E. Bobillier, 1798–1840. French geometer.
Crater (6.5 km/1230 m).

Bowen (Manilius A) [17.6°N, 9.1°E] Ira Sprague Bowen, 1898–1973. American astronomer.
Crater with a flat floor (9 km).

Buckland, Dorsum [21°N, 12°E] William Buckland, 1784–1856. British naturalist.
Wrinkle ridge, length 150 km.

Daubrée (Menelaus S) [15.7°N, 14.7°E] Gabriel-Auguste Daubrée, 1814–1896. French geologist.
Crater with a flooded floor (14 km/1590 m).

Doloris, Lacus (Lake of Suffering) [17°N, 9°E]
Diameter 110 km.

Gast, Dorsum [24°N, 9°E] Paul Werner Gast, 1930–1973. American geochemist.
Wrinkle ridge, length 60 km.

Gaudii, Lacus (Lake of Joy) [17°N, 13°E]
Diameter 100 km.

Haemus, Montes (Haemus Mountains) [17°N, 13°E] An old name for a Balkan mountain range given by Hevelius ('*Haemus, mons Thraciae*').
Length 400 km.

Hiemalis, Lacus (Winter Lake) [15°N, 14°E]
Diameter 50 km.

Hornsby [23.8°N, 12.5°E] Thomas Hornsby, 1733–1810. British astronomer.
Crater (3 km).

Joy (Hadley A) [25.0°N, 6.6°E] Alfred H. Joy, 1882–1973. American astronomer.
Crater (6 km/1000 m).

Krishna [24.5°N, 11.3°E] Indian male name.
Crater (2.8 km).

Lenitatis, Lacus (Lake of Tenderness) [14°N, 12°E]
Diameter 80 km.

Linné [27.7°N, 11.8°E] Carl von Linné (Linnaeus), 1707–1778. Swedish botanist, physician, traveller.
Small circular crater with a sharp rim (2.4 km/600 m), *surrounded by light material; under high illumination it resembles a brightly shining white patch; in the past numerous observers recorded alleged changes in the size and appearance of the formation.*

Manilius [14.5°N, 9.1°E] Manilius, first century BC. Roman poet, author of the poem *Astronomicon,* which contains a description of well-known constellations.
Very prominent crater with terraces and central peaks (39 km/3050 m).

Menelaus [16.3°N, 16.0°E] Menelaus, *c.* 100 AD. Greek geometer and astronomer of Alexandria. Author of '*Spherica*', which deals with spherical trigonometry.
Prominent crater with a sharp rim and central peaks (27 km/3010 m).

Odii, Lacus (Lake of Hate) [19°N, 7°E]
Diameter 70 km.

Owen, Dorsum [25°N, 11°E] George Owen, 1552–1613. British naturalist.
Wrinkle ridge, length 50 km.

Serenitatis, Mare (Sea of Serenity). See p. 52.

Sulpicius Gallus [19.6°N, 11.6°E] Sulpicius Gallus, *c.* 168 BC. Roman consul, orator and scholar. Foretold lunar eclipse on eve of battle of Pydna, Macedonia.
Circular crater with a sharp rim (12.2 km/2160 m).

Sulpicius Gallus, Rimae [21°N, 10°E]
System of prominent rilles, length 90 km.

von Cotta, Dorsum [24°N, 12°E] Carl Bernhard von Cotta, 1808–1879. German naturalist.
Wrinkle ridge, length 220 km.

Note. Small craters **Bessel F** (0.5 km/50 m) and **Bessel G** (1 km/70 m) when observed from the Earth appear as mere bright patches. Their size was determined from photographs taken by *Lunar Orbiter 4*.

13

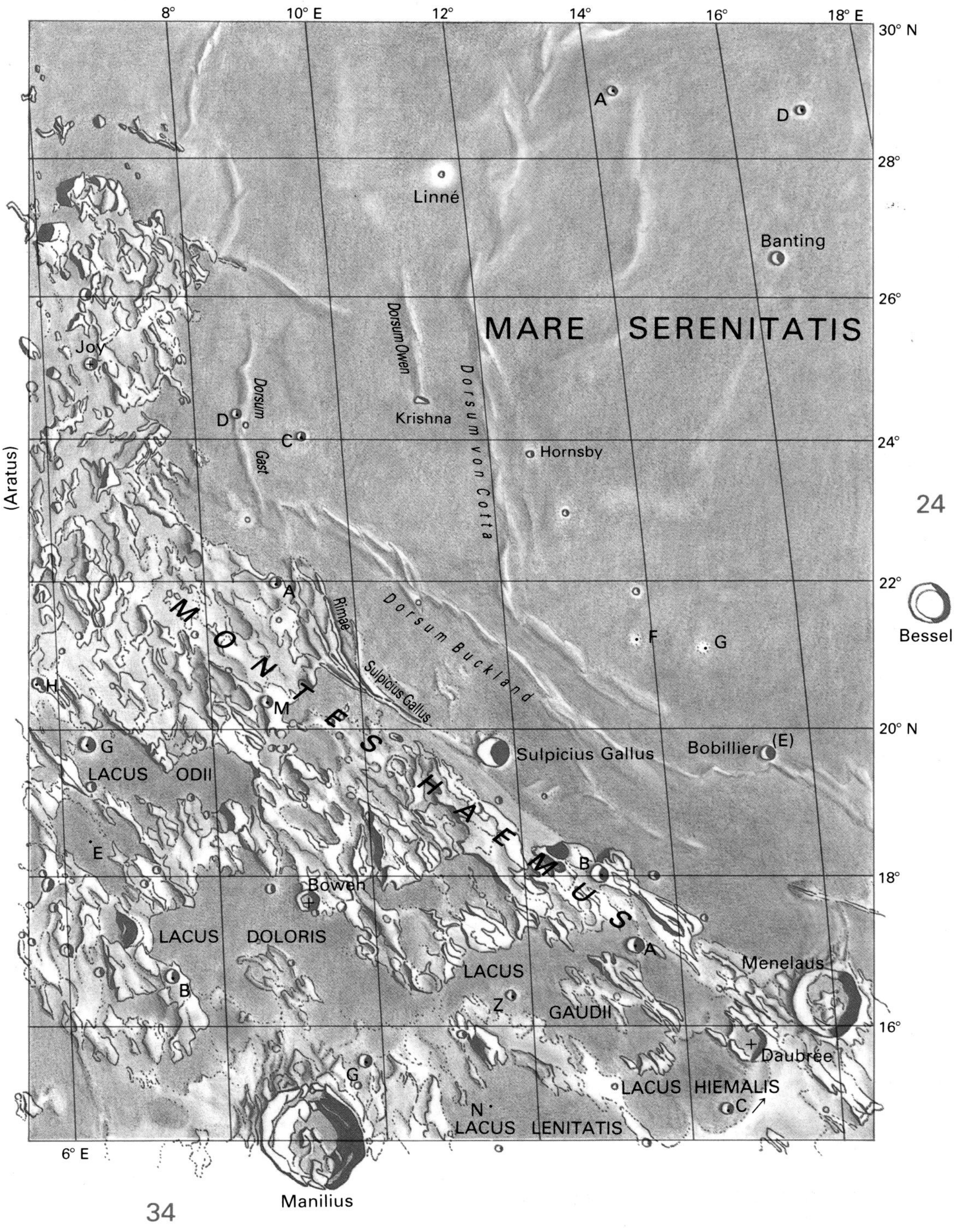

8° 10° E 12° 14° 16° 18° E

30° N
28°
26°
24°
22°
20° N
18°
16°

6° E

MARE SERENITATIS

MONTES HAEMUS

A
D
Linné
Banting
Joy
Dorsum Owen
Dorsum von Cotta
Dorsum Gast
D
C
Krishna
Hornsby
(Aratus)
24
22
A
Rimae Sulpicius Gallus
Dorsum Buckland
F
G
Bessel
H
M
G
LACUS ODII
Sulpicius Gallus
Bobillier (E)
E
B
Bowen
LACUS DOLORIS
A
LACUS
Menelaus
B
Z
GAUDII
Daubrée
G
LACUS HIEMALIS
N
C
LACUS LENITATIS
Manilius

34

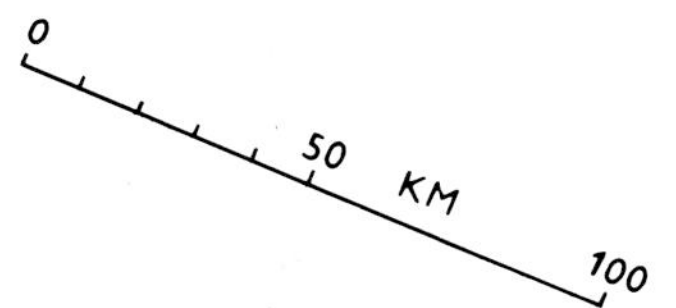

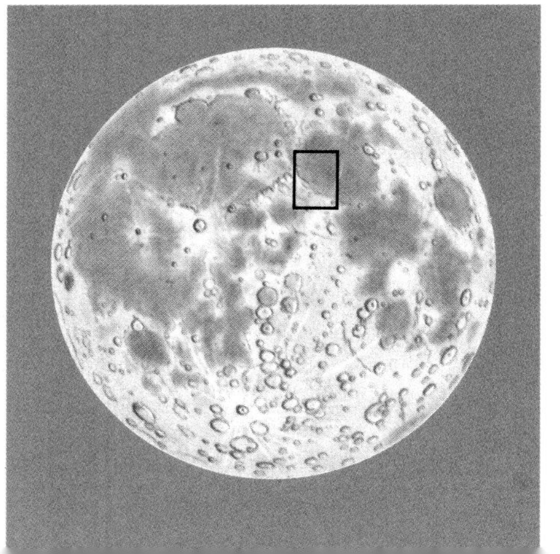

24. BESSEL

The eastern part of Mare Serenitatis is intersected in many places by wrinkle ridges, the largest of which approximately follows the line of the meridian 25 °E and has a serpentine shape (formerly called Serpentine Ridge). The edge of Mare Serenitatis has a dark border and a series of shallow rilles. The summit of Posidonius Gamma (old name) has a crater pit 2 km in diameter.

Abetti [19.9°N, 27.7°E] (1) Antonio Abetti, 1846–1928; (2) Antonio Abetti, 1882–1982. Italian astronomers.
Flooded, not very prominent crater (7 km).

Al-Bakri (Tacquet A). See map 35.

Aldrovandi, Dorsa [24°N, 29°E] Ulisse Aldrovandi, 1522–1605. Italian naturalist.
System of wrinkle ridges, length 120 km.

Archerusia, Promontorium [17°N, 22°E] Old name for a cape on the southern edge of Pontus Euxinus (Black Sea), which on Hevelius' map was marked in place of M. Serenitatis and M. Tranquillitatis.

Argaeus, Mons [19°N, 29°E] Old name for a mountain in Cappadocia.
Mountain massif, length 50 km.

Auwers [15.1°N, 17.2°E] Arthur von Auwers, 1838–1915. German astronomer, Director of Gotha observatory.
Flooded crater, open to the north (20 km/1680 m).

Azara, Dorsum [26°N, 20°E] Felix de Azara, 1746–1811. Spanish naturalist.
Wrinkle ridge, length 110 km.

Bessel [21.8°N, 17.9°E] Friedrich Wilhelm Bessel, 1784–1846. Prominent German astronomer. One of the first to measure the parallax of a star (61 Cygni, in 1838).
Prominent crater (16 km/1740 m).

Borel (le Monnier C) [22.3°N, 26.4°E] Felix E. E. Borel, 1871–1956. French mathematician.
Crater (5 km/950 m).

Brackett [17.9°N, 23.6°E] Frederick Sumner Brackett, 1896–1972. American physicist.
Flooded, not very prominent crater (9 km).

Dawes [17.2°N, 26.4°E] William R. Dawes, 1799–1868. English physician and astronomer, gave his name to a rule for the testing of the resolving power of a telescope ('Dawes' Limit').
Crater with a sharp rim (18 km/2330 m).

Deseilligny [21.1°N, 20.6°E] Jules A. P. Deseilligny, 1868–1918. French selenographer.
Circular crater (6.6 km/1190 m).

Finsch [23.6°N, 21.3°E] O. F. H. Finsch, 1839–1917. German zoologist.
Flooded, not very prominent crater (4 km).

Lister, Dorsa [19°N, 22°E] Martin Lister, 1638–1712. British zoologist.
System of wrinkle ridges, length 290 km.

Littrow, Catena [22°N, 29°E] J. Littrow. See map 25.
Small crater chain, length 10 km.

Menelaus, Rimae [17°N, 17°E] Menelaus. See map 23.
System of rilles, length 140 km.

Nicol, Dorsum [18°N, 23°E] William Nicol, 1768–1851. Scottish physicist, inventor of the Nicol prism.
Wrinkle ridge, length 50 km.

Plinius [15.4°N, 23.7°E] Gaius Plinius Secundus (senior), 'Pliny' the Elder', 23–79 AD. Author of the encyclopedia *Historia Naturalis* in 37 books. Died during the destruction of Pompeii.
Prominent crater with a sharp rim, terraces and central peaks (43 km/2320 m).

Plinius, Rimae [17°N, 24°E].
System of prominent rilles, length 120 km.

Sarabhai (Bessel A) [24.7°N, 21.0°E] Vikram Ambalal Sarabhai, 1919–1971. Indian astrophysicist.
Crater (7.6 km/1660 m).

Serenitatis, Mare (Sea of Serenity). See map 13.

Smirnov, Dorsa [25°N, 25°E] Sergei S. Smirnov, 1895–1947. Soviet naturalist.
System of sizeable wrinkle ridges, length 130 km.

Tacquet [16.6°N, 19.2°E] André Tacquet, 1612–1660. Belgian mathematician.
Circular crater (6.9 km/1260 m).

Very (le Monnier B) [25.6°N, 25.3°E] Frank W. Very, 1852–1927. American astronomer.
Crater (5.1 km/950 m).

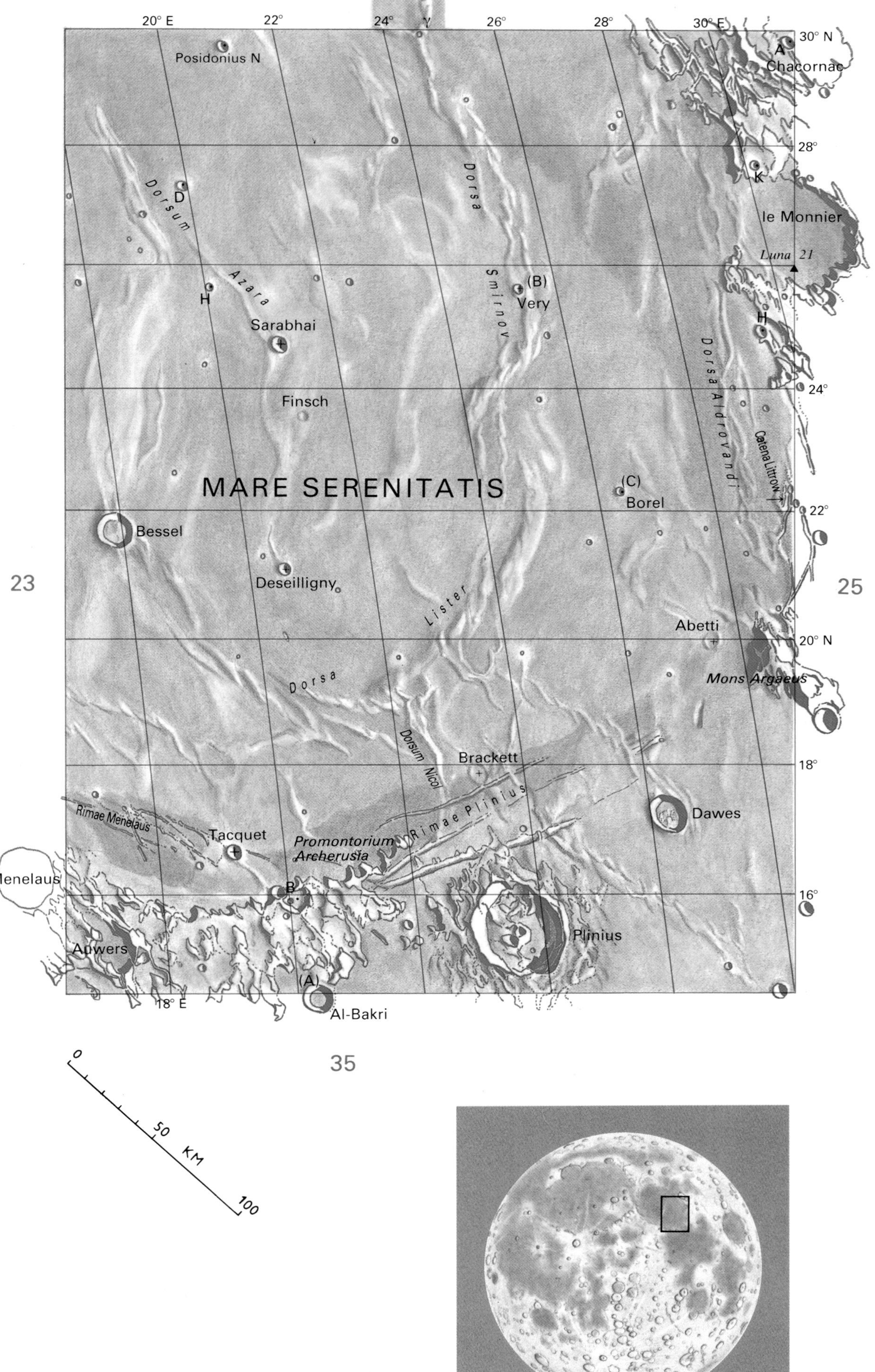
14
20° E
22°
24°
26°
28°
30° E
30° N
28°
24°
22°
20° N
18°
16°
18° E
Posidonius N
A
Chacornac
K
le Monnier
Luna 21
D
Dorsum
H
Azara
Sarabhai
Dorsa
Smirnov
(B)
Very
H
Dorsa Aldrovandi
Finsch
Catena Littrow
MARE SERENITATIS
(C)
Borel
Bessel
Deseilligny
23
25
Lister
Abetti
Mons Argaeus
Dorsa
Dorsum Nicol
Brackett
Rimae Menelaus
Rimae Plinius
Dawes
Tacquet
Promontorium Archerusia
Menelaus
B
Plinius
Auwers
(A)
Al-Bakri
35
0
50
KM
100

25. RÖMER

Mountainous area adjoining Mare Serenitatis and Mare Tranquillitatis. Römer is surrounded by a dense field of craters with disintegrated walls. The region of Römer, Chacornac and Littrow has a system of conspicuous rilles. The *Apollo 17* expedition landed in a valley between mountains to the north of Littrow. *Lunokhod 2* was active in le Monnier.

Amoris, Sinus (Bay of Love) [19°N, 38°E]
Reaches 250 km *from the border of M. Tranquillitatis.*
Argaeus, Mons. See map 24.

Barlow, Dorsa. See map 36.

Beketov (Jansen C) [16.3°N, 29.2°E] N. N. Beketov, 1827–1911. Russian chemist.
Crater (8.4 km/1000 m).

Bonitatis, Lacus (Lake of Good) [23°N, 44°E]
Diameter 130 km.

Brewster (Römer L) [23.3°N, 34.7°E] David Brewster, 1781–1868. Scottish optician, experimented with polarized light.
Crater (11 km/2130 m).

Carmichael (Macrobius A) [19.6°N, 40.4°E] Leonard Carmichael, 1898–1973. American psychologist.
Crater (20 km/3640 m).

Chacornac. See map 14.
Chacornac, Rimae [29°N, 32°E]
System of rilles, length 120 km.

Ching-te [20.0°N, 30.0°E] Chinese male name.
Crater (3.9 km).

Clerke (Littrow B) [21.7°N, 29.8°E] Agnes Mary Clerke, 1842–1907. British astronomer.
Crater (7 km/1430 m).

Esclangon (Macrobius L) [21.5°N, 42.1°E] Ernest B. Esclangon, 1876–1954. French astronomer.
Flooded crater with a low wall (16 km).

Fabbroni (Vitruvius E) [18.7°N, 29.2°E] Giovanni V. M. Fabbroni, 1752–1822. Italian chemist.
Crater (11.1 km/2090 m).

Franck (Römer K) [22.6°N, 35.5°E] James Franck, 1882–1964. German physicist and Nobel Laureate.
Crater (12 km/2510 m).

Franz [16.6°N, 40.2°E] Julius H. Franz, 1847–1913. German astronomer, selenographer.
Flooded, considerably eroded crater (26 km/590 m).

Gardner (Vitruvius A) [17.7°N, 33.8°E] Irvine Clifton Gardner, 1889–1972. American physicist.
Crater with a flat floor (18.4 km/3000 m).

Hill (Macrobius B) [20.9°N, 40.8°E] George W. Hill, 1838–1914. American astronomer and mathematician.
Crater (16 km/3340 m).

Jansen, Rima [15°N, 29°E]. Jansen, see map 36.
Narrow rille, length 35 km.

le Monnier [26.6°N, 30.6°E] Pierre Charles le Monnier, 1715–1799. French astronomer and physicist.
Flooded crater with a very dark floor, which forms a small bay in M. Serenitatis (61 km/2400 m); see also map 24.

Littrow [21.5°N, 31.4°E] Johann J. von Littrow, 1781–1840. Austrian astronomer.
Flooded crater, with damaged southern wall (31 km).

Lucian (Maraldi B). See map 36.

Maraldi [19.4°N, 34.9°E] Giovanni D. Maraldi, 1709–1788. Italian astronomer, assistant to Giovanni-Domenico Cassini.
Flooded crater with a very dark floor (40 km).
Maraldi, Mons [20°N, 35°E]
Mountain massif to the north-east of Maraldi, length 15 km.

Newcomb [29.9°N, 43.8°E] Simon Newcomb, 1835–1909. Canadian-born American mathematician and astronomer.
Crater (39 km/2180 m).

Römer [25.4°N, 36.4°E] Olaus Römer, 1644–1710. Danish astronomer. The first to determine the velocity of light by observations of Jupiter's satellites.
Prominent crater with a sharp rim, terraces and a central peak (40 km).
Römer, Rimae [27°N, 35°E]
System of pronounced rilles, length 110 km.

Taurus, Montes (Taurus Mountains) [26°N, 36°E] Name given by Hevelius to the mountainous region north of Römer. *Mountainous region, diameter* 500 km.

Theophrastus (Maraldi M) [17.5°N, 39.0°E] Theophrastus, 372–287 BC. Greek philosopher and botanist.
Crater (9 km/1700 m).

Vitruvius [17.6°N, 31.3°E] Marcus Pollio Vitruvius, first century BC. Roman architect, author of *De Architectura*. Interests: astronomy, physics, water docks and sun dials.
Flooded crater (30 km/1550 m).
Vitruvius, Mons [19°N, 31°E]
Mountain base, diameter 15 km.

25

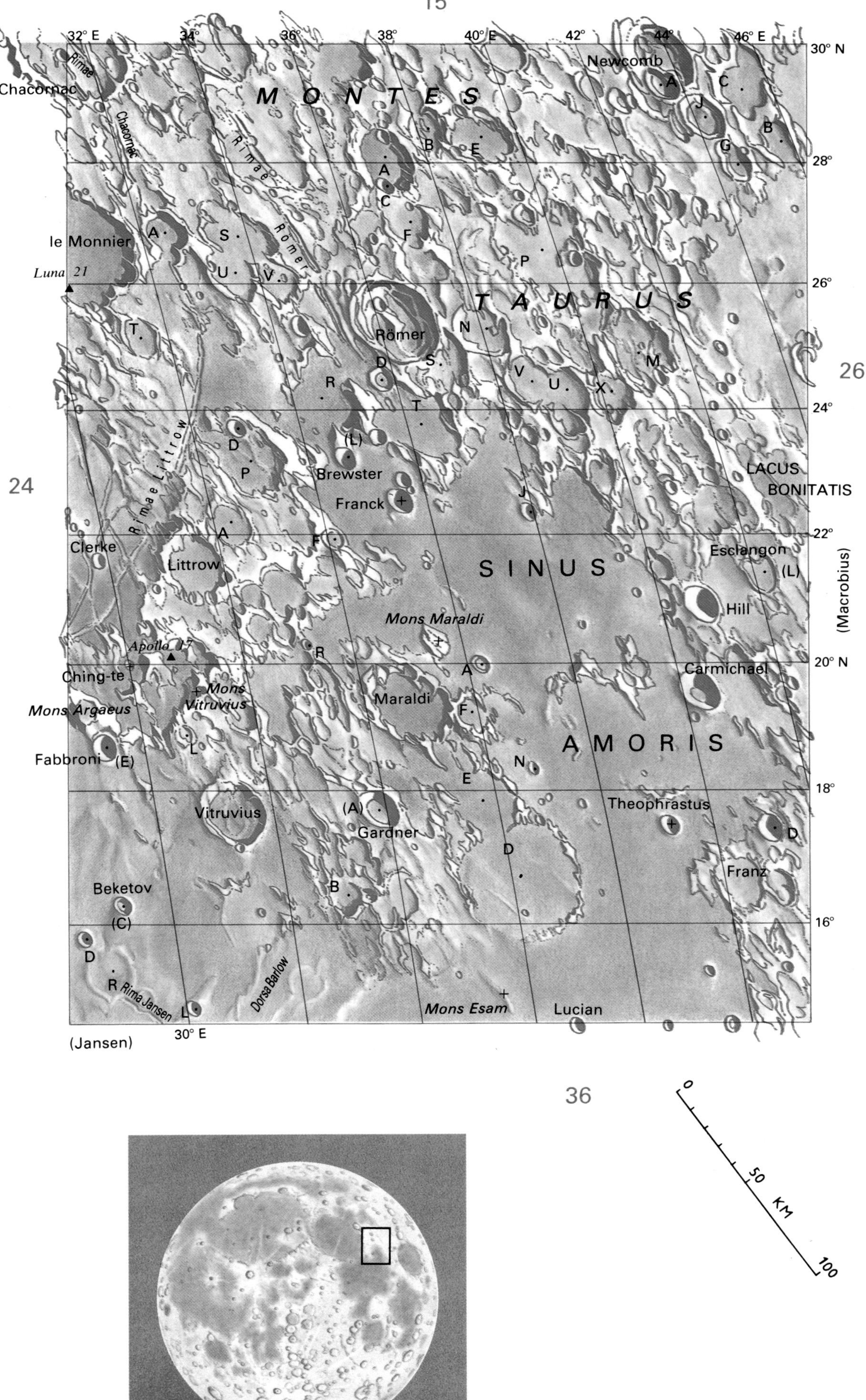

15
24
26
36
32° E
34°
36°
38°
40° E
42°
44°
46° E
30° N
28°
26°
24°
22°
20° N
18°
16°
30° E
MONTES
TAURUS
SINUS
AMORIS
LACUS
BONITATIS
Chacornac
Rimae Chacornac
Rimae Römer
Newcomb
le Monnier
Luna 21
Römer
Rimae Littrow
Brewster
Franck
Clerke
Littrow
Esclangon
Hill
(Macrobius)
Mons Maraldi
Apollo 17
Ching-te
Mons Vitruvius
Mons Argaeus
Maraldi
Carmichael
Fabbroni
Vitruvius
Gardner
Theophrastus
Franz
Beketov
Rima Jansen
Dorsa Barlow
Mons Esam
Lucian
(Jansen)
0
50
KM
100

26. CLEOMEDES

The north-western part of Mare Crisium is encircled by large mountain massifs. Under oblique illumination numerous wrinkle ridges are prominent on the surface of the mare. Under high illumination the rays from the crater Proclus are prominent.

Bonitatis, Lacus (Lake of Good). See map 25.

Cleomedes [27.7°N, 55.5°E] Cleomedes, first century BC or later. Greek astronomer.
Very prominent crater (126 km), *with clefts on the floor and a central peak.*

Cleomedes, Rima
Rille inside Cleomedes, length 30 km.

Crisium, Mare (Sea of Crises). See p. 80.

Curtis (Picard Z) [14.6°N, 56.6°E] Heber Doust Curtis, 1872–1942. American astronomer.
Crater (3 km).

Debes [29.5°N, 51.7°E] Ernest Debes, 1840–1923. German cartographer, prepared lunar maps and atlases.
Crater (31 km) *connected with the crater Debes A.*

Delmotte [27.1°N, 60.2°E] Gabriel Delmotte, 1876–1950. Prominent French selenographer.
Crater (33 km).

Eckert [17.3°N, 58.3°E] Wallace J. Eckert, 1902–1971. American astronomer.
Crater (3 km).

Fredholm (Macrobius D) [18.4°N, 46.5°E] Erik Ivar Fredholm, 1866–1927. Swedish mathematician.
Crater (15 km).

Lavinium, Promontorium; Olivium, Promontorium [15°N, 49°E]. Old, unofficial names, now little used. Two sharp 'capes' situated opposite each other on the western edge of Mare Crisium (named by Birt). These two promontories are separated by two eroded craters (not a bridge, which was once claimed).

Macrobius [21.3°N, 46.0°E] Ambrosius T. Macrobius, fourth century AD. Possibly Roman grammarian, author of a commentary to Cicero's *'Scipion's Dream'* (vision of a flight to the stars).
Prominent crater with terraces (64 km).

Oppel, Dorsum [19°N, 52°E] Albert Oppel, 1831–1865. German palaeontologist.
Prominent wrinkle ridge, length 300 km.

Peirce [18.3°N, 53.5°E] Benjamin Peirce, 1809–1880. American mathematician and astronomer.
Crater (18.5 km).

Picard [14.6°N, 54.7°E] Jean Picard, 1620–1682. French astronomer, founder of the ephemeris *Connaissance des Temps.*
Prominent crater with a sharp rim (23 km).

Proclus [16.1°N, 46.8°E]. Proclus Diadochos, 410–485 AD. Athenian philosopher and mathematician.
Polygonal crater with sharp contours, centre of a bright ray system (28 km/2400 m).

Somnii, Palus (Marsh of Sleep). See map 37.

Swift (Peirce B) [19.3°N, 53.4°E] Lewis Swift, 1820–1913. American astronomer.
Crater (11 km).

Tisserand [21.4°N, 48.2°E] François F. Tisserand, 1845–1896. French astronomer; celestial mechanics, orbits of comets.
Crater (37 km).

Tralles [28.4°N, 52.8°E] Johann G. Tralles, 1763–1822. German physicist.
Crater (43 km).

Yerkes [14.6°N, 51.7°E] Charles T. Yerkes, 1837–1905. Chicago millionaire, financed the construction of an observatory with the largest refractor in the world, 1 m diameter (Yerkes Observatory, opened in 1897).
Flooded crater (36 km).

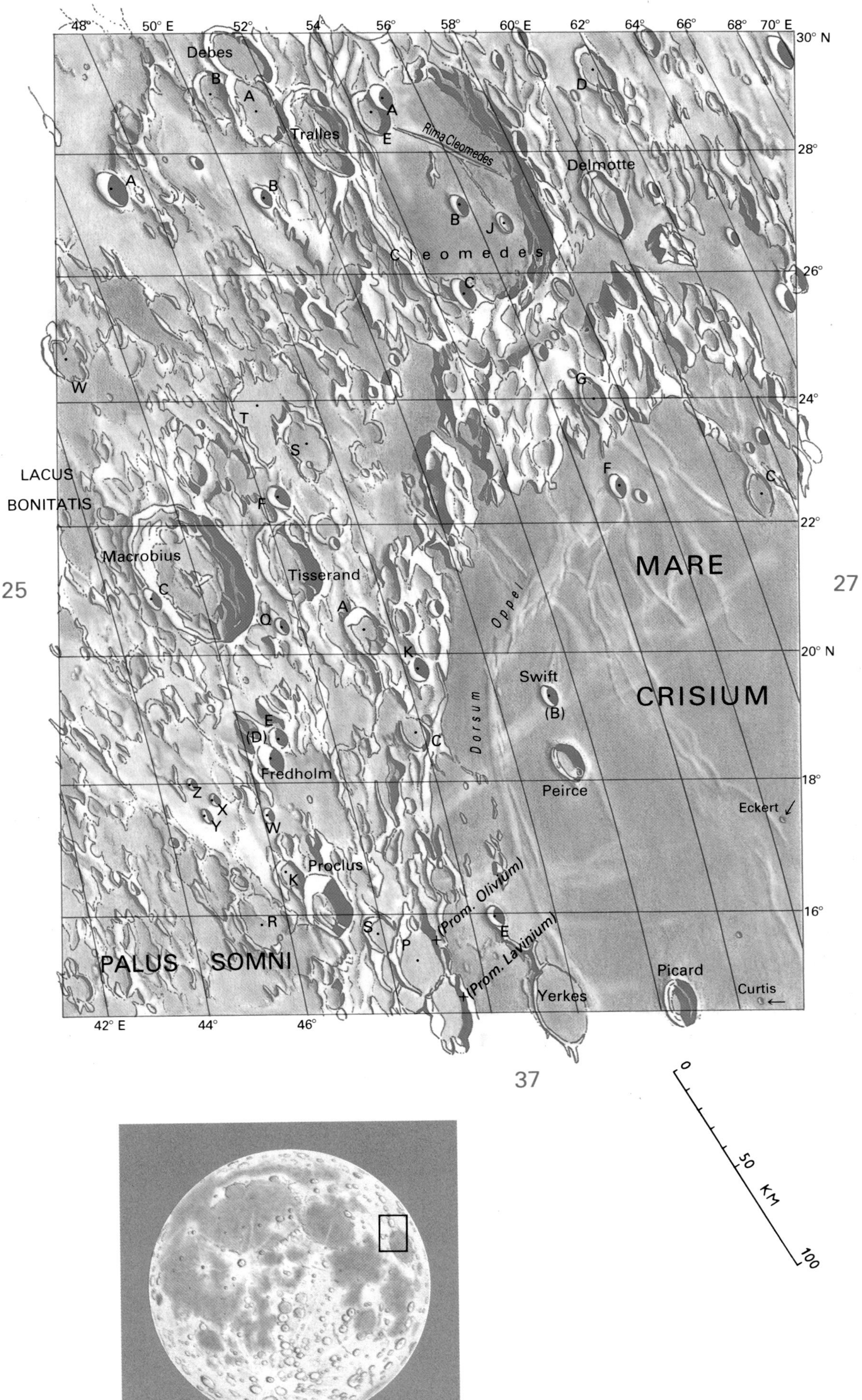
16
48°
50° E
52°
54°
56°
58°
60° E
62°
64°
66°
68°
70° E
30° N
28°
26°
24°
22°
20° N
18°
16°
42° E
44°
46°
Debes
B
A
Tralles
A
E
Rima Cleomedes
D
Delmotte
A
B
B
J
C l e o m e d e s
C
W
G
T
S
LACUS
BONITATIS
F
F
C
Macrobius
C
Tisserand
MARE
CRISIUM
Oppel
Dorsum
O
A
K
Swift
(B)
E
(D)
C
Fredholm
Peirce
Eckert
Z
X
Y
W
Proclus
K
R
S
P
(Prom. Olivium)
(Prom. Lavinium)
E
PALUS SOMNI
Picard
Curtis
Yerkes
25
27
37
0
50
KM
100

27. PLUTARCH

The eastern part of Mare Crisium and eastern margin of the Moon. Convenient points of orientation are the craters Eimmart, Alhazen and Plutarch. During a favourable libration the crater Goddard, with its dark floor, and the Mare Marginis are clearly visible.

Alhazen [15.9°N, 71.8°E] Abu Ali al-Hasan, 987–1038. Arabian mathematician at the court of Caliph Hakem in Cairo.
Prominent crater with a sharp rim wall (33 km).

Anguis, Mare (The Serpent Sea) [22°N, 67°E] Franz's name for a narrow sinuous lowland area east of Mare Crisium.
Surface area 10 000 sq km, *length about* 130 km.

Cannon [19.9°N, 81.4°E] Annie J. Cannon, 1863–1941. American astronomer; worked on the classification of stellar spectra.
Flooded crater with a lighter floor (57 km).

Crisium, Mare (Sea of Crises) [17°N, 59°E] Oval mare, with major axis running east–west, and surrounded by a mountainous wall.
Surface area 176 000 sq km *(comparable with that of Great Britain), diameter* 570 km.

Eimmart [24.0°N, 64.8°E] Georg Christoph Eimmart, 1638–1705. German engraver and amateur astronomer, author of a map of the Moon.
Crater (46 km).

Goddard [14.8°N, 89.0°E] Robert H. Goddard, 1882–1945. American physicist, pioneer of rocket-technology.
Flooded crater with a dark floor (89 km).

Harker, Dorsa [14°N, 64°E] Alfred Harker, 1859–1939. British petrologist.
System of wrinkle ridges, length 200 km.

Hubble [22.1°N, 86.9°E] Edwin P. Hubble, 1889–1953. American astronomer. Investigation into galaxies and how they move away from each other, with velocities directly proportional to their distances apart.
Partly flooded crater (81 km).

Liapunov [26.3°N, 89.3°E] Alexander M. Liapunov, 1857–1918. Russian mathematician.
Crater (66 km).

Marginis, Mare (The Border Sea) [12°N, 88°E]. Name given by Franz to a small, irregularly shaped mare along the eastern edge of the Moon. Extends beyond the normally visible hemisphere on to the lunar farside.
Surface area 62 000 sq km, *diameter* 360 km.

Plutarch [24.1°N, 79.0°E] Plutarchos, *c.* 50–120 AD. Greek philosopher and writer. In his dialogue *De facie in orbe lunae* he developed some early theories on the nature of the Moon.
Prominent crater (68 km).

Rayleigh [29.0°N, 89.2°E] (Lord) John W. Strutt Rayleigh, 1842–1919. British physicist, awarded Nobel Prize for Physics in 1904 for research in optics.
Walled plain (107 km) *along the eastern limb of the Moon.*

Seneca [26.6°N, 80.2°E] Lucius A. Seneca, *c.* 4 BC–65 AD. Roman philosopher and writer, teacher of Nero. In his work *Quaestiones Naturales* he concluded that comets are celestial bodies.
Inconspicuous, irregularly shaped crater (53 km).

Tetyaev, Dorsa [19°N, 65°E] Mikhail M. Tetyaev, 1882–1956. Soviet geologist.
System of wrinkle ridges, length 150 km.

Urey (Rayleigh A) [27.9°N, 87.4°E] Harold C. Urey, 1893–1981. American chemist, Nobel Laureate.
Crater (38 km).

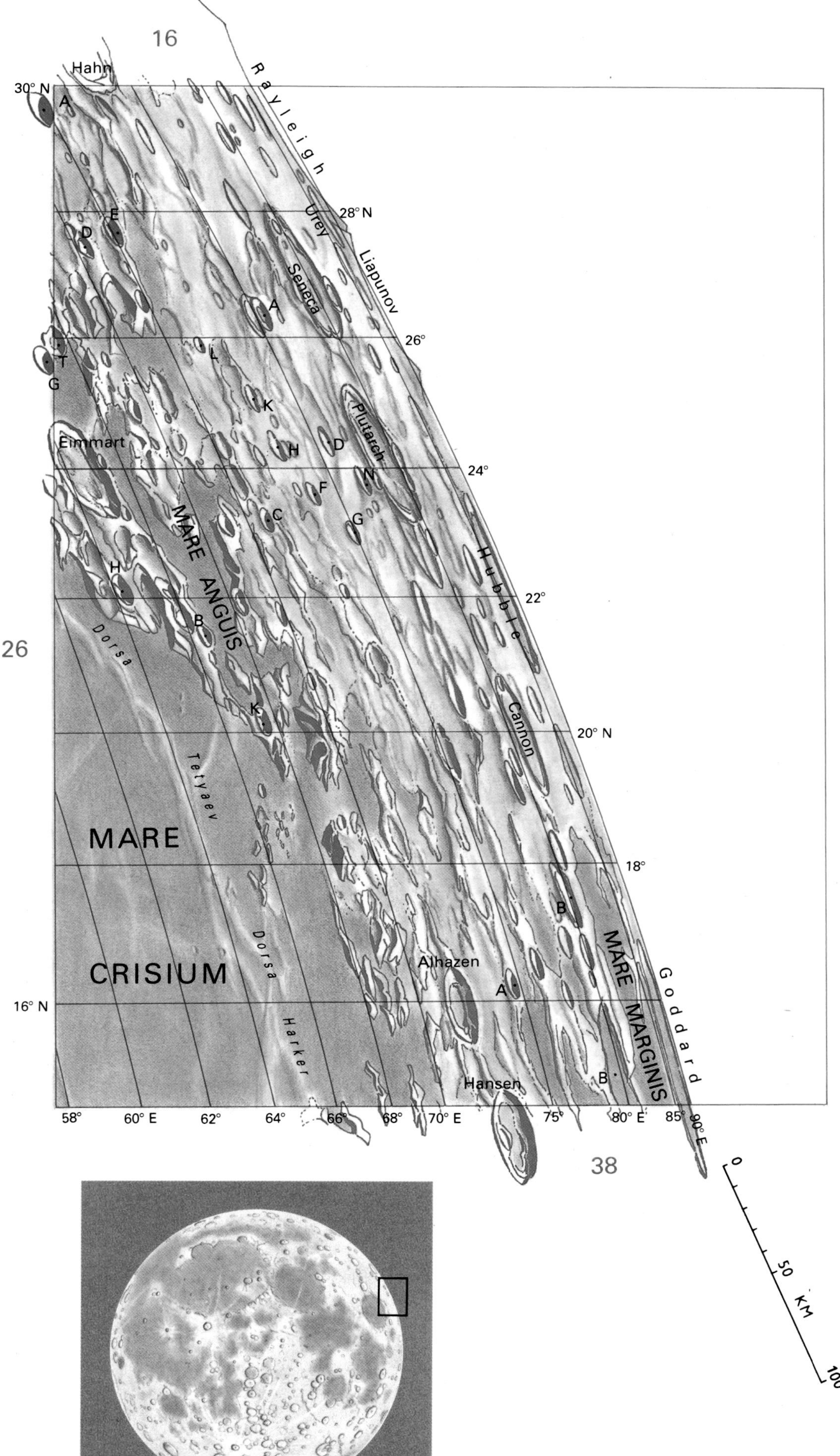
16
Hahn
30° N
A
Rayleigh
Urey
28° N
E
D
Seneca
Liapunov
A
26°
L
T
G
K
Plutarch
Eimmart
H
D
N
24°
F
C
G
MARE
ANGUIS
H
Hubble
22°
B
26
Dorsa
K
Cannon
20° N
Tetyaev
MARE
18°
B
Dorsa
Alhazen
MARE
MARGINIS
CRISIUM
A
Goddard
16° N
Harker
Hansen
B
58°
60° E
62°
64°
66°
68°
70° E
75°
80° E
85°
90° E
38
0
50
KM
100

28. GALILAEI

The western limb of the Moon and western edge of Oceanus Procellarum. The observer using a large telescope will be attracted by the clefts in the floor of the crater Hevelius. Here also lies the unique formation Reiner Gamma. The region also contains the landing site of the probe *Luna 9* (which made the first soft-landing on the Moon).

Bohr [12.8°N, 86.4°W] Niels H. D. Bohr, 1885–1962. Danish physicist. (Bohr's model of the atom.) Nobel Prize for Physics in 1922.
Crater in a libration zone (71 km).

Cardanus [13.2°N, 72.4°W] Girolamo Cardano, 1501–1576. Italian mathematician, astrologer and physician.
Crater (50 km) *connected by a crater chain with the crater Krafft* (see map 17).

Cavalerius [5.1°N, 66.8°W] Buonaventura Cavalieri, 1598–1647. Italian mathematician, pupil of Galileo.
Prominent crater (58 km).

Galilaei [10.5°N, 62.7°W] Galileo Galilei, 1564–1642. Famous Italian physicist and astronomer. First telescopic observation of the Moon and planets, advocate of Copernican theories; observed sunspots and discovered Jupiter's satellites.
Crater with a sharp rim (15.5 km).

Galilaei, Rima [13°N, 59°W]
Sinuous rille, length 180 km.

Hedin [2.9°N, 76.5°W] Sven A. Hedin, 1865–1952. Swedish explorer and traveller, expeditions to Central Asia.
Remains of a walled plain (143 km).

Hevelius [2.2°N, 67.3°W] Johann Hewelcke (Hevel), 1611–1687. Polish astronomer and selenographer. Proposed a new nomenclature for the Moon, of which only six names have been preserved.
Walled plain (106 km).

Hevelius, Rimae [2°N, 66°W]
System of rilles inside and to the south of Hevelius, length 190 km.

Krafft, Catena. See map 17.

Olbers [7.4°N, 75.9°W] Heinrich W. H. Olbers, 1758–1840. German physician and astronomer, discovered and observed comets.
Crater (75 km).

Planitia Descensus (Plain of Descent) [7°N, 64°W]. Site of the first soft-landing on the Moon, by *Luna 9.* It is situated among low hills at the edge of Oceanus Procellarum.

Procellarum, Oceanus (Ocean of Storms). See p. 84.

Reiner Gamma [8°N, 59°W] (Reiner, see p. 84).
Entirely flat feature formed of bright material.

Vasco da Gama [13.9°N, 83.8°W] Vasco da Gama, 1469–1524. Portuguese navigator. Made the first voyage to India around the Cape of Good Hope (1498).
Crater (96 km).

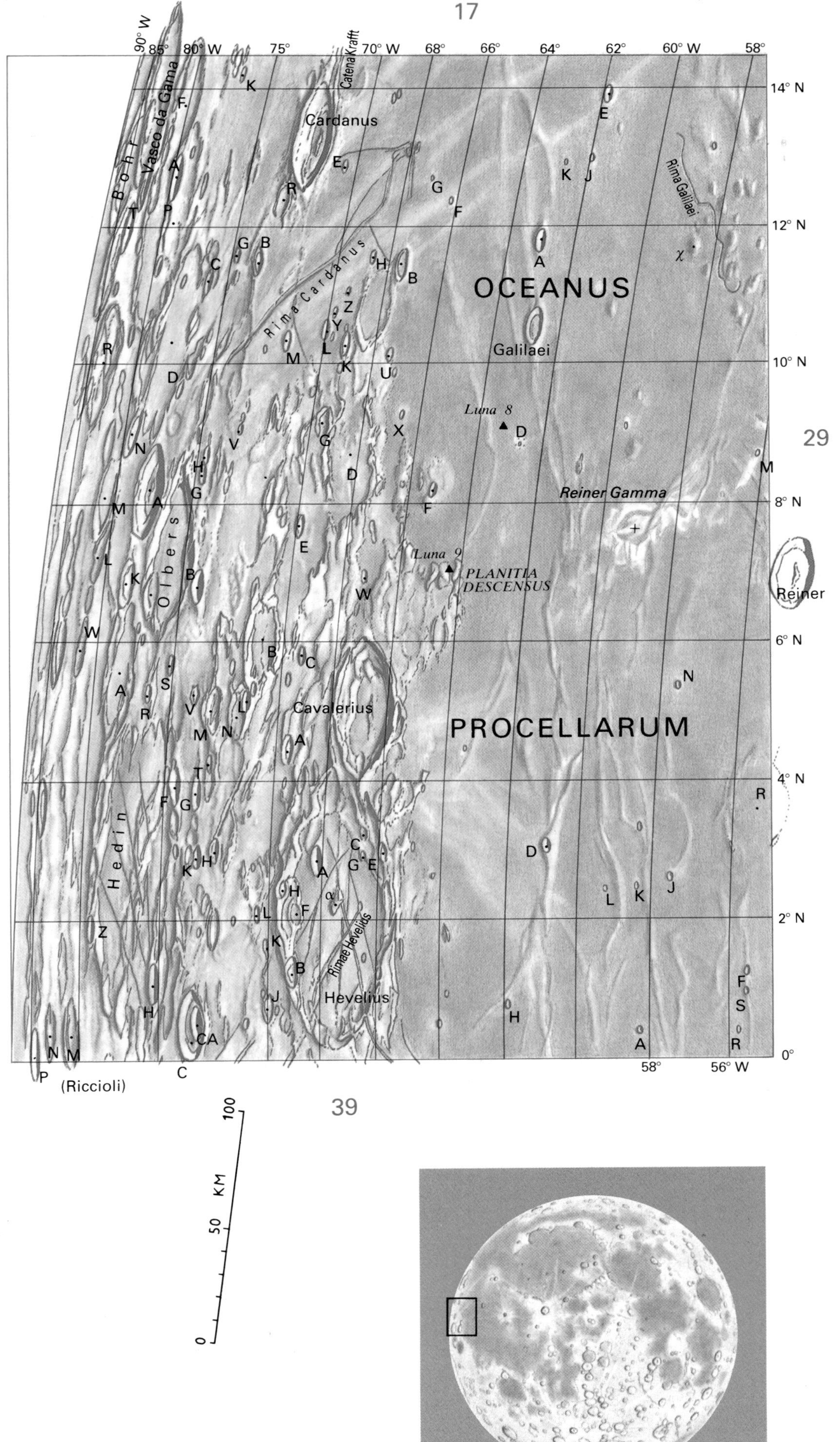
17
90° W
85°
80° W
75°
Catena Krafft
70° W
68°
66°
64°
62°
60° W
58°
14° N
12° N
10° N
8° N
6° N
4° N
2° N
0°
58°
56° W
Bohr
Vasco da Gama
Cardanus
Rima Cardanus
Rima Galilaei
OCEANUS
Galilaei
Luna 8
29
Reiner Gamma
Luna 9
PLANITIA DESCENSUS
Reiner
Olbers
Cavalerius
PROCELLARUM
Hedin
Rimae Hevelius
Hevelius
(Riccioli)
39
0
50
100
KM

29. MARIUS

The western part of Oceanus Procellarum, poor in large craters, but very rich in lunar domes, especially near the crater Marius. This area, geologically very interesting with distinct traces of volcanic activity, was selected as a target for one of the *Apollo* missions, subsequently cancelled.

Maestlin [4.9°N, 40.6°W] Michael Möstlin (Maestlin), 1550–1631. German mathematician and astronomer, teacher of Johannes Kepler, whom he introduced to the Copernican heliocentric system.
Small crater (7.1 km/1650 m) *located to the north of the remains of the walled plain* **Maestlin R** (60 km).

Maestlin, Rimae [2°N, 40°W]
System of short, straight rilles, length 80 km.

Marius [11.9°N, 50.8°W] Simon Mayer (Marius), 1570–1624. German astronomer; independently discovered Jupiter's satellites.
Regular flooded crater (41 km); *a small crater* **Marius G** (3.3 km) *is situated on its flat floor.*

Procellarum, Oceanus (Ocean of Storms).
The largest lunar 'mare'. Surface area 2 102 000 sq km, *i.e. less than the Mediterranean Sea on Earth. The western, northern and southern borders are relatively distinct, but the eastern edge is indefinite. The surface is furrowed by numerous wrinkle ridges. The laser altimeter of* Apollo 15 *discovered some exceptionally flat places in Oceanus Procellarum with height differences of* ±80 m *over a distance of* 200 km. *Some of the bright rays radiating from the crater Kepler are visible in this area.*

Reiner [7.0°N, 54.9°W] Vincentio Reinieri, died in 1648. Italian mathematician and astronomer, pupil and friend of Galileo.
Prominent crater (30 km).

Suess [4.4°N, 47.6°W] Edward Suess, 1831–1914. Austrian geologist and selenographer, advocated cosmic origin of tektites.
Small crater (9.2 km).

Suess, Rima [6°N, 47°W]
Narrow, sinuous rille, not observable with small telescopes, length 200 km.

Small craters: **Kepler C** (12.2 km/2170 m)
Kepler D (10.0 km/350 m)
Kepler E (5.2 km/1000 m)
Maestlin G (2.8 km/670 m)
Maestlin H (7.1 km/1370 m)

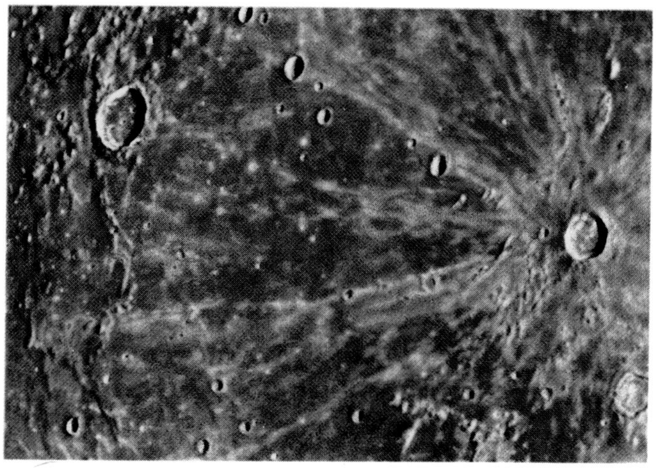

Bright rays radiating from the crater ***Kepler*** *are conspicuous on the dark area of the Oceanus Procellarum. They reach up to the vicinity of the crater* ***Marius.***

18

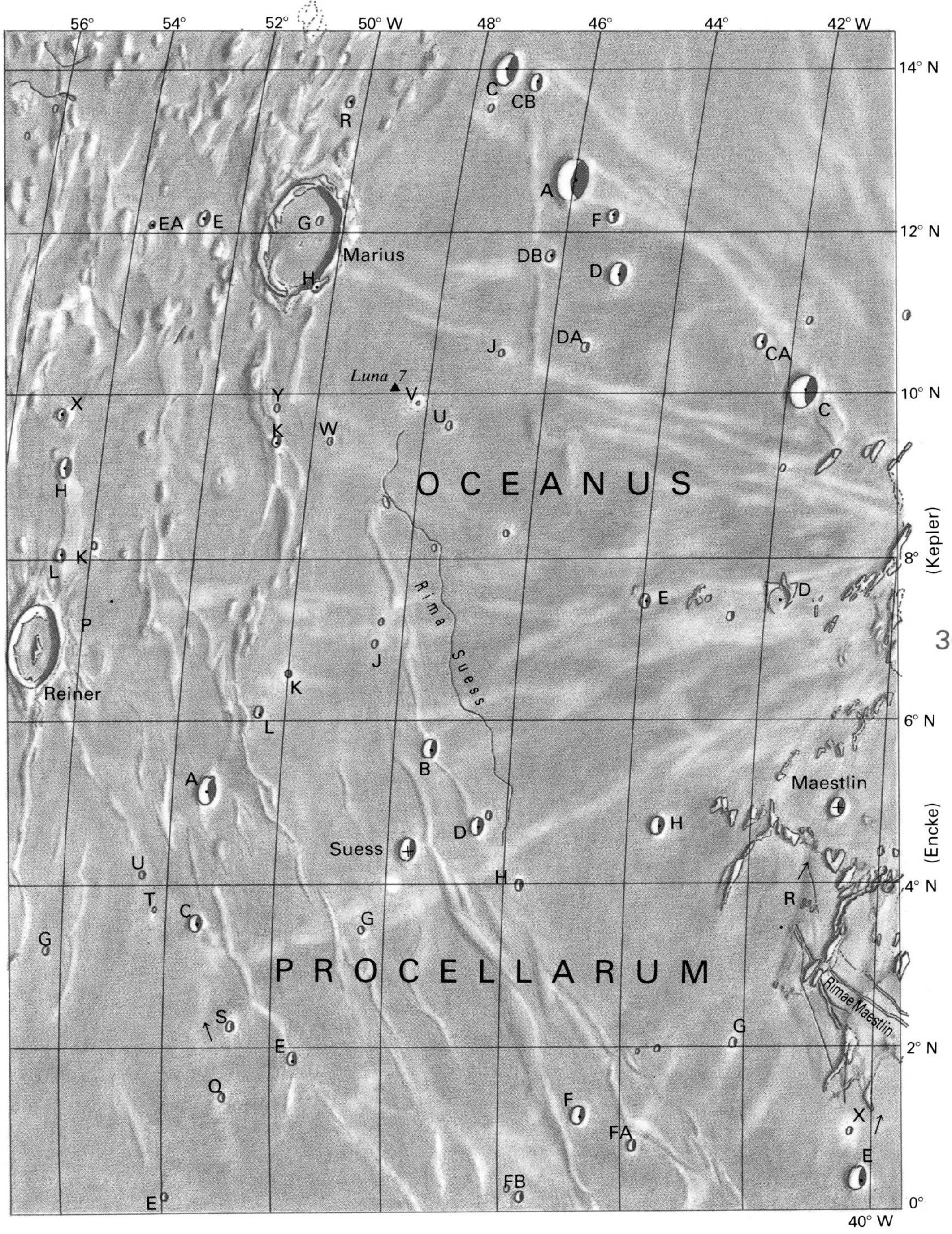

56° 54° 52° 50° W 48° 46° 44° 42° W

14° N
12° N
10° N
8° N
6° N
4° N
2° N
0°

40° W

C CB
R
A
EA E G Marius
F
DB
H
D
J DA
CA
Luna 7
X Y V U C
K W
OCEANUS
H
(Kepler)
L K
Rima Suess
E D
28 P
J
Reiner
K
30
L
B
A Maestlin
H
D
Suess (Encke)
U
R H
T
R
C G
G
PROCELLARUM
Rimae Maestlin
S
G
E
O
F X
FA
E
FB
E

40

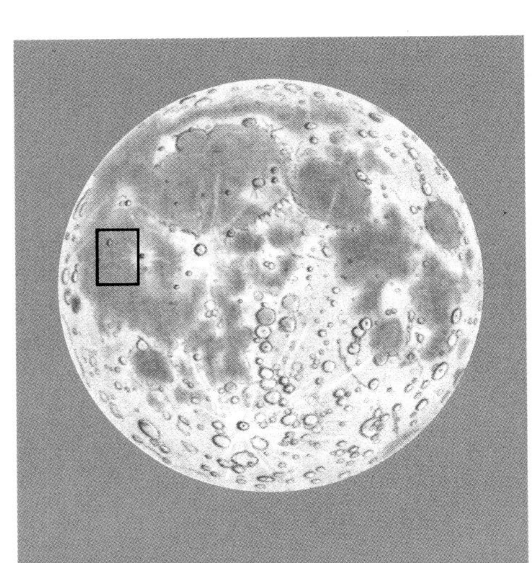

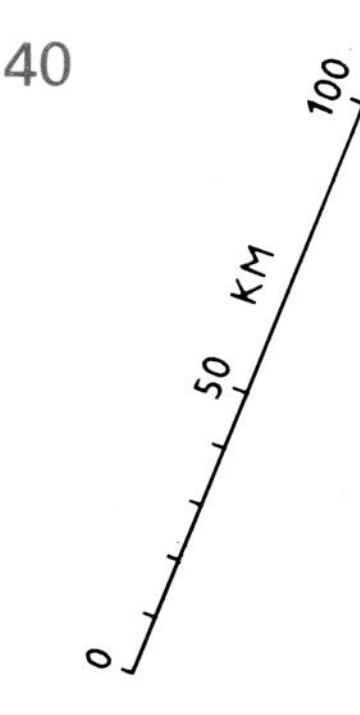

30. KEPLER

The dark background of Mare Insularum contrasts with bright ray systems from two main centres: the crater Copernicus (in the east) and, especially, the crater Kepler which dominates this area. Perhaps the best-known group of lunar domes is that situated north of the crater Hortensius. Other domes that can be seen through even quite a small telescope lie to the west of the crater Milichius and to the south of the crater T. Mayer (see map 19).

Encke [4.6°N, 36.6°W] Johann Franz Encke, 1791–1865. German astronomer. Calculated the elements of the orbit of 'Encke's Comet' (the first known short-period comet).

Crater with uneven floor (29 km/750 m); *a small crater* **Encke N** (3.5 km/590 m) *is situated on its western wall.*

Hortensius [6.5°N, 28.0°W] Martin van den Hove, 1605–1639. Dutch astronomer, professor of mathematics at Amsterdam.

Small crater with a sharp rim (14.6 km/2860 m); *to the north there is a group of domes, most of them having summit craterlets.*

Insularum, Mare (Sea of Isles) [7°N, 22°W] Name approved by the IAU in 1976.

An area of about 900 km in diameter, approximately between the craters Kepler and Encke in the west and the Sinus Aestuum in the east. The northern boundary is formed by the Carpathian Mountains, the southern border is indefinite, linking with Mare Cognitum. The largest 'isle' here is the crater Copernicus, which is surrounded by light ejecta, visible under high illumination. (See also maps 31 and 32.)

Kepler [8.1°N, 38.0°W] Johannes Kepler, 1571–1630. German astronomer and ingenious theoretician, who on the basis of Tycho Brahe's observations formulated the three Laws of Planetary Motion that bear his name and which describe the motions of the planets around the Sun.

Very prominent crater (32 km/2570 m) *with uneven floor; centre of a bright ray system.*

Kunowsky [3.2°N, 32.5°W] Georg K. F. Kunowsky, 1786–1846. German lawyer and amateur astronomer, observer of the Moon and planets.

Flooded crater (18 km/850 m).

Milichius [10.0°N, 30.2°W] Jacob Milich, 1501–1559. German physician, philosopher and mathematician.

Circular crater with a sharp rim (13 km/2510 m).

Milichius Pi *is a typical dome with a summit craterlet.*

Milichius, Rima [8°N, 33°W].

Sinuous, narrow rille, unobservable with small telescopes, length 110 km.

Small craters: **Encke B** (11.5 km/2230 m)
Hortensius A (10.2 km/1850 m)
Hortensius B (6.7 km/1170 m).

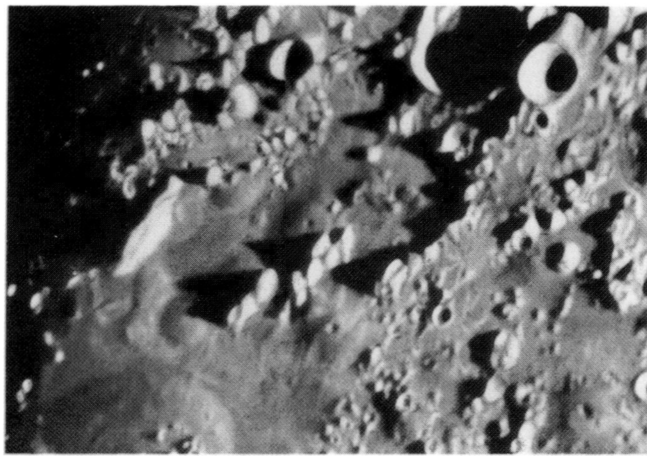

A group of numerous ***lunar domes*** *is situated in a broad valley between the mountain massifs T. Mayer Alpha and T. Mayer Zeta (as shown in the top right-hand corner of map 30). These very low formations are observable only in the vicinity of the terminator.*

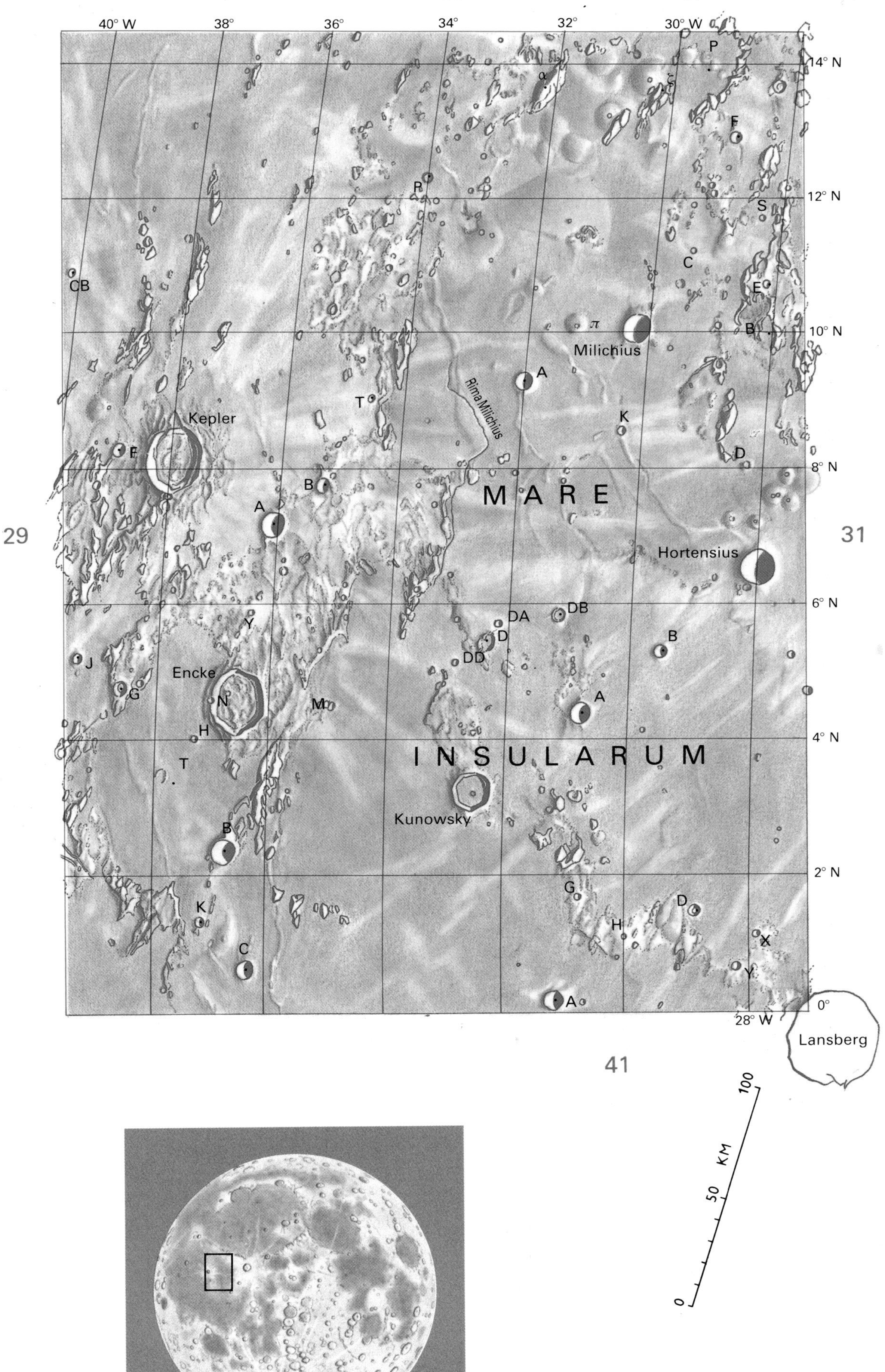
19
40° W
38°
36°
34°
32°
30° W
14° N
12° N
10° N
8° N
6° N
4° N
2° N
0°
28° W
29
31
41
Kepler
Milichius
Rima Milichius
Hortensius
Encke
Kunowsky
Lansberg
MARE
INSULARUM
CB
P
F
R
S
C
E
B
A
T
K
D
π
α
ζ
Y
J
G
N
M
H
DA
DB
DD
X
0
50
100
KM

31. COPERNICUS

The crater Copernicus is undoubtedly one of the best-known and most typical of lunar formations; it is also a centre of bright rays which can be traced across the surface of Mare Imbrium (see maps 20 and 21). To the west of Copernicus is a group of scattered solitary hills, which rise to a height of several hundred metres.

Carpatus, Montes (Carpathian Mountains) [15°N, 25°W] Mädler's name.
Mountain range at the southern margin of Mare Imbrium. It stretches approximately east–west and its length is about 400 km. *It consists of individual hills and mountain massifs, whose heights are between* 1000 *and* 2000 m *(see also maps 19 and 20).*

Copernicus [9.7°N, 20.0°W] Nikłas (Nicholas) Copernicus, 1473–1543. Renowned Polish astronomer, one of the founders of modern astronomy. His heliocentric system was explained in his main work *De Revolutionibus Orbium Celestium* (1543).
Ring mountain, 93 km *in diameter and* 3760 m *deep; terraced walls, relatively flat floor with a group of central peaks (heights up to* 1200 m*); the height of the wall above the surrounding terrain is* 900 m.

Fauth [6.3°N, 20.1°W] Philipp J. H. Fauth, 1867–1941. Renowned German selenographer and observer of the planets, author of lunar maps.
Double-crater **Fauth** *and* **Fauth A** *is shaped like a keyhole; Fauth has a diameter of* 12.1 km *and a depth of* 1960 m, *Fauth A is* 9.6 km *in diameter and* 1540 m *deep.*

Gay-Lussac [13.9°N, 20.8°W] Louis Joseph Gay-Lussac, 1778–1850. French physicist and chemist (Gay-Lussac laws).
Crater at the southern edge of the Carpathian Mountains, slightly disintegrated wall (26 km/830 m).

Gay-Lussac, Rima [13°N, 22°W]
Wide, distinct rille, length 40 km.

Insularum, Mare (Sea of Isles). See map 30.

Reinhold [3.3°N, 22.8°W] Erasmus Reinhold, 1511–1553. German mathematician and astronomer.
Prominent crater with terraced walls (48 km/3260 m).

Small craters: **Copernicus H** (4.6 km/870 m) – *dark 'halo' crater*
Gambart A (12 km/2440 m)
Gay-Lussac A (14 km/2550 m)
T. Mayer C (15.6 km/2510 m)
T. Mayer D (8.6 km/1470 m).

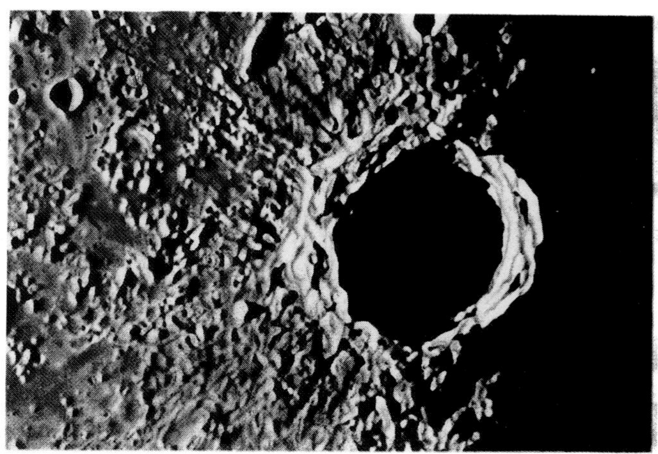

The crater ***Copernicus*** *is a magnificent sight shortly before sunset, when the whole floor disappears in shadow and the narrowing part of the eastern wall appears as a crescent Moon shining brightly at the terminator.*

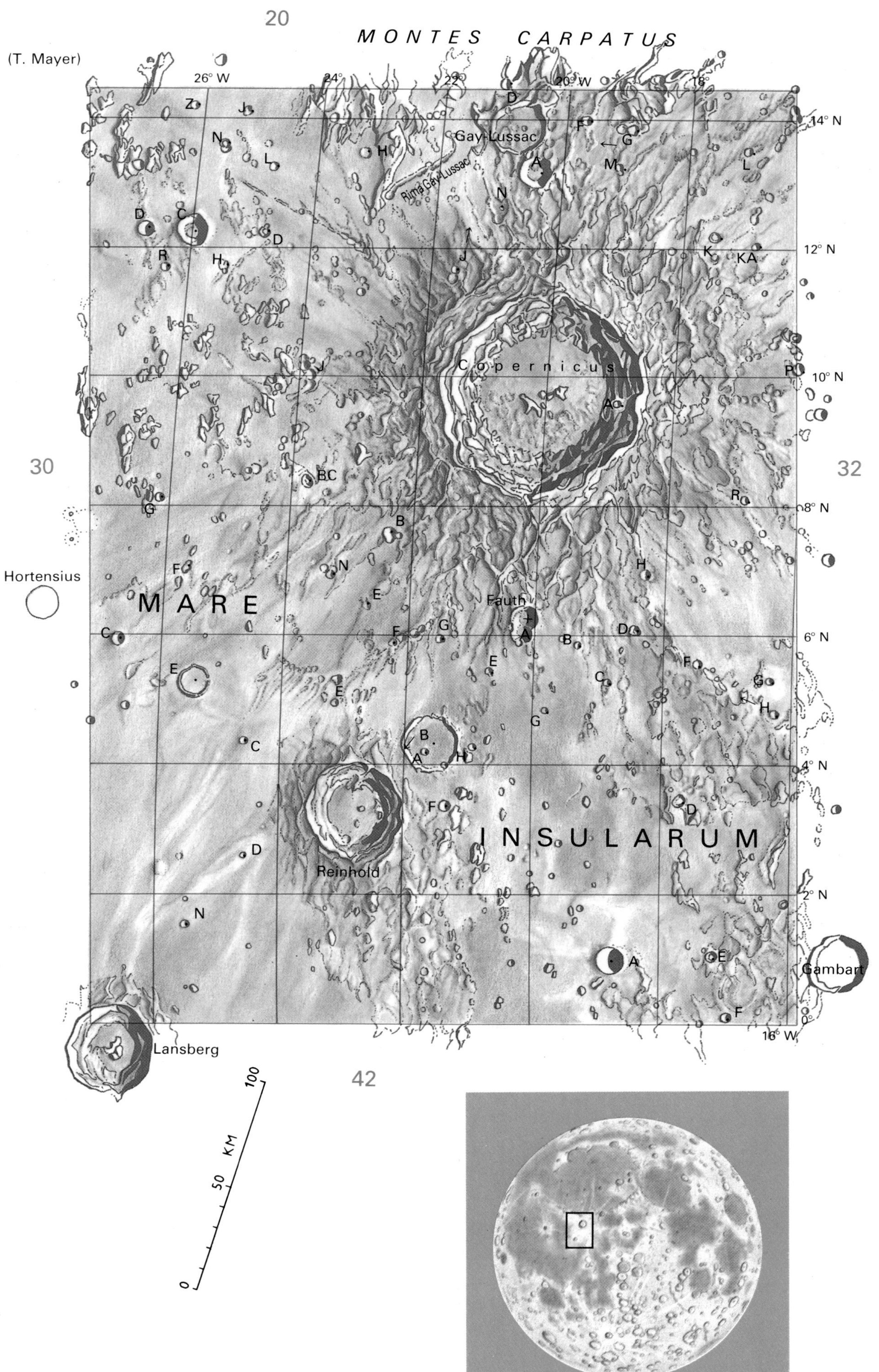

20
MONTES CARPATUS
(T. Mayer)
26° W
24°
22°
20° W
18°
14° N
12° N
10° N
8° N
6° N
4° N
2° N
0°
16° W
Gay-Lussac
Rima Gay-Lussac
Copernicus
30
32
Hortensius
MARE
INSULARUM
Fauth
Reinhold
Lansberg
Gambart
42
0
50
KM
100

32. STADIUS

The dark monotonous area of Sinus Aestuum is surrounded by several interesting formations, the most striking being the beautiful crater Eratosthenes. A mountain range projects south-west of Eratosthenes towards the 'submerged crater' Stadius. North-west of Stadius is a crater chain which continues further north (see map 20). A dome is clearly visible close to the crater Gambart C. Under high illumination large, prominent dark patches can be seen to the south of the crater Copernicus C and to the north of the crater Schröter D. The area near Gambart C was the landing site of *Surveyor 2.*

Aestuum, Sinus (Bay of Billows) [12°N, 8°W]
A very flat mare-like area, partly disintegrated by inconspicuous wrinkle ridges and crater pits; total surface area 40 000 sq km, *diameter about* 230 km.

Eratosthenes (see map 21). In contrast with nearby remains of Stadius, Eratosthenes offers examples of two completely different aspects. Whereas under low illumination it appears as a prominent feature, at Full Moon it seems almost to disappear and is as faint as Stadius (see also p. 14, fig. 11).

Gambart [1.0°N, 15.2°W] Jean F. Gambart, 1800–1836. French astronomer, discovered thirteen comets.
Flooded crater with a single wall (25 km/1050 m).

Insularum, Mare (Sea of Isles). See map 30.

Schröter [2.6°N, 7.0°W] Johann H. Schröter, 1745–1816. German selenographer, experienced observer, author of *Selenotopographische Fragmente,* discovered numerous clefts and rilles on the Moon.
Crater with considerably disintegrated wall, open to the south (34.5 km).

Schröter, Rima [1°N, 6°W]
Rille, length 40 km.

Sömmering [0.1°N, 7.5°W] Samuel T. Sömmering, 1755–1830. German surgeon and naturalist.
Crater with considerably disintegrated wall (28 km).

Stadius [10.5°N, 13.7°W] Jan Stade, 1527–1579. Belgian mathematician and astronomer, author of planetary tables *Tabulae Bergenses.*
Circular depression, edged with incomplete low walls and crater pits; diameter 69 km, *height of northeast wall* 650 m.

Small craters: **Gambart B** (11.5 km/2170 m)
Gambart C (12.2 km/2300 m)
Schröter A (4.2 km/620 m)
Schröter W (10.1 km/610 m).

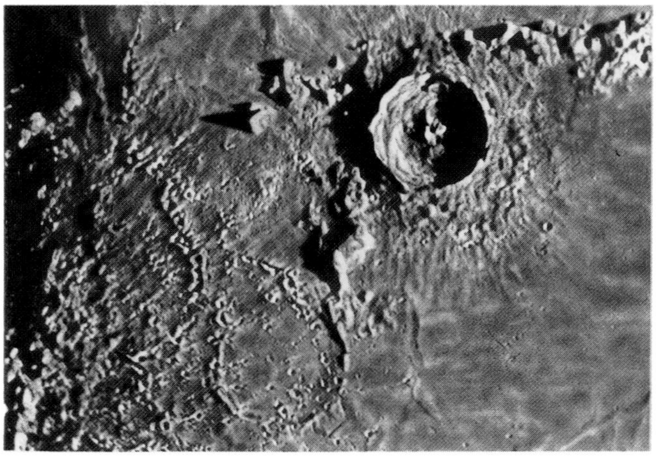

A considerable number of secondary craters, which originated from matter ejected during the formation of the crater Copernicus, is found in the vicinity of the lunar formation ***Stadius.*** *The prominent crater at the top is* ***Eratosthenes.***

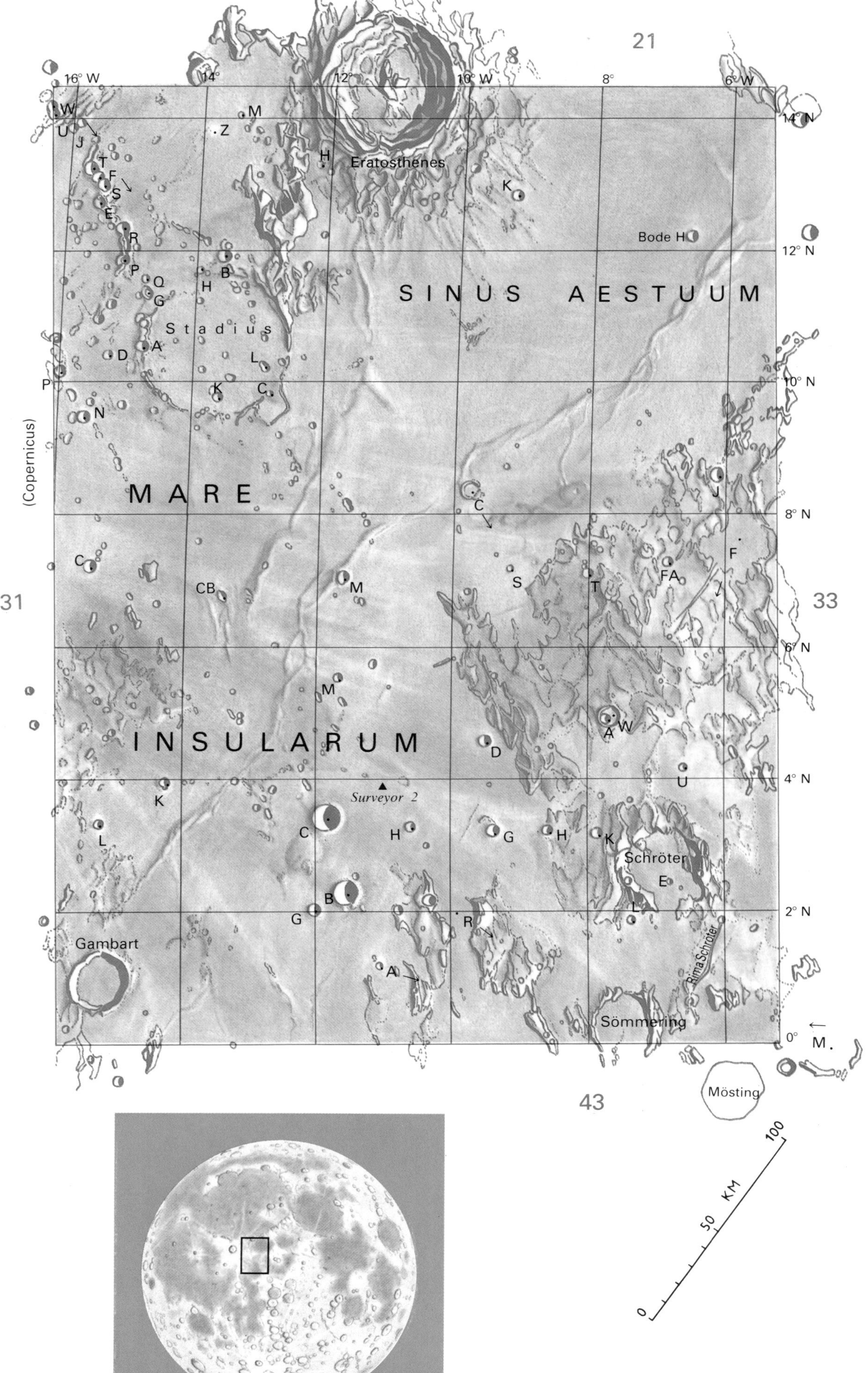
21
16° W
14°
12°
10° W
8°
6° W
14° N
12° N
10° N
8° N
6° N
4° N
2° N
0°
Eratosthenes
Bode H
SINUS AESTUUM
Stadius
MARE
INSULARUM
(Copernicus)
31
33
Surveyor 2
Schröter
Rima Schröter
Gambart
Sömmering
M.
Mösting
43
0
50 KM
100

33. TRIESNECKER

A 'continental' area, surrounded by Mare Vaporum, Sinus Medii and Sinus Aestuum. There is a complex system of clefts in the vicinity of the crater Triesnecker. The region also contains the landing sites of the probes *Surveyor 4* and *Surveyor 6*.

Aestuum, Sinus (Bay of Billows). See map 32.

Blagg [1.3°N, 1.5°E] Mary Adela Blagg, 1858–1944. English selenographer who played an important part in preparing the modern lunar nomenclature adopted by the IAU in 1935.
Small crater (5.4 km/910 m).

Bode [6.7°N, 2.4°W] Johann E. Bode, 1747–1826. German astronomer. 'Bode's Law' relates the distances of individual planets from the Sun.
Crater (18.6 km/3480 m).

Bode, Rimae [10°N, 4°W]
System of rilles visible through larger telescopes.

Bruce [1.1°N, 0.4°E] Catherine W. Bruce, 1816–1900. American patron of art and science who supported astronomers and astronomical institutions both at home and abroad.
Small crater (6.7 km/1260 m).

Chladni [4.0°N, 1.1°E] Ernst F. F. Chladni, 1756–1827. German physicist. In 1794 he was the first to demonstrate that meteorites are of cosmic origin. 'Chladni's figures' relate to the vibrations of sound.
Crater (13.6 km/2630 m).

Medii, Sinus (Central Bay). Mädler's name for a small mare at the centre of the near-side of the Moon.
Surface area 52 000 sq km, *diameter* 350 km.

Murchison [5.1°N, 0.1°W] Sir Roderick I. Murchison, 1792–1871. Scottish soldier, geologist and geographer.
Crater with a considerably disintegrated wall (58 km).

Pallas [5.5°N, 1.6°W] Peter Simon Pallas, 1741–1811. German naturalist and explorer. Discovered the 'Pallas meteorite' close to Krasnoyarsk.
Crater (50 km/1260 m).

Rhaeticus [0°N, 4.9°E] Georg Joachim von Lauchen (Rhaeticus), 1514–1576. German mathematician and astronomer, pupil of Copernicus.
Irregular crater with disintegrated wall (43 × 49 km).

Triesnecker [4.2°N, 3.6°E] Franz de Paula Triesnecker, 1745–1817. Austrian mathematician and astronomer.
Prominent crater with central peaks (26 km/2760 m).

Triesnecker, Rimae [5°N, 5°E]
The richest and best-known system of rilles, visible through even quite a small telescope. Length, measured from north to south, about 200 km.

Ukert [7.8°N, 1.4°E] Friedrich A. Ukert, 1780–1851. German historian and philologist.
Crater (23 km/2800 m).

Vaporum, Mare (Sea of Vapours) [13°N, 3°E] Riccioli's name for a circular mare situated to the south-east of the Apennines.
Surface area 55 000 sq km, *diameter about* 230 km.

Small craters: **Bode A** (12.3 km/2820 m)
Bode B (10.2 km/1780 m)
Bode C (7.0 km/1300 m)
Pallas A (10.6 km/2080 m).

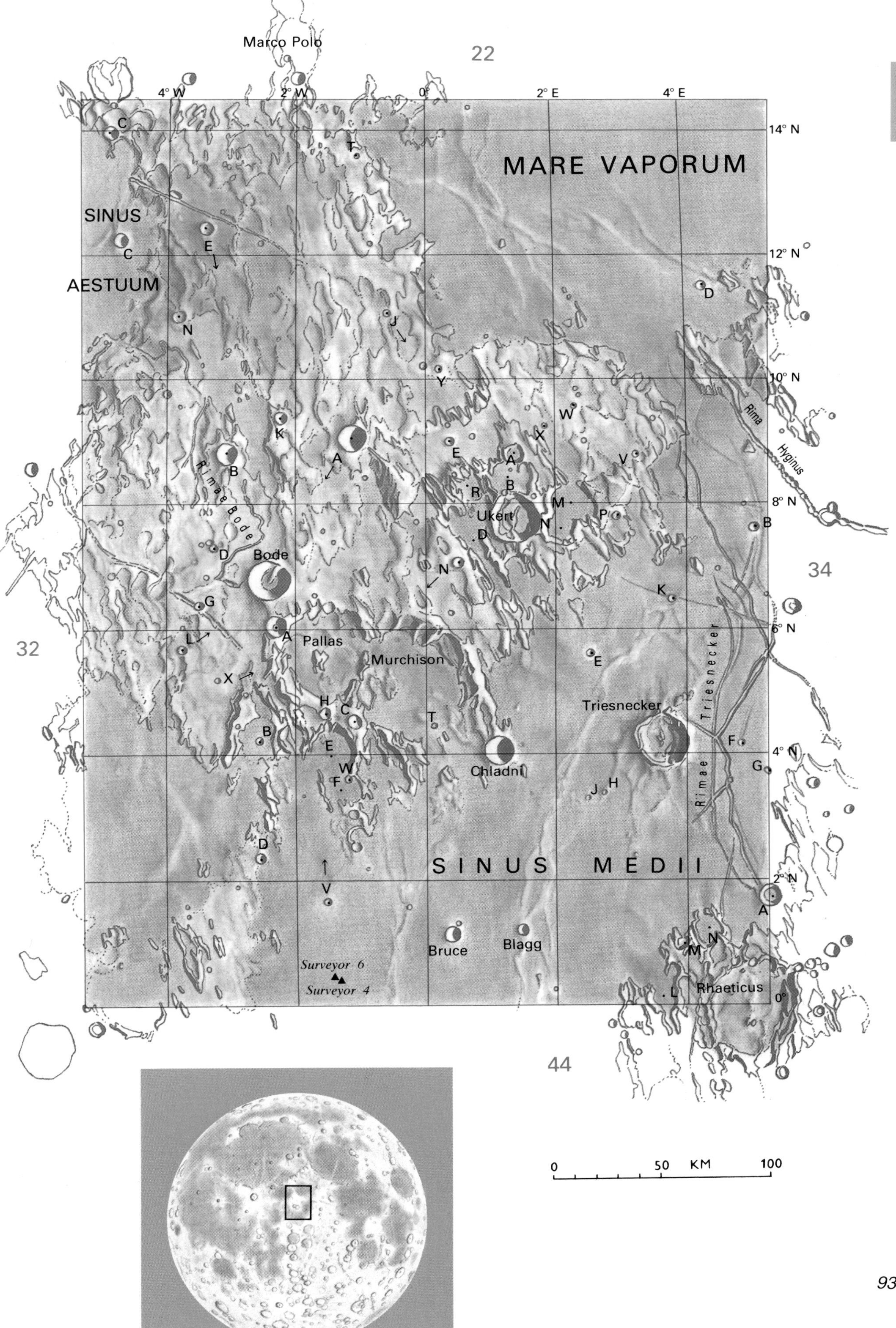
Marco Polo
22
4° W
2° W
0°
2° E
4° E
14° N
12° N
10° N
8° N
6° N
4° N
2° N
0°
MARE VAPORUM
SINUS
AESTUUM
Rimae Bode
Bode
Pallas
Murchison
Ukert
Triesnecker
Rimae Triesnecker
Rima Hyginus
Chladni
SINUS MEDII
Bruce
Blagg
Rhaeticus
Surveyor 6
Surveyor 4
32
34
44
0
50 KM
100

34. HYGINUS

A region with a prominent radial structure, directed towards the basin of Mare Imbrium, containing the rilles Rima Hyginus and Rima Ariadaeus which are readily visible. An interesting plateau is situated about 10 km north of the crater Godin C.

Ariadaeus, Rima [7°N, 13°E] (Ariadaeus, see map 35).
Wide rille, length 220 km.

Agrippa [4.1°N, 10.5°E] Agrippa, *c.* 92 AD. Greek astronomer. In 92 AD observed an occultation of the Pleiades by the Moon.
Crater (46 km/3070 m).

Boscovich [9.8°N, 11.1°E] Ruggiero G. Boscovich, 1711–1787. Croatian from Dubrovnik. Mathematician, physicist and astronomer.
Crater with a considerably disintegrated wall (46 km/1770 m).

Boscovich, Rimae
Rilles in Boscovich, length 40 km.

Cayley [4.0°N, 15.1°E] Arthur Cayley, 1821–1895. English mathematician and astronomer.
Circular crater (14.3 km/3130 m).

d'Arrest [2.3°N, 14.7°E] Heinrich L. d'Arrest, 1822–1875. German astronomer. Studied comets and asteroids.
Crater with a disintegrated wall (30 km/1490 m).

Dembowski [2.9°N, 7.2°E] Ercole Dembowski, 1812–1881. Italian astronomer. Measured 20 000 positions of double stars.
Crater with open wall to the east (26 km).

de Morgan [3.3°N, 14.9°E] Augustus de Morgan, 1806–1871. English mathematician.
Circular crater (10 km/1860 m).

Godin [1.8°N, 10.2°E] Louis Godin, 1704–1760. French mathematician and geodesist.
Crater (35 km/3200 m).

Hyginus [7.8°N, 6.3°E] Caius Julius Hyginus, first century AD. Spanish by origin, friend of Ovid. Described constellations and their mythology.
Crater (10.6 km/770 m).

Hyginus, Rima [7.8°N, 6.3°E]
Shallow valley, partially formed by a chain of craters. Length 220 km.

Julius Caesar [9.0°N, 15.4°E] Julius Caesar, 100–44 BC. Roman Emperor, honoured by Riccioli because of his reform of the calendar.
Flooded crater with a wide wall and a dark floor (90 km).

Lenitatis, Lacus (Lake of Tenderness). See map 23.

Manilius, see p. 72.

Silberschlag [6.2°N, 12.5°E] Johann E. Silberschlag, 1721–1791. German theologian and astronomer.
Circular crater (13.4 km/2530 m).

Tempel [3.9°N, 11.9°E] Ernest W. L. Tempel, 1821–1889. German astronomer, discovered six asteroids and many comets.
Crater with a disintegrated wall (48 km/1250 m).

Triesnecker, Rimae See map 33.

Vaporum, Mare (Sea of Vapours). See p. 92.

Whewell [4.2°N, 13.7°E] William Whewell, 1794–1866. English philosopher and historian of science.
Circular crater (14 km/2260 m).

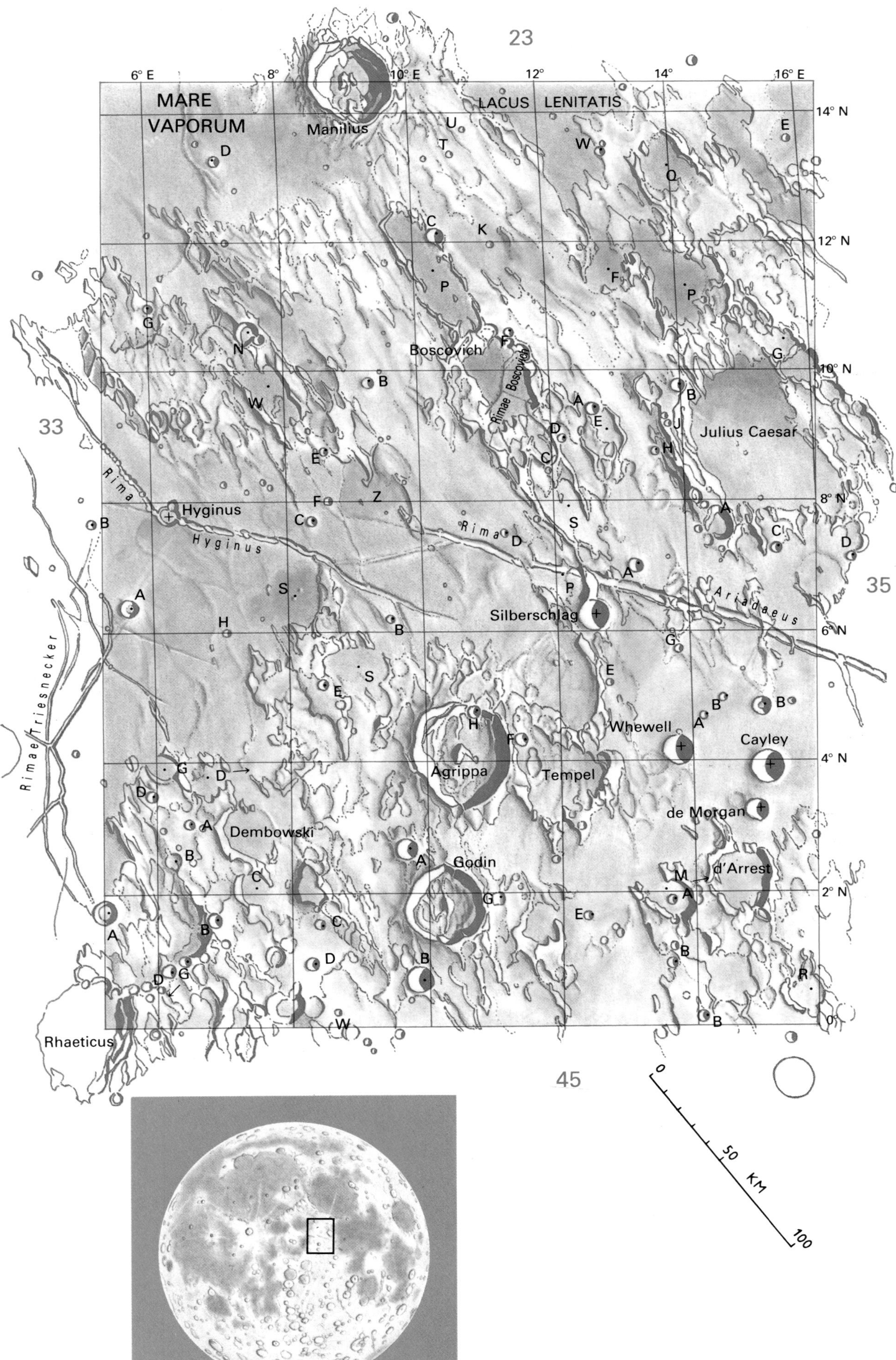
23
6° E
8°
10° E
12°
14°
16° E
14° N
12° N
10° N
8° N
6° N
4° N
2° N
0°
MARE VAPORUM
Manilius
LACUS LENITATIS
Boscovich
Rimae Boscovich
Julius Caesar
33
35
Rima Hyginus
Hyginus
Rima Ariadaeus
Silberschlag
Rimae Triesnecker
Whewell
Cayley
Agrippa
Tempel
de Morgan
Dembowski
Godin
d'Arrest
Rhaeticus
45
0
50
KM
100

35. ARAGO

The western part of Mare Tranquillitatis. Note the remarkable system of rilles running down the western edge of the mare, and the prominent network of wrinkle ridges surrounding the formation Lamont. Also of interest are the large domes Arago Alpha and Beta, and a group of four small domes north-west of Arago Alpha. This area contains the crash-landing sites of *Ranger 6* and *Ranger 8;* and the soft-landing sites of *Surveyor 5* and the *Apollo 11* mission.

Al-Bakri (Tacquet A) [14.3°N, 20.2°E] A. A. Al-Bakri, 1010–1094. Arabian geographer from Spain.
Crater with a flat floor (12 km/1000 m).

Aldrin (Sabine B) [1.4°N, 22.1°E] Edwin E. Aldrin, Jr., b. 1930. American astronaut (*Apollo 11*).
Crater (3.4 km/600 m).

Arago [6.2°N, 21.4°E] Dominique F. J. Arago, 1786–1853. French astronomer.
Crater (26 km).

Ariadaeus [4.6°N, 17.3°E] Philippus Arrhidaeus, d. 317 BC; Macedonian king, whose name was entered in the Babylonian list of eclipses.
Crater (11.2 km/1830 m).

Armstrong (Sabine E) [1.4°N, 25.0°E] Neil A. Armstrong, b. 1930. American astronaut (*Apollo 11*). First man to set foot on the Moon (1969).
Crater (4.6 km/670 m).

Carrel (Jansen B) [10.7°N, 26.7°E] Alexis Carrel, 1873–1944. French physician and physiologist, Nobel Laureate.
Crater (16 km).

Collins (Sabine D) [1.3°N, 23.7°E] Michael Collins, b. 1930. American astronaut (*Apollo 11*).
Crater (2.4 km/560 m).

Dionysius [2.8°N, 17.3°E] St Dionysius, 9–120 AD. According to Riccioli, he observed a solar eclipse when Christ was crucified.
Circular crater (17.6 km), *very bright at Full Moon.*

Honoris, Sinus (Bay of Honour) [12°N, 18°E].
Promontory of Mare Tranquillitatis, about 100 km *long.*

Lamont [5.0°N, 23.2°E] John Lamont, 1805–1879. Scottish-born German astronomer.
Inconspicuous formation 75 km *in diameter, whose wall is outlined by wrinkle ridges.*

Maclear [10.5°N, 20.1°E] Sir Thomas Maclear, 1794–1879. Irish astronomer.
Flooded crater (20 km/610 m).

Maclear, Rimae [13°N, 20°E]
Parallel rilles, length 100 km.

Manners [4.6°N, 20.0°E] Russell Henry Manners, 1800–1870. English admiral and astronomer.
Crater (15 km/1710 m).

Ritter [2.0°N, 19.2°E] (1) Karl Ritter, 1779–1859, German geographer. (2) August Ritter, 1826–1908, German astrophysicist.
Crater with an uneven floor (31 km).

Ritter, Rimae [3°N, 18°E]
Parallel rilles, length about 100 km.

Ross [11.7°N, 21.7°E] (1) Sir James C. Ross, 1800–1862, English polar explorer, gave his name to the Ross Sea. (2) Frank E. Ross, 1874–1966, American astronomer, studied ultra-violet radiation.
Crater with an elongated shape (26 km).

Sabine [1.4°N, 20.1°E] Sir Edward Sabine, 1788–1883. Irish astronomer.
Crater (30 km).

Schmidt [1.0°N, 18.8°E] (1) Johann F. J. Schmidt, 1825–1884, German selenographer. (2) Bernhard Schmidt, 1879–1935, German optician, inventor of the Schmidt camera. (3) Otto J. Schmidt, 1891–1956. Soviet naturalist.
Circular crater (11.4 km/2300 m).

Sosigenes [8.7°N, 17.6°E] Sosigenes, first century BC. Greek astronomer. Julius Caesar's adviser, helped to introduce reformed Julian calendar, 46 BC.
Crater (18 km/1730 m).

Sosigenes, Rimae [7°N, 19°E]
Parallel rilles, length about 150 km.

Tranquillitatis, Mare (Sea of Tranquillity). See p. 98. Between the craters Carrel and Lamont there is a remarkable transition from dark to light mare material.

Tranquillitatis, Statio (Tranquillity Base) [0.7°N, 23.5°E]. Landing site of *Apollo 11* (1969).

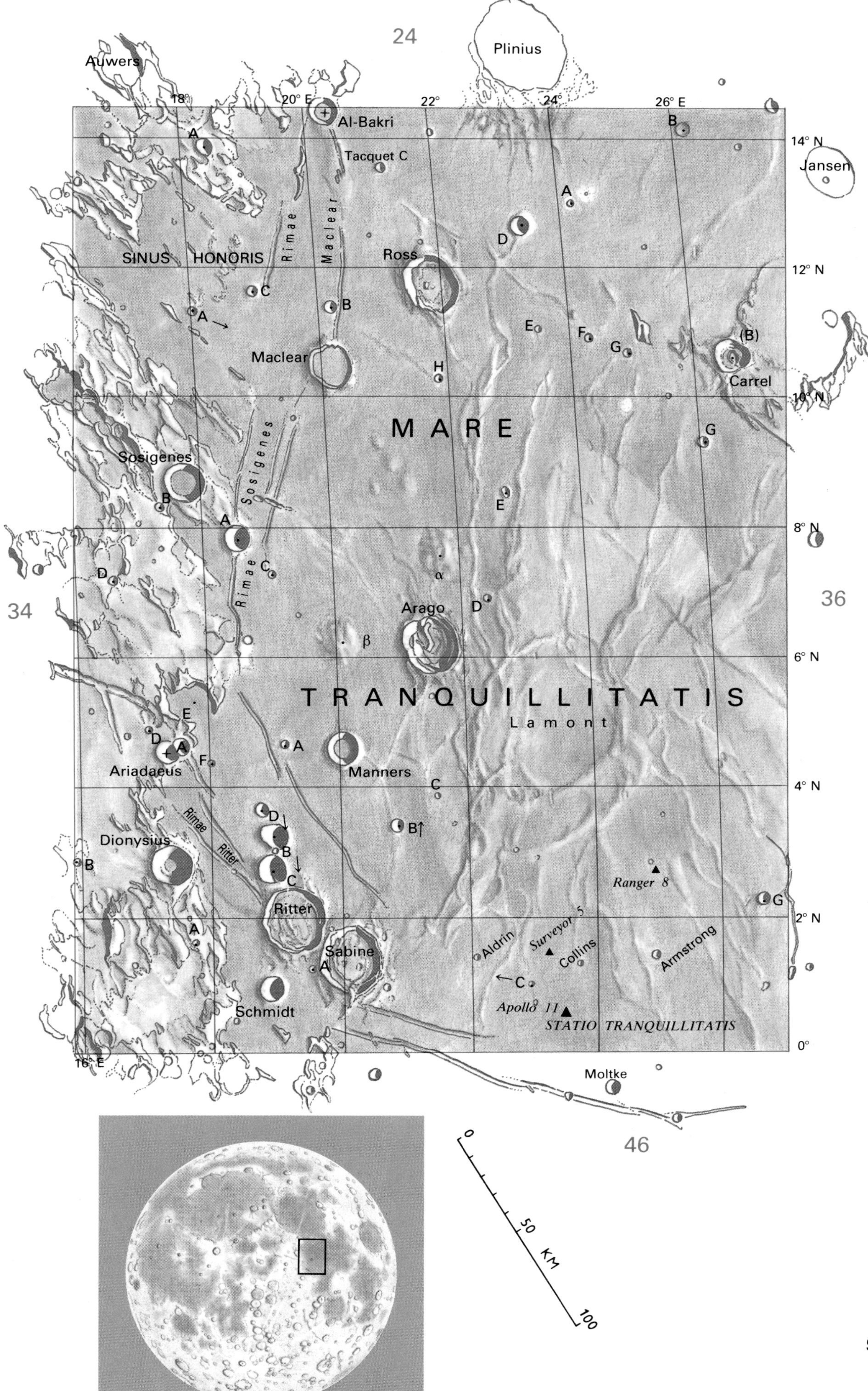
24
Plinius
Auwers
18°
20° E
22°
24°
26° E
Al-Bakri
Tacquet C
14° N
Jansen
SINUS HONORIS
Rimae Maclear
Ross
12° N
Maclear
Carrel
(B)
10° N
MARE
Sosigenes
Rimae Sosigenes
8° N
α
Arago
β
34
36
6° N
TRANQUILLITATIS
Lamont
Ariadaeus
Manners
4° N
Rimae Ritter
Dionysius
Ranger 8
Ritter
2° N
Sabine
Aldrin
Surveyor 5
Collins
Armstrong
Apollo 11
STATIO TRANQUILLITATIS
Schmidt
0°
16° E
Moltke
46
0
50
KM
100

36. CAUCHY

The central part of Mare Tranquillitatis. The most interesting area is in the neighbourhood of the crater Cauchy, where, under low illumination, a modest telescope will show rilles, a striking fault Rupes Cauchy and the typical domes Omega Cauchy and Tau Cauchy: on the summit of Omega Cauchy is a small craterlet, which is visible with a large telescope. A narrow cleft is present to the west of the crater Maskelyne (lower left-hand edge of the map).

Aryabhata (Maskelyne E) [6.2°N, 35.1°E] Aryabhata, 476–550 AD. Indian astronomer and mathematician.
Flooded crater with part of its eastern wall intact (22 km).

Barlow, Dorsa [15°N, 31°E] William Barlow, 1845–1934. British crystallographer.
System of wrinkle ridges, length 120 km.

Cajal (Jansen F) [12.6°N, 31.1°E] Santiago Ramón y Cajal. Spanish histologist, Nobel Laureate.
Crater (9 km/1800 m).

Cauchy [9.6°N, 38.6°E] Augustin L. Cauchy, 1789–1857. French mathematician.
Circular crater, bright at Full Moon (12.4 km/2610 m).

Cauchy, Rima [10°N, 39°E]
Clearly visible wide rille, length 210 km.

Cauchy, Rupes [9°N, 37°E]
A fault, about 120 km *long, changing over into a rille. At sunrise the north-eastern wall casts a striking shadow, but it is bright under the setting sun: compare with the similar Rupes Recta* (map 54).

Esam, Mons [14.6°N, 35.7°E] Arabian male name.
Hill, base diameter 8 km.

Jansen [13.5°N, 28.7°E] Zacharias Janszoon, 1580–1638. Dutch optician from Middleburg, one of the first telescope makers.
Flooded crater with a low wall (23 km/620 m).

Jansen, Rima [15°N, 29°E]
Rille, length 35 km. (See also map 25.)

Lucian (Maraldi B) [14.3°N, 36.7°E] Lucian of Samosata, 120–180 AD. Greek writer.
Crater (7 km/1490 m).

Lyell [13.6°N, 40.6°E] Sir Charles Lyell, 1797–1875. Scottish geologist and explorer.
Crater with an irregular, disintegrated wall and a dark floor (32 km).

Maskelyne [2.2°N, 30.1°E] Nevil Maskelyne, 1732–1811. Englishman, the fifth Astronomer Royal.
Crater with terraced wall and central peak (24 km).

Menzel [3.4°N, 36.9°E] Donald H. Menzel, 1901–1976. American astrophysicist.
Crater (3 km).

Sinas [8.8°N, 31.6°E] Simon Sinas, 1810–1876. Greek merchant, patron of astronomers, bequeathed Athens Observatory.
Circular crater (12.4 km/2260 m).

Tranquillitatis, Mare (Sea of Tranquillity). Named by Riccioli.
The surface area of M. Tranquillitatis is 421 000 sq km *and therefore can be compared with that of the Black Sea. There are numerous wrinkle ridges and domes, especially in the western part.*

Wallach (Maskelyne H) [4.9°N, 32.3°E] Otto Wallach, 1847–1931. German chemist.
Crater (6 km/1140 m).

Small craters: **Jansen Y** (3.6 km/690 m)
Maskelyne B (9.2 km/1910 m)
Sinas A (5.8 km/1140 m)
Sinas E (9.2 km/1700 m).

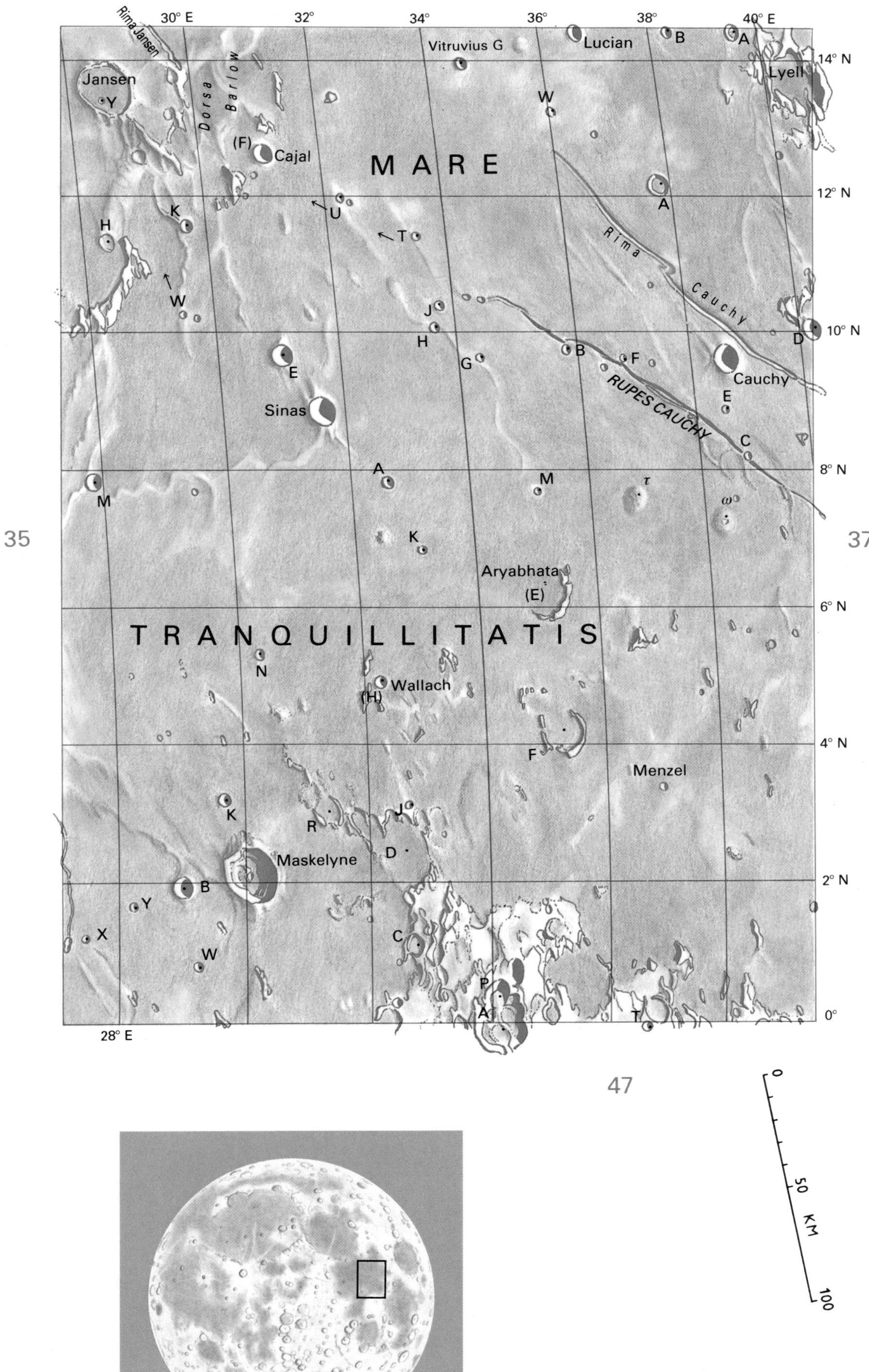
25
35
37
47
30° E
32°
34°
36°
38°
40° E
28° E
14° N
12° N
10° N
8° N
6° N
4° N
2° N
0°
Rima Jansen
Jansen
Y
Dorsa Barlow
(F)
Cajal
MARE
TRANQUILLITATIS
Vitruvius G
Lucian
B
A
Lyell
W
A
Rima Cauchy
U
T
K
H
W
J
H
G
B
F
D
E
Sinas
Cauchy
E
RUPES CAUCHY
C
A
M
M
τ
ω
K
Aryabhata
(E)
N
Wallach
(H)
F
Menzel
K
R
J
Maskelyne
D
B
Y
X
W
C
P
A
T
0
50
KM
100

37. TARUNTIUS

A narrow 'continental' strip separates Mare Tranquillitatis from Mare Fecunditatis. To the northeast is the dark area of the Mare Crisium, bordered by a mountain massif and numerous craters; under low illumination the edge of Mare Crisium offers a beautiful view. This area is intersected from the west by the rille Rima Cauchy.

Abbot (Apollonius K) [5.6°N, 54.8°E] Charles G. Abbot, 1872–1973. American astrophysicist. *Crater* (10 km).

Anville (Taruntius G) [1.9°N, 49.5°E] Jean-Baptiste d'Anville, 1697–1782. French cartographer.
Crater (11 km).

Asada (Taruntius A) [7.3°N, 49.9°E] Goryu Asada, 1734–1799. Japanese astronomer. *Crater* (12 km).

Cameron (Taruntius C) [6.2°N, 45.9°E] Robert C. Cameron, 1925–1972. American astronomer.
Crater (11 km).

Cato, Dorsa [1°N, 47°E] Marcus Porcius Cato, 234–149 BC. Roman architect.
System of wrinkle ridges, length 140 km.

Cayeux, Dorsum [1°N, 51°E] Lucien Cayeux, 1864–1944. French geologist. *Wrinkle ridge, length* 130 km.

Concordiae, Sinus (Bay of Concord) [11°N, 43°E]
Bay, length about 160 km.

Crile (Proclus F) [14.2°N, 46.0°E] G. Crile, 1864–1943. American physician. *Crater* (9 km).

Crisium, Mare (Sea of Crises). See also maps 26, 27 and 38.

Cushman, Dorsum [1°N, 49°E] J. A. Cushman, 1881–1949. American micropalaeontologist.
Wrinkle ridge, length 80 km.

da Vinci [9.1°N, 45.0°E] Leonardo da Vinci, 1452–1519. Famous Florentine artist, sculptor, mathematician, architect and engineer. The first to explain 'earthshine' on the Moon.
Inconspicuous crater with a disintegrated wall (38 km).

Fecunditatis, Mare (Sea of Fertility). See also maps 48, 49 and 59.

Glaisher [13.2°N, 49.5°E] James Glaisher, 1809–1903. English meteorologist.
Prominent crater at edge of Mare Crisium (16 km).

Greaves (Lick D) [13.2°N, 52.7°E] William M. H. Greaves, 1897–1955. British astronomer. Astronomer Royal for Scotland. *Crater* (14 km).

Lawrence (Taruntius M) [7.4°N, 43.2°E] Ernest O. Lawrence, 1901–1958. American physicist, Nobel Laureate.
Flooded crater (24 km/1000 m).

Lick [12.4°N, 52.7°E] James Lick, 1796–1876. American financier and philanthropist, endowed the Lick Observatory in California.
Flooded crater (31 km).

Secchi [2.4°N, 43.5°E] Pietro Angelo Secchi, 1818–1878. Italian astronomer, pioneer of stellar spectroscopy.
Inconspicuous crater with an open wall (24.5 km/1910 m).

Secchi, Montes [3°N, 43°E]
Inconspicuous mountain range, length about 50 km.

Secchi, Rimae [1°N, 44°E]
Pair of rilles occupying a 40 km *diameter area.*

Smithson (Taruntius N) [2.4°N, 53.6°E] James Smithson, 1765–1829. British chemist and mineralogist.
Crater (6 km).

Somni, Palus (Marsh of Sleep).
A 'continental' region stretching into Mare Tranquillitatis and separated from the remaining 'continental region' in the vicinity of Mare Crisium by bright rays from the crater Proclus. It stands out as a grey surface area at Full Moon.

Taruntius [5.6°N, 46.5°E] Lucius Taruntius Firmanus, *c.* 86 BC. Roman mathematician, philosopher and astrologer.
Crater with an outlined concentric wall and a central peak (56 km).

Tebbutt (Picard G) [9.6°N, 53.6°E] John Tebbutt, 1834–1916. Australian astronomer.
Flooded crater (32 km).

Watts (Taruntius D) [8.9°N, 46.3°E] Chester B. Watts, 1889–1971. American astronomer.
Flooded crater (15 km).

Zähringer (Taruntius E) [5.6°N, 40.2°E] Joseph Zähringer, 1929–1970. German physicist.
Crater (11.3 km/2110 m).

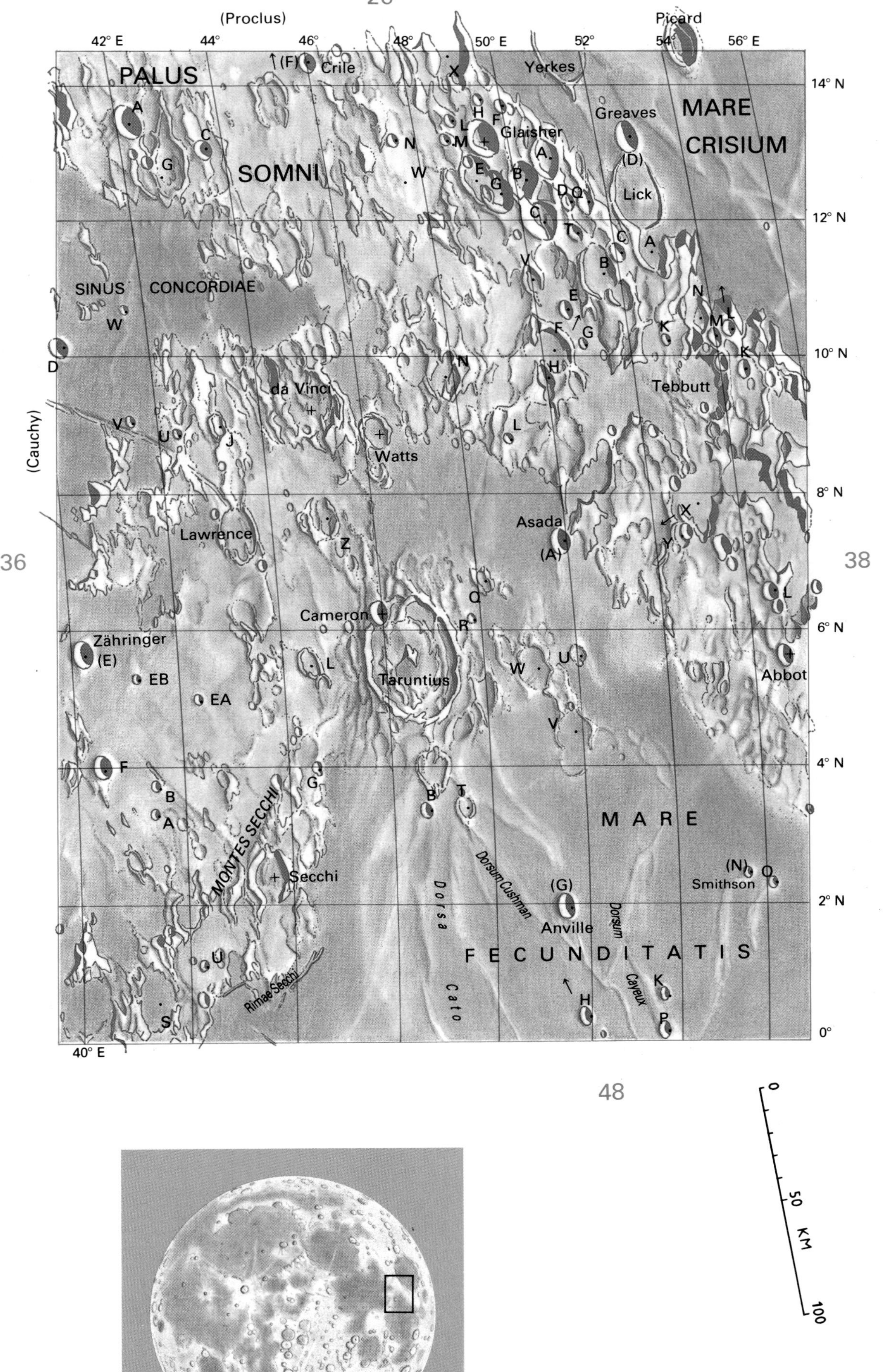
26
(Proclus)
Picard
42° E
44°
46°
48°
50° E
52°
54°
56° E
14° N
12° N
10° N
8° N
6° N
4° N
2° N
0°
40° E
PALUS
SOMNI
MARE
CRISIUM
SINUS
CONCORDIAE
Crile
Yerkes
Greaves
Glaisher
Lick
(D)
Tebbutt
da Vinci
Watts
(Cauchy)
Lawrence
Asada
(A)
36
38
Cameron
Taruntius
Zähringer
(E)
EB
EA
Abbot
MONTES SECCHI
Secchi
Rimae Secchi
Dorsa
Cato
Dorsum Cushman
Dorsum
Cayeux
(G)
Anville
MARE
FECUNDITATIS
(N)
Smithson
48
0
50
KM
100

38. NEPER

The eastern limb of the Moon, with a group of small lunar maria: M. Marginis, M. Smythii, M. Spumans and M. Undarum. This region appeared in its true form for the first time on the historic photograph taken by *Luna 3*. The landing sites of the probes *Luna 15, 18, 20* and *24* are shown on the map.

Agarum, Promontorium (Cape Agarum) [14°N, 66°E] Named after a cape in the Azov Sea, near the Crimea in Russia.

Ameghino (Apollonius C) [3.3°N, 57.0°E] Florentino Ameghino, 1854–1911. Italian historian of natural sciences.
Crater (9 km).

Apollonius [4.5°N, 61.1°E] Apollónios, second half of the third century BC. Greek mathematician, the 'Great Geometer' of Alexandria.
Flooded crater with a dark floor (53 km).

Auzout [10.3°N, 64.1°E] Adrien Auzout, 1622–1691. French astronomer, inventor of the filar micrometer.
Crater with peaks rising from its floor (33 km)

Back (Schubert B) [1.1°N, 80.7°E] Ernst E. A. Back, 1881–1959. German physicist. *Crater* (35 km).

Banachiewicz [5.2°N, 80.1°E] Tadeusz Banachiewicz, 1882–1954. Polish astronomer and mathematician.
Walled plain (92 km).

Boethius (Dubiago U) [5.6°N, 72.3°E] Boethius, 480–524 AD. Greek physicist. *Crater* (10 km).

Bombelli (Apollonius T) [5.3°N, 56.2°E] R. Bombelli, 1526–1572. Italian mathematician. *Crater* (10 km).

Cartan (Apollonius D) [4.2°N, 59.3°E] E. J. Cartan, 1869–1951. French mathematician. *Crater* (16 km).

Condon (Webb R) [1.9°N, 60.4°E] Edward W. Condon, 1902–1974. American physicist.
Flooded crater (35 km).

Condorcet [12.1°N, 69.6°E] Jean A. de Condorcet, 1743–1794. French mathematician and philosopher.
Flooded crater with a dark floor (74 km).

Crisium, Mare (Sea of Crises). See p. 80.

Daly (Apollonius P) [5.7°N, 59.6°E] Reginald A. Daly, 1871–1957. Canadian geologist. *Crater* (17 km).

Dubiago [4.4°N, 70.0°E] (1) Dmitri Dubiago, 1849–1918; (2) Alexander D. Dubiago, 1903–1959. Russian astronomers.
Flooded crater with a dark floor (51 km).

Fahrenheit (Picard X) [13.1°N, 61.7°E] Daniel G. Fahrenheit, 1686–1736. German physicist. *Crater* (6 km).

Firmicus [7.3°N, 63.4°E] Firmicus Maternus, *c.* 330 AD. Astrologer, Sicilian by origin.
Flooded crater with a dark floor (56 km).

Hansen [14.0°N, 72.5°E] Peter Andreas Hansen, 1795–1874. Danish astronomer. *Crater with a central peak* (40 km).

Harker, Dorsa See map 27.

Jansky [8.5°N, 89.5°E] Karl Jansky, 1905–1950. American radio-physicist and radio-astronomer. *Crater* (73 km).

Jenkins (Schubert Z) [0.3°N, 78.1°E] Louise F. Jenkins, 1888–1970. American astronomer. *Crater* (38 km).

Knox-Shaw (Banachiewicz F) [5.3°N, 80.2°E] Harold Knox-Shaw, 1885–1970. British astronomer. *Crater* (12 km).

Krogh (Auzout B) [9.4°N, 65.7°E] Schack A. S. Krogh, 1874–1949. Danish zoologist. *Crater* (20 km).

Liouville (Dubiago S) [2.6°N, 73.5°E] Joseph Liouville, 1809–1882. French mathematician. *Crater* (16 km).

Marginis, Mare (The Border Sea) [12°N, 88°E].
Surface area 62 000 sq km.

Neper [8.8°N, 84.5°E] John Napier, 1550–1617. Scottish mathematician, inventor of logarithms in 1614.
Very prominent crater with a central massif and a dark floor (137 km).

Nobili (Schubert Y) [0.2°N, 75.9°E] Leopoldo Nobili, 1784–1835. Italian physicist. *Crater* (42 km).

Peek [2.6°N, 86.9°E] Bertrand M. Peek, 1891–1965. British astronomer. *Crater* (13 km).

Perseverantiae, Lacus (Lake of Persistence – marked on the map as LAC. PER.) [8°N, 62°E]
Diameter 70 km.

Petit (Apollonius W) [2.3°N, 63.5°E] Alexis T. Petit, 1771–1820. French physicist. *Crater* (5 km).

Pomortsev (Dubiago P) [0.7°N, 66.9°E] Mikhail M. Pomortsev, 1851–1916. Russian pioneer of rocket propulsion.
Crater (23 km).

Respighi (Dubiago C) [2.8°N, 71.9°E] Lorenzo Respighi, 1824–1890. Italian astronomer. *Crater* (19 km).

Sabatier [13.2°N, 79.0°E] Paul Sabatier, 1854–1941. French chemist, Nobel Laureate. *Crater* (10 km).

Schubert [2.8°N, 81.0°E] Theodor F. von Schubert, 1789–1865. Russian cartographer. *Crater* (54 km).

Shapley (Picard H) [9.4°N, 56.9°E] Harlow Shapley, 1885–1972. American astronomer. *Crater* (23 km).

Smythii, Mare (Smyth's Sea) [2°S, 87°E]. See also map 49. William Henry Smyth, 1788–1865. British astronomer, writer, and admiral.
Circular mare; surface area 104 000 sq km.

Spumans, Mare (Foaming Sea) [1°N, 65°E]
Surface area 16 000 sq km.

Stewart (Dubiago Q) [2.2°N, 67.0°E] John Q. Stewart, 1894–1972. American astrophysicist. *Crater* (13 km).

Successus, Sinus (Bay of Success) [1°N, 58°E]
Diameter 100 km.

Tacchini (Neper K) [4.9°N, 85.8°E] Pietro Tacchini, 1838–1905. Italian astronomer. *Crater* (40 km).

Termier, Dorsum [11°N, 58°E] Pierre-Marie Termier, 1859–1930. French geologist.
Wrinkle ridge, length 90 km.

Theiler [13.4°N, 83.3°E] Max Theiler, 1899–1972. South-African bacteriologist, Nobel Laureate. *Crater* (8 km).

Townley (Apollonius G) [3.4°N, 63.3°E] Sidney D. Townley, 1867–1946. American astronomer. *Crater* (19 km).

Undarum, Mare (Sea of Waves) [7°N, 69°E]
Surface area 21 000 sq km.

Usov, Mons [12°N, 63°E] Mikhail A. Usov, 1883–1933. Soviet geologist. *Mountain, diameter* 15 km.

van Albada (Auzout A) [9.4°N, 64.3°E] Gale B. van Albada, 1912–1972. Dutch astronomer. *Crater* (22 km).

Virchow (Neper G) [9.8°N, 83.7°E] Rudolph L. K. Virchow, 1821–1902. German physician and pathologist.
Crater (17 km).

Wildt (Condorcet K) [9.0°N, 75.8°E] Rupert Wildt, 1905–1976. German-born American astronomer. Studied the constitution of the planets. *Crater* (11 km).

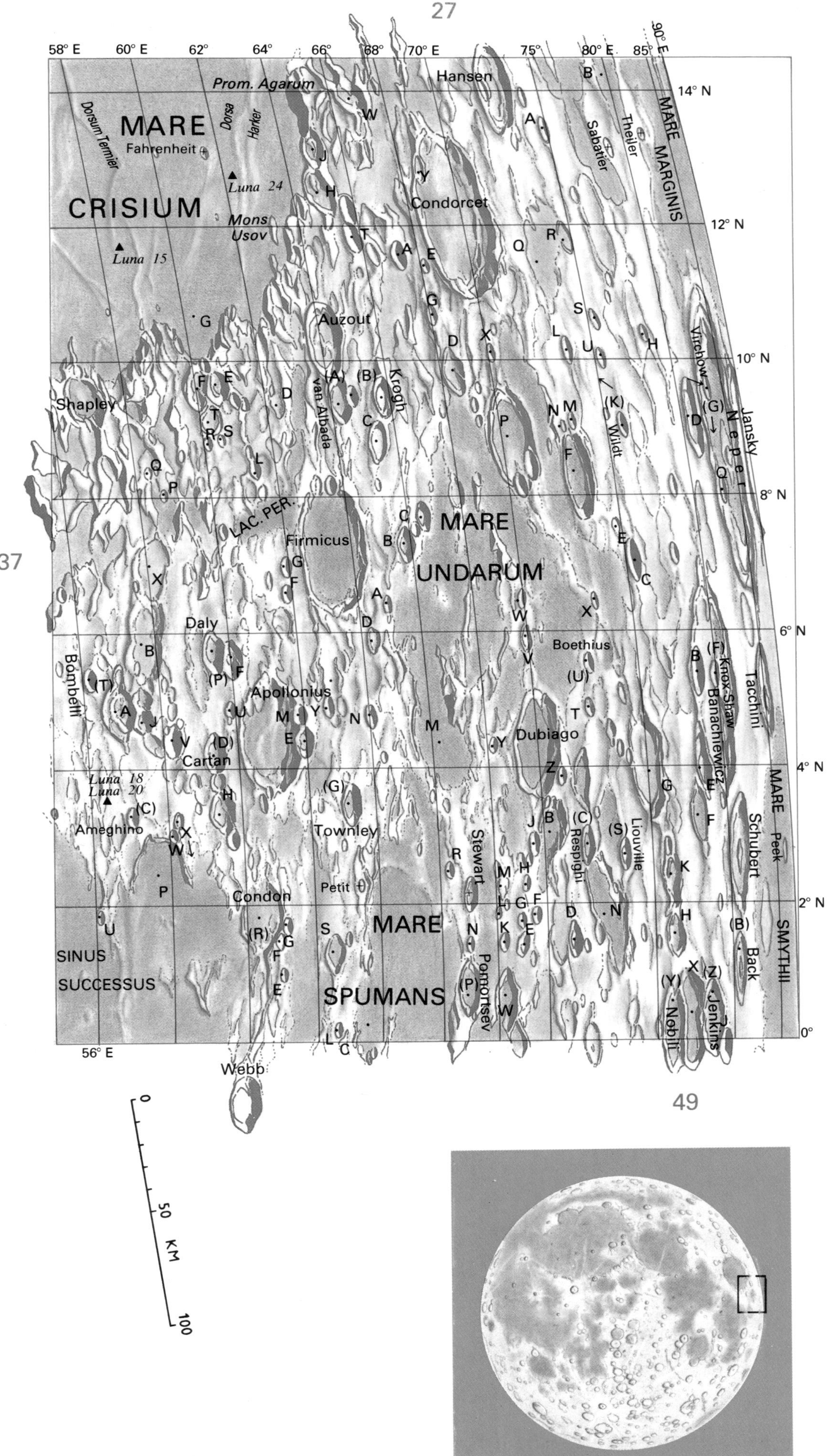
27
58° E
60° E
62°
64°
66°
68°
70° E
75°
80° E
85°
90° E
14° N
12° N
10° N
8° N
6° N
4° N
2° N
0°
56° E
Prom. Agarum
Hansen
MARE
CRISIUM
Dorsum Termier
Dorsa Harker
Fahrenheit
Luna 24
Mons Usov
Luna 15
Sabatier
Theiler
MARE MARGINIS
Condorcet
Auzout
Shapley
van Albada
Krogh
Virchow
Wildt
Neper
Jansky
LAC. PER.
Firmicus
MARE
UNDARUM
37
Daly
Boethius
Bombelli
Apollonius
Knox-Shaw
Banachiewicz
Tacchini
Dubiago
Cartan
Luna 18
Luna 20
Ameghino
Townley
Stewart
Respighi
Liouville
Schubert
Peek
MARE SMYTHII
Petit
Condon
MARE
SPUMANS
SINUS
SUCCESSUS
Pomortsev
Back
Nobili
Jenkins
Webb
49
0
50
KM
100

39. GRIMALDI

The western limb of the Moon, the south-western edge of Oceanus Procellarum. The north-eastern edge of the basin Mare Orientale. A small lunar basin Grimaldi.

Aestatis, Lacus (Summer Lake) [15°S, 69°W]
Two elongated dark patches north of the crater Crüger (see also map 50); total surface area about 1000 sq km.

Autumni, Lacus (Autumn Lake) [14°S, 82°W]
Dark patches situated on the inside of the Cordillera Mountains; total surface area 3000 sq km, *the longest distance across* 240 km.

Cordillera, Montes (Cordillera Mountains) [20°S, 80°W]
Ring mountain formation 900 km *in diameter, which forms the outer wall of the basin Mare Orientale (continued on map 50). Length about* 1500 km. *From Earth only part of the mountains is visible, during favourable librations.*

Damoiseau [4.8°N, 61.1°W] Marie Charles T. de Damoiseau, 1768–1846. French astronomer.
Crater (37 km).

Grimaldi [5.2°S, 68.6°W] Francesco M. Grimaldi, 1618–1663. Italian physicist and astronomer, author of a map of the Moon which was used by Riccioli as a basis for his nomenclature.
Basin whose flooded centre is surrounded by an inner wall 222 km *in diameter; the damaged external wall has a diameter of* 430 km.

Grimaldi, Rimae [9°S, 64°W]
System of rilles, length about 230 km.

Hartwig [6.1°S, 80.5°W] Karl E. Hartwig, 1851–1923. German astronomer.
Crater (80 km).

Hermann [0.9°S, 57.3°W] Jacob Hermann, 1678–1733. Swiss mathematician.
Circular crater (15.5 km).

Lallemand (Koppf A) [14.3°S, 84.1°W] André Lallemand, 1904–1978. French astronomer.
Crater (18 km).

Lohrmann [0.5°S, 67.2°W] Wilhelm G. Lohrmann, 1796–1840. German geodesist and selenographer.
Crater (31 km).

Procellarum, Oceanus (Ocean of Storms). See p. 84.

Riccioli [3.0°S, 74.3°W] Giovanni Baptista Riccioli, 1598–1671. Italian philosopher, theologian and astronomer, author of *Almagestum Novum* in which he introduced the system of lunar nomenclature still in use today.
Walled plain (146 km).

Riccioli, Rimae [2°S, 74°W]
System of rilles, length up to 390 km.

Rocca [12.7°S, 72.8°W] Giovanni A. Rocca, 1607–1656. Italian mathematician.
Disintegrated crater (90 km).

Schlüter [5.9°S, 83.3°W] Heinrich Schlüter, 1815–1844. German astronomer, assistant to Bessel.
Prominent crater with terraced walls (89 km).

Sirsalis [12.5°S, 60.4°W] Gerolamo Sirsalis (Sersale), 1584–1654. Italian Jesuit selenographer.
Crater (42 km).

Sirsalis, Rimae [14°S, 60°W]
System of rilles, clearly visible through even a small telescope; the largest is called Rima Sirsalis (see also map 50).

Veris, Lacus (Spring Lake). See map 50.

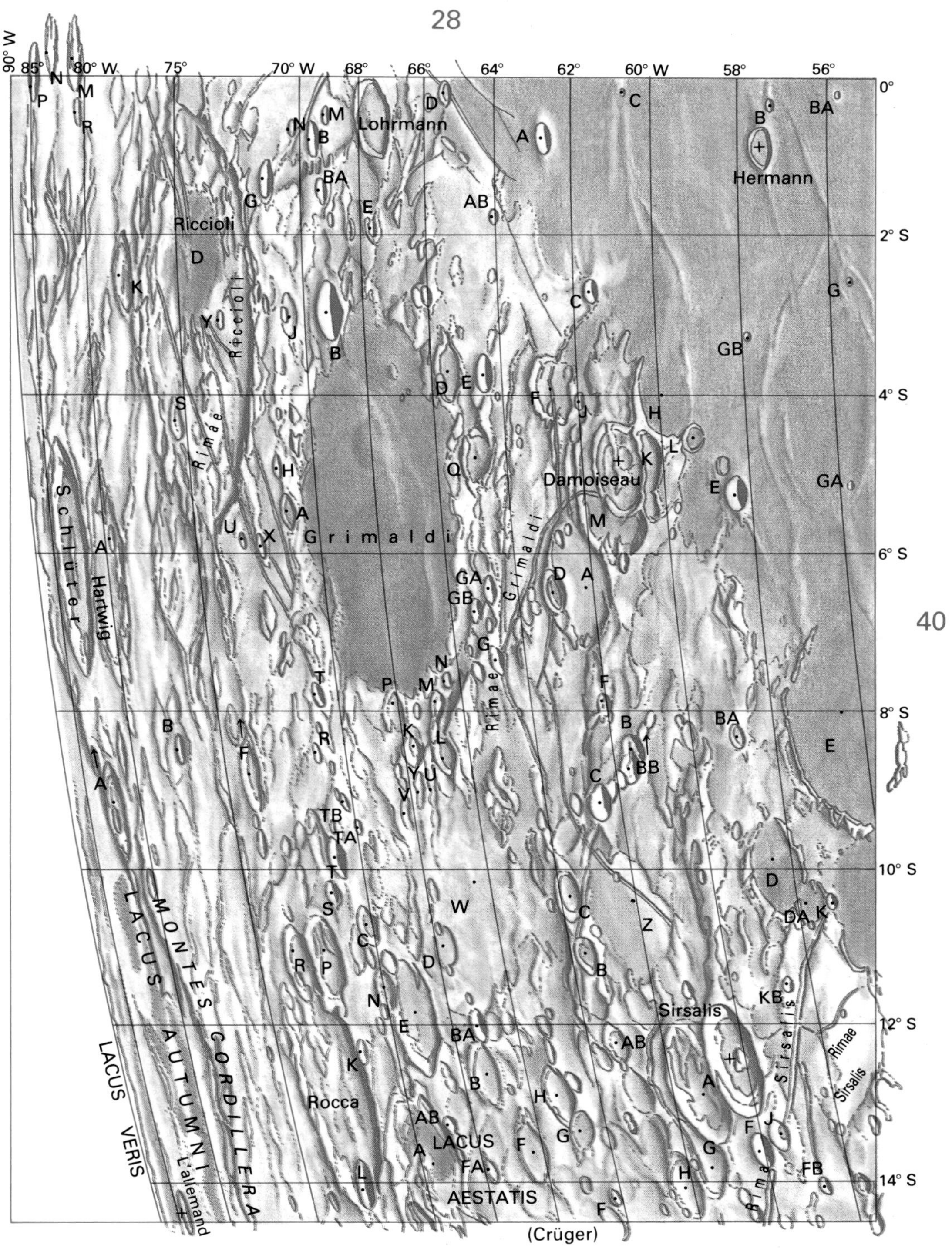

90° W 85° 80° W 75° 70° W 68° 66° 64° 62° 60° W 58° 56°

0° 2° S 4° S 6° S 8° S 10° S 12° S 14° S

Lohrmann

Hermann

Riccioli

Rimae Riccioli

Damoiseau

Grimaldi

Rimae Grimaldi

Schlüter

Hartwig

MONTES CORDILLERA

LACUS AUTUMNI

LACUS VERIS

Lallemand

Rocca

LACUS AESTATIS

Sirsalis

Rima Sirsalis

Rimae Sirsalis

(Crüger)

40

50

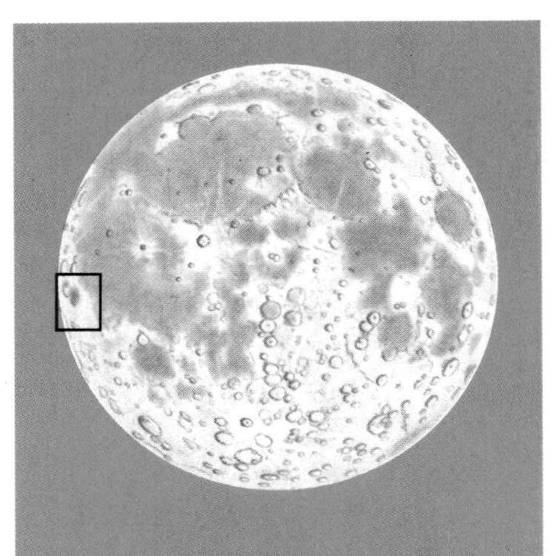

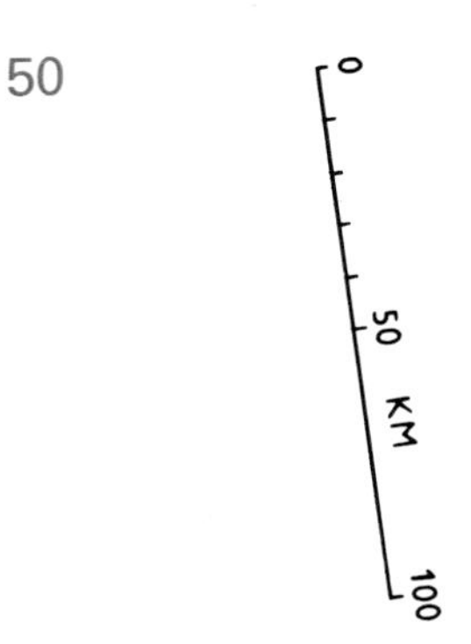

40. FLAMSTEED

The southern part of Oceanus Procellarum. The two craters Billy and Hansteen are useful points of orientation: they are separated by the massif Mons Hansteen. *Surveyor 1* landed close to the remains of the wall of the flooded crater Flamsteed P.

Billy [13.8°S, 50.1°W] Jacques de Billy, 1602–1679. French Jesuit mathematician and astronomer; rejected astrology and superstitious ideas about comets.
Flooded crater with a very dark floor (46 km).

Flamsteed [4.5°S, 44.3°W] John Flamsteed, 1646–1719. Englishman, the first Astronomer Royal. Author of the first star catalogue since that of Tycho Brahe. Flamsteed's system of numbering the stars is still in use today.
Crater (21 km/2160 m).

Hansteen [11.5°S, 52.0°W] Christopher Hansteen, 1784–1873. Norwegian geophysicist, discovered the position of the north geomagnetic pole.
Crater with hills on its floor (45 km).

Hansteen, Mons (earlier Hansteen Alpha) [12°S, 50°W]
Mountain massif of triangular contour, which appears very bright under high illumination; diameter 30 km.

Hansteen, Rima [12°S, 53°W]
Inconspicuous rille, length 25 km.

Letronne [10.6°S, 42.4°W] Jean Antoine Letronne, 1787–1848. French archaeologist, in his time an authority on the civilization of ancient Egypt.
Remains of a flooded walled plain 119 km *in diameter; it resembles a semi-circular bay in Oceanus Procellarum.*

Procellarum, Oceanus (Ocean of Storms). See p. 84.
This part of Oceanus Procellarum contains numerous wrinkle ridges, remains of the walls of flooded craters and many small hills. The adjacent area on map 41 is of similar character.

Rubey, Dorsa [10°S, 42°W] William Malden Rubey, 1898–1974. American geologist.
System of wrinkle ridges, length 100 km.

Winthrop (Letronne P) [10.7°S, 44.4°W] John Winthrop, 1714–1779. American astronomer.
Remains of a flooded crater, diameter 18 km.

Small craters: **Flamsteed F** (5.4 km/1050 m)
Letronne B (5.2 km/1000 m)
Letronne T (3.0 km/620 m)

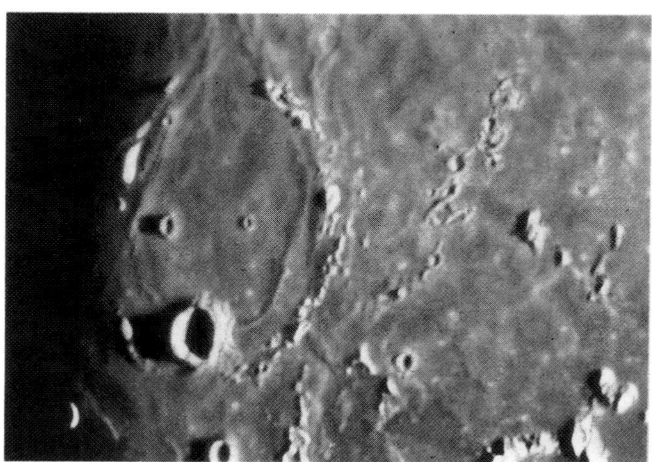

*The crater **Flamsteed** lies on the southern edge of the circular formation Flamsteed P, which is a flooded crater 110 km in diameter.*

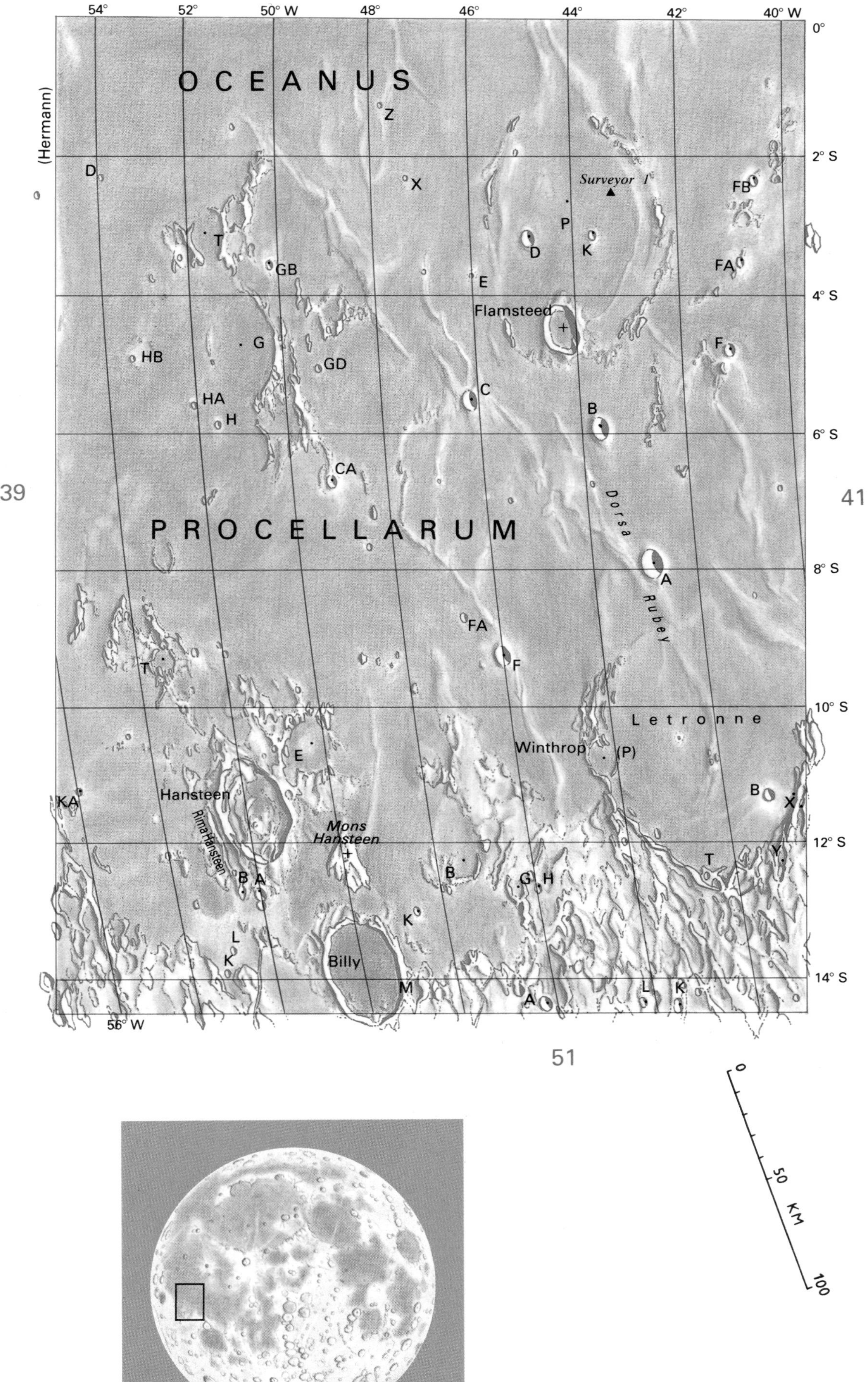
29
39
41
51
54°
52°
50° W
48°
46°
44°
42°
40° W
0°
2° S
4° S
6° S
8° S
10° S
12° S
14° S
56° W
OCEANUS
PROCELLARUM
(Hermann)
Z
X
D
T
GB
Surveyor 1
FB
P
D
K
FA
E
Flamsteed
G
HB
GD
F
C
HA
H
B
CA
Dorsa
Rubey
A
FA
F
T
Letronne
E
Winthrop
(P)
KA
Hansteen
Rima Hansteen
Mons Hansteen
B
X
Y
T
B
A
B
G
H
K
L
K
Billy
M
A
L
K
0
50
KM
100

41. EUCLIDES

The south-eastern part of Oceanus Procellarum. The Riphaeus Mountains stretch along the eastern edge of the map. To the south-east of the crater Lansberg D there is a large dome. This part of Oceanus Procellarum contains numerous wrinkle ridges and small isolated hills, and is interesting for telescopic observation, especially under low illumination. Larger telescopes will reveal the narrow sinuous rille Rima Herigonius.

Euclides [7.4°S, 29.5°W] Euclid, *c.* 300 BC. Famous Greek mathematician, founder of the Alexandrian mathematical school; 'Euclidean geometry'.
Very prominent and bright crater (12 km).

Ewing, Dorsa [11°S, 38°W] William M. Ewing, 1906–1974. American geophysicist.
System of wrinkle ridges, total length 320 km.

Herigonius [13.3°S, 33.9°W] Pierre Herigone, *c.* 1644. French mathematician. His six-volume *Cursus Mathematicus* includes spherical astronomy and the theory of the motion of the planets.
Crater (15 km/2100 m).

Herigonius, Rima [13°S, 37°W]
Sinuous rille with numerous 'meanders', length about 100 km.

Norman (Euclides B) [11.8°S, 30.4°W] Robert Norman, *c.* 1590. British naturalist.
Crater (10.3 km/2000 m).

Procellarum, Oceanus (Ocean of Storms). See p. 84.
The areas north of the crater Euclides F (5.2 km/1090 m) and in the neighbourhood of the crater Herigonius contain one of the richest systems of wrinkle ridges on the Moon.

Riphaeus, Montes (Riphaeus Mountains) [7°S, 28°W] According to ancient Greek geographers this was the mountain range from which north winds used to blow (today the Ural Mountains). (See also map 42.)
Range 150 km.

Scheele (Letronne D) [9.4°S, 37.8°W] Carl Wilhelm Scheele, 1742–1786. Swedish chemist.
Crater (5 km/750 m).

Wichmann [7.5°S, 38.1°W] Moritz L. G. Wichmann, 1821–1859. German astronomer, determined the inclination of the lunar equator and was one of the first to confirm the existence of the Moon's physical librations.
Circular crater (10.6 km).

Small craters: **Lansberg B** (9.9 km/2030 m)
Lansberg C (19.8 km/810 m)
Lansberg G (9.9 km/270 m)
Wichmann C (2.8 km/490 m).

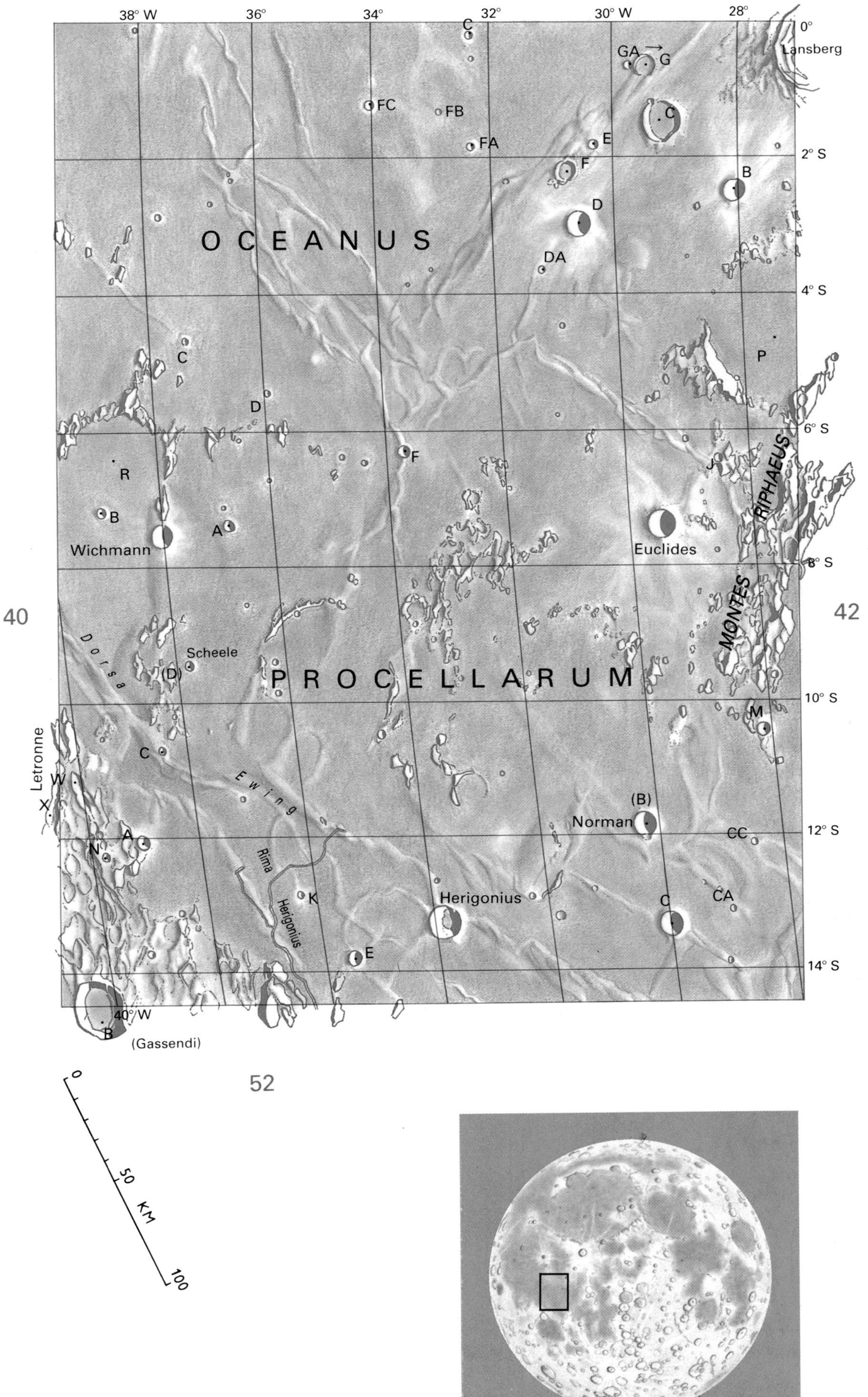
30
38° W
36°
34°
32°
30° W
28°
0°
Lansberg
GA
G
C
FC
FB
FA
E
F
2° S
B
D
OCEANUS
DA
4° S
C
P
D
6° S
F
R
J
B
A
Wichmann
Euclides
RIPHAEUS
8° S
40
42
Dorsa
MONTES
Scheele
(D)
PROCELLARUM
10° S
M
Letronne
C
W
X
Ewing
(B)
Norman
12° S
CC
A
N
Rima
Herigonius
K
Herigonius
C
CA
E
14° S
40° W
B
(Gassendi)
52
0
50
KM
100

42. FRA MAURO

The southern edge of Mare Insularum extends to between the craters Lansberg and Fra Mauro. The lower half of the map is occupied by Mare Cognitum. Although this area is seemingly uninteresting, it contains a number of geologically important locations and therefore was one of the most visited regions of the Moon. This is the area where the probe *Ranger 7* crash landed and where two *Apollo* expeditions landed: *Apollo 12* close to *Surveyor 3,* and *Apollo 14* in the hills at the edge of the crater Fra Mauro.

Bonpland [8.3°S, 17.4°W] Aimé Bonpland, 1773–1858. French botanist who accompanied Humboldt on his expeditions to Mexico and Colombia.
Remains of a walled plain with narrow clefts on its floor (60 km).

Cognitum, Mare [10°S, 23°W] (the 'Known' Sea). Named in 1964 after the successful flight of the probe Ranger 7, which transmitted to Earth the first detailed television pictures of the lunar surface. The seemingly smooth, flat surface of the mare is pitted with numerous craters, some of which are a thousand times smaller than the smallest visible through large Earth-based telescopes. A new stage in the exploration of the Moon began here.

Darney [14.5°S, 23.5°W] Maurice Darney, 1882–1958. French, observer of the Moon.
Circular crater (15 km/2620 m); *adjacent crater* **Darney C** (13.3 km/2330 m).

Eppinger (Euclides D) [9.4°S, 25.7°W] H. Eppinger, 1879–1946. Austrian physician.
Crater (6 km/1250 m).

Fra Mauro [6.0°S, 17.0°W] Fra Mauro, d. 1459. Venetian geographer; prepared a map of the World (1457).
Remains of a walled plain whose centre is intersected by longitudinal clefts (95 km).

Guettard, Dorsum [10°S, 18°W] Jean-Etienne Guettard, 1715–1786. French geologist.
Wrinkle ridge, length 40 km.

Insularum, Mare (Sea of Isles). See maps 30, 31.

Kuiper (Bonpland E) [9.8°S, 22.7°W] Gerard P. Kuiper, 1905–1973. Dutch-born American astronomer, made significant discoveries in the solar system; Director of Lunar and Planetary Laboratory, University of Arizona.
Crater (6.8 km/1330 m).

Lansberg [0.3°S, 26.6°W] Philippe van Lansberge, 1561–1632. Belgian physician and astronomer, author of a treatise on the use of astrolabes and gnomons.
Prominent crater (39 km/3110 m).

Moro, Mons [12°S, 20°W] Antonio L. Moro, 1687–1764. Italian naturalist.
Elevation about 10 km *long, situated on a wrinkle ridge.*

Opelt, Rimae [13°S, 18°W]. Named after the crater Opelt (map 53).
System of rilles, length 70 km.

Parry, Rimae [8°S, 17°W]. Named after the crater Parry (map 43).
System of rilles, visible through even quite a small telescope, length 300 km.

Riphaeus, Montes. See map 41.

Tolansky, see map 43.

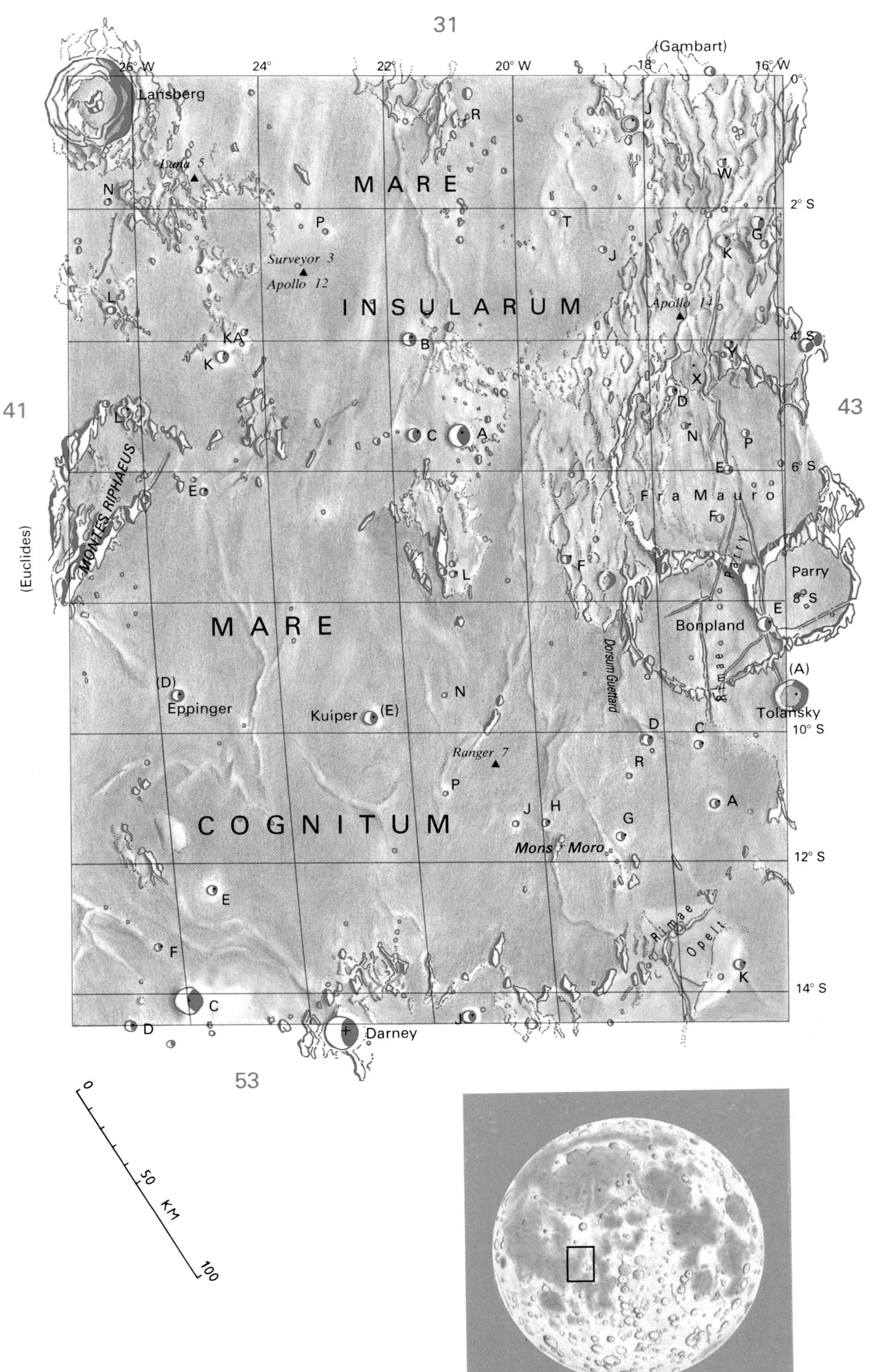

31
41
43
53
(Gambart)
(Euclides)
26° W
24°
22°
20° W
18°
16° W
0°
2° S
4° S
6° S
8° S
10° S
12° S
14° S
Lansberg
Luna 5
MARE
INSULARUM
Surveyor 3
Apollo 12
Apollo 14
MONTES RIPHAEUS
Fra Mauro
Parry
Bonpland
Dorsum Guettard
Rimae Parry
(A)
Tolansky
MARE
COGNITUM
(D)
Eppinger
Kuiper
(E)
Ranger 7
Mons Moro
Rimae Opelt
Darney
0
50
100
KM

43. LALANDE

Predominantly mare area, which is penetrated by Mare Nubium from the south. In the west it adjoins the group of three craters Bonpland, Parry and Fra Mauro. The edge of a vast 'continental' area, which shows traces of the development of the basin of Mare Imbrium (faults, valleys, clefts), protrudes into this region from the east. The crater Davy Y contains an interesting chain of crater pits.

Davy [11.8°S, 8.1°W] Sir Humphry Davy, 1778–1829. English physicist and chemist, inventor of the miners' safety-lamp.
Crater (35 km), *the wall of which is intersected by the crater* **Davy A** (15 km).

Davy, Catena [11°S, 7°W]
Typical crater chain, length about 50 km.

Guericke [11.5°S, 14.1°W] Otto von Guericke, 1602–1686. German physicist. In 1654 demonstrated the existence of atmospheric pressure by the well-known experiment with the so-called 'Magdeburg hemispheres'.
Remains of a walled plain (58 km).

Kundt (Guericke C) [11.5°S, 11.5°W] August Kundt, 1839–1894. German physicist.
Crater (11 km).

Lalande [4.4°S, 8.6°W] Joseph J. le François de Lalande, 1732–1807. French astronomer, Director of the Paris Observatory.
Prominent terraced crater (24 km/2590 m), *centre of a bright ray system.*

Mösting [0.7°S, 5.9°W] Johann S. von Mösting, 1759–1843. Danish patron of astronomers, one of the founders of the journal *Astronomische Nachrichten.*
Crater with terraced walls (26 km/2760 m).

Mösting A
Small bright ring crater (13 km/2700 m), *the fundamental point in the selenographical network of coordinates. Position:* 3°12′43.2″S, 5°12′39.6″W (Davies, 1987).

Palisa [9.4°S, 7.2°W] Johann Palisa, 1848–1925. Austrian astronomer. Discovered 127 asteroids.
Crater with disintegrated wall, open to south-west (33 km).

Parry [7.9°S, 15.8°W] Sir William E. Parry, 1790–1855. English admiral and Arctic explorer.
Crater with a flooded floor (48 km/560 m).

Parry, Rimae. See map 42.

Tolansky (Parry A) [9.5°S, 16.0°W] Samuel Tolansky, 1907–1973. British physicist.
Crater with a flat floor (13 km/880 m).

Turner [1.4°S, 13.2°W] Herbert H. Turner, 1861–1930. English astronomer. Participated in the formulation of international lunar nomenclature. In 1903 he discovered a nova in the constellation Gemini.
Ring crater (11.8 km/2630 m).

Small craters: **Davy C** (3.4 km/540 m)
Fra Mauro R (3.4 km/650 m)
Guericke D (7.6 km/1500 m)*
Lalande A (13.2 km/2600 m)
Parry D (2.8 km/330 m)

* This crater is situated on the top of the hill Fra Mauro Eta.

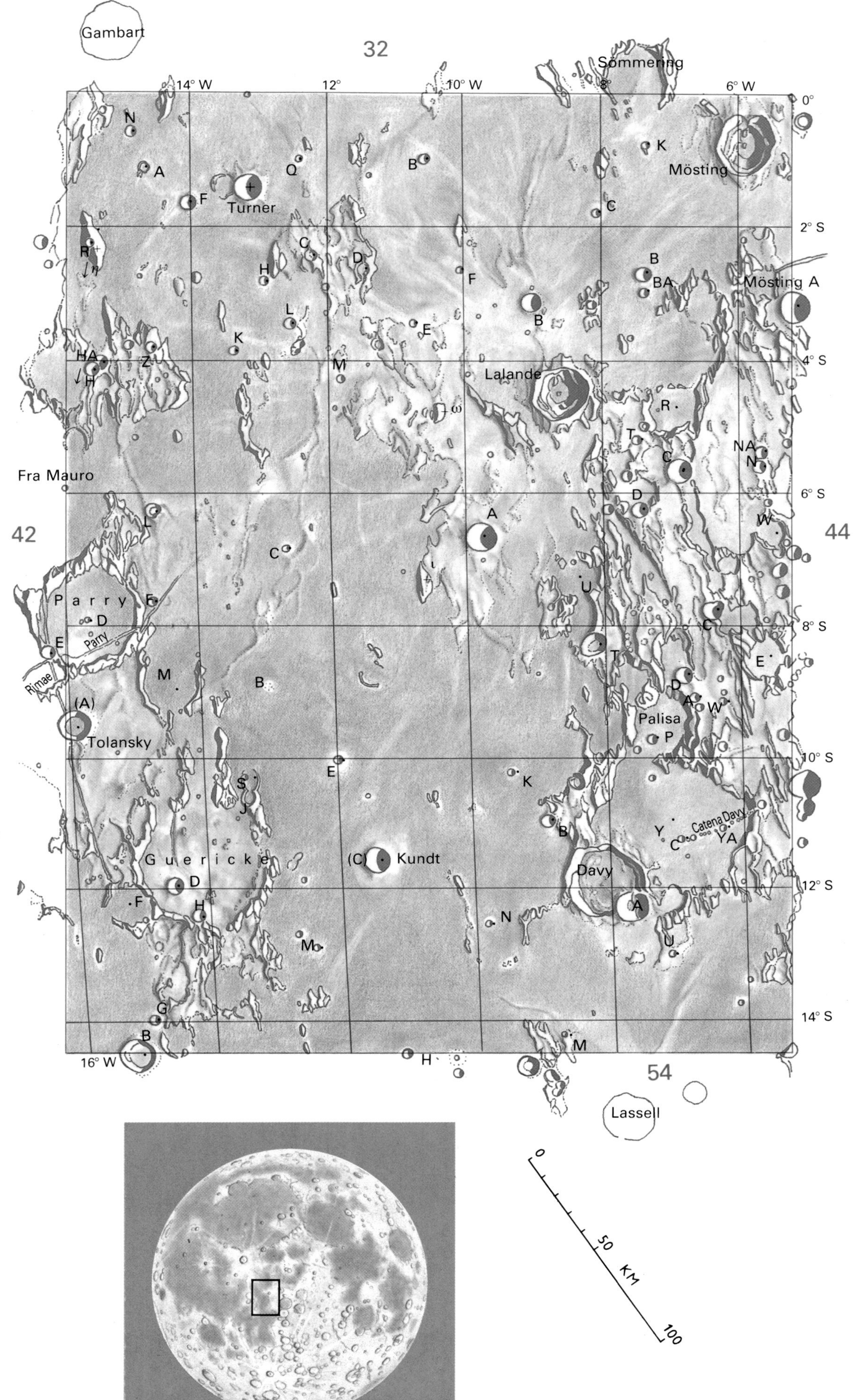
Gambart
32
Sömmering
14° W
12°
10° W
8°
6° W
0°
N
K
A
Q
B
Mösting
F
Turner
C
2° S
R
C
D
B
H
F
BA
Mösting A
L
B
E
K
HA
Z
M
4° S
H
Lalande
ω
R
T
NA
C
N
Fra Mauro
6° S
D
L
A
W
42
44
C
U
Parry
F
D
C
8° S
E
Parry
T
Rimae
M
B
E
D
A
W
(A)
Palisa
Tolansky
P
10° S
E
K
S
J
B
Catena Davy
Y
C
YA
Guericke
(C)
Kundt
Davy
D
12° S
F
H
A
N
M
U
G
14° S
B
M
16° W
H
54
Lassell
0
50
KM
100

44. PTOLEMAEUS

The central part of the near-side of the Moon, with large walled plains. The dark surface of Sinus Medii, bordered by clefts, extends into this area from the north. This region, like that shown on map 43, contains a network of valleys and clefts radiating from Mare Imbrium. The floor of the crater Alphonsus was the crash-landing site of the probe *Ranger 9*.

Albategnius [11.2°S, 4.1°E] Muhammed ben Geber al Batani, 852–929 AD. Arabian prince and astronomer.
Ring mountain (136 km).

Alphonsus [13.4°S, 2.8°W] Alfonso X, 'El Sabio' (The Wise), 1221–1284 AD. King of Castile, astronomer. *Alphonsine Tables.*
Ring mountain with a central peak, on its floor rilles and crater pits with dark haloes.

Alphonsus, Rimae
System of rilles inside Alphonsus.

Ammonius (Ptolemaeus A) [8.5°S, 0.8°W] Ammonius, *c.* 517 AD. Greek philosopher.
Crater (9 km/1850 m).

Flammarion [3.4°S, 3.7°W] Camille Flammarion, 1842–1925. French astronomer, famous popularizer of astronomy.
Walled plain (75 km); **Mösting A** *rises from its western wall.* See also map 43.

Flammarion, Rima [2°S, 5°W]
Rille, length 80 km.

Gyldén [5.3°S, 0.3°E] Hugo Gyldén, 1841–1896. Finnish astronomer, Director of Stockholm Observatory.
Disintegrated crater (47 km).

Herschel [5.7°S, 2.1°W] William Herschel, 1738–1822. German-born English astronomer, discoverer of Uranus; pioneered stellar astronomy, discovered 2500 nebulae and galaxies.
Prominent terraced crater (41 km/3770 m).

Hipparchus [5.5°S, 4.8°E] Hipparchos, *c.* 190–125 BC. Renowned Greek astronomer, author of the first star catalogue.
Considerably disintegrated walled plain (150 km/3320 m).

Klein [12.0°S, 2.6°E] Hermann J. Klein, 1844–1914. German selenographer, popularizer of astronomy.
Crater (44 km/1460 m).

Medii, Sinus (Central Bay). See also map 33.

Müller [7.6°S, 2.1°E] Karl Müller, 1866–1942. Austrian selenographer.
Elongated crater (24 × 20 km).

Oppolzer [1.5°S, 0.5°W] Theodor E. von Oppolzer, 1841–1886. Austrian astronomer, author of tables of solar and lunar eclipses to 2163 AD.
Remains of a crater (43 km), *cleft on floor.*

Oppolzer, Rima [1°S, 2°E]
Rille, 110 km *long.*

Ptolemaeus [9.2°S, 1.8°W] Claudius Ptolemaeus, *c.* 90–160 AD. Greek astronomer, author of the *Almagest.* Geocentric model of the Universe.
Very prominent walled plain with numerous pits and depressions on its floor (153 km/2400 m).

Réaumur [2.4°S, 0.7°E] René A. F. de Réaumur, 1683–1757. French physicist.
Remains of a crater (53 km).

Réaumur, Rima [3°S, 3°E]
Rille, length 45 km.

Seeliger [2.2°S, 3.0°E] Hugo von Seeliger, 1849–1924. German astronomer.
Circular crater (8.5 km/1800 m).

Spörer [4.3°S, 1.8°W] Friedrich W. G. Spörer, 1822–1895. German astronomer. The study of solar activity (law of distribution of sunspots).
Indistinct, shallow crater (28 km/310 m).

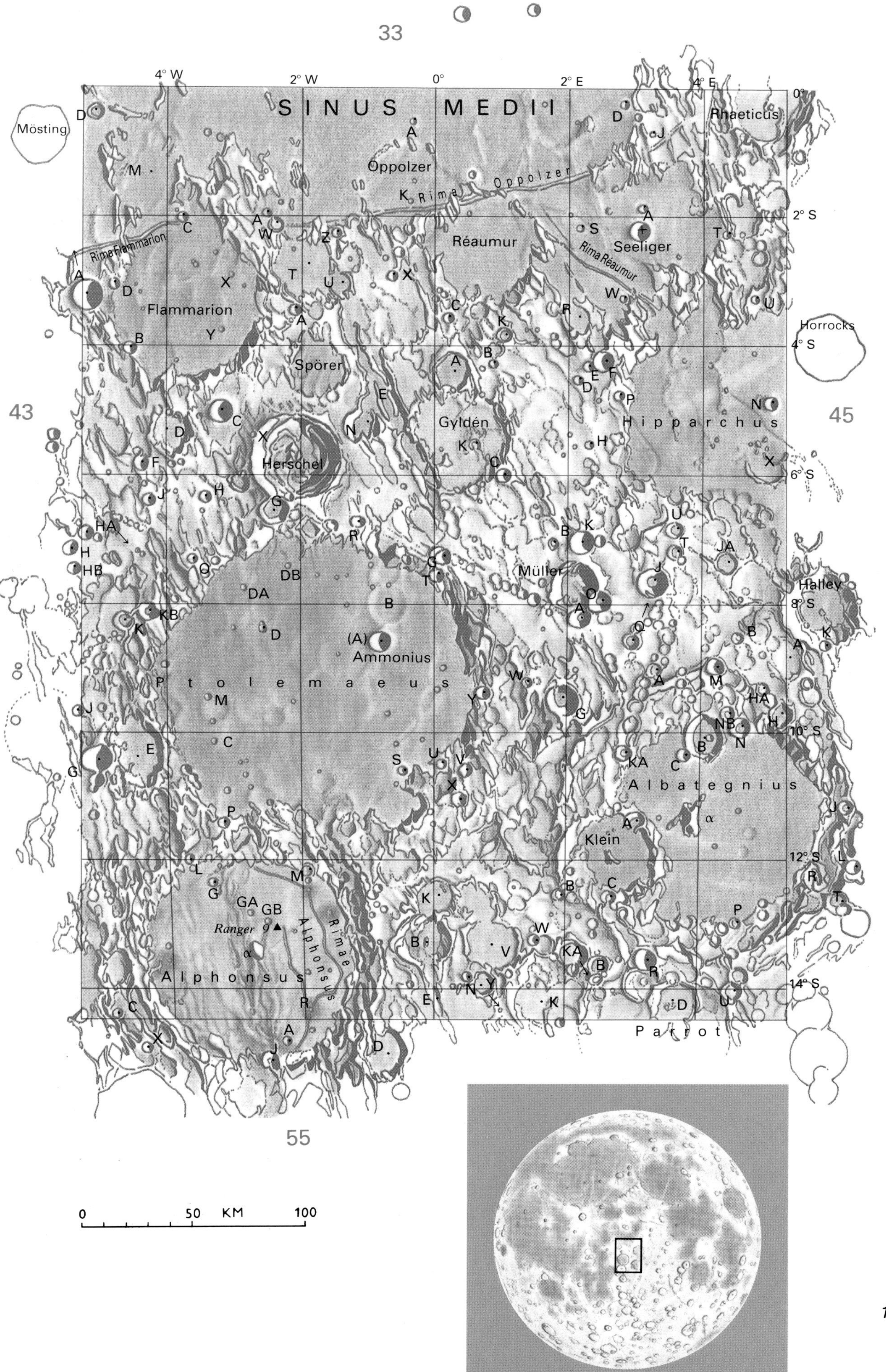
33
43
45
55
SINUS MEDII
Mösting
Rhaeticus
Oppolzer
Rima Oppolzer
Réaumur
Seeliger
Rima Réaumur
Rima Flammarion
Flammarion
Spörer
Herschel
Gyldén
Hipparchus
Horrocks
Müller
Halley
Ammonius
Ptolemaeus
Albategnius
Klein
Alphonsus
Rimae Alphonsus
Ranger 9
Parrot
4° W
2° W
0°
2° E
4° E
0°
2° S
4° S
6° S
8° S
10° S
12° S
14° S
0 50 KM 100

45. ANDĚL

A 'continental' area with several badly damaged craters. The effects of the cataclysm that created Mare Imbrium can be seen crossing this area. There is an interesting row of 'diminishing' craters: Halley, Hind, Hipparchus C and Hipparchus L. The *Apollo 16* mission landed some distance north of the crater Descartes.

Abulfeda [13.8°S, 13.9°E] Ismail Abu'l Fida, 1273–1331. Syrian prince, geographer and astronomer.
Crater (62 km/3110 m).

Anděl [10.4°S, 12.4°E] Karel Anděl, 1884–1947. Czech teacher and selenographer. *Mappa Selenographica* (1926).
Polygonal crater whose wall is open to the south (35 km).

Burnham [13.9°S, 7.3°E] Sherburne W. Burnham, 1838–1921. American amateur astronomer, discovered over 1300 double stars.
Inconspicuous crater with a disintegrated wall (25 km).

Descartes [11.7°S, 15.7°E] René Descartes, 1596–1650. Great French philosopher and mathematician.
Crater (48 km).

Dollond [10.4°S, 14.4°E] John Dollond, 1706–1761. English optician, deviced the achromatic telescope objective.
Circular crater (11.1 km/1580 m).

Halley [8.0°S, 5.7°E] Edmond Halley, 1656–1742. English astronomer. On the basis of Newton's laws of gravitation he proved the periodicity of the orbit of the comet that bears his name.
Crater (36 km/2510 m).

Hind [7.9°S, 7.4°E] John Russell Hind, 1823–1895. English astronomer.
Crater (29 km/2980 m) *in line with the craters* **Hipparchus C** (17 km/2940 m) *and* **Hipparchus L** (13 km/2630 m).

Hipparchus. See map 44.

Horrocks [4.0°S, 5.9°E] Jeremiah Horrocks, 1619–1641. English astronomer; the first to observe a transit of Venus across the Sun, in 1639.
Prominent crater (30 km/2980 m).

Lade [1.3°S, 10.1°E] Heinrich E. von Lade, 1817–1904. German banker and amateur astronomer.
Remains of a flooded crater (56 km); *the crater* **Lade B,** *to the north, is filled to the brim.*

Lindsay (Dollond C) [7.0°S, 13.0°E] Eric M. Lindsay, 1907–1974. Irish astronomer.
Crater (32 km/1550 m).

Pickering [2.9°S, 7.0°E] Edward C. Pickering, 1846–1919. American astronomer.
Circular crater (15 km/2740 m).

Ritchey [11.1°S, 8.5°E] George W. Ritchey, 1864–1945. American astronomer and optician.
Crater (25 km/1300 m).

Saunder [4.2°S, 8.8°E] Samuel A. Saunder, 1852–1912. English selenographer. Catalogued the positions of 3000 points on the Moon.
Crater with a low, irregular wall (45 km).

Theon Junior [2.3°S, 15.8°E] Theon of Alexandria, *c.* 380 AD. Last astronomer of the Alexandrian school.
Prominent circular crater (18.6 km/3580 m).

Theon Senior [0.8°S, 15.4°E] Theon of Smyrna, *c.* 100 AD. Greek mathematician and astronomer.
Crater (18.2 km/ 3470 m).

34
44
46
56
Rhaeticus
Theon Senior
Theon Junior
Lade
Pickering
Horrocks
Saunder
Taylor
Hipparchus
Lindsay
Halley
Hind
Apollo 16
Andel
Dollond
Albategnius
Ritchey
Descartes
Burnham
Abulfeda
Vogel
0 50 KM 100

46. THEOPHILUS

From the north, this area is penetrated by the edge of Mare Tranquillitatis with the shallow rilles Rimae Hypatia. The large crater Theophilus is one of the most spectacular formations on the Moon; it forms a prominent trio with the craters Cyrillus and Catharina (map 57). Craters Kant B and Zöllner E are situated on the opposite edges of an extensive plateau which resembles a filled crater similar to Wargentin (map 70).

Alfraganus [5.4°S, 19.0°E] Muhammed ebn Ketir al Fargani, *c.* 840 AD. Persian astronomer.
Irregular crater (21 km/3830 m).

Asperitatis, Sinus (Bay of Asperity) [6°S, 25°E] The name corresponds to the very uneven character of this mare area.
Diameter about 180 km.

Cyrillus [13.2°S, 24.0°E] St Cyril, d. 444 AD. Bishop of Alexandria, after Theophilus.
Ring mountain with a considerably disintegrated wall (98 km).

Delambre [1.9°S, 17.5°E] Jean B. J. Delambre, 1749–1822. French astronomer; collaborated in trigonometrical surveys which led to the derivation of the 'metre'.
Prominent terraced crater (52 km).

Hypatia [4.3°S, 22.6°E] Hypatia, d. 415 AD. Daughter of Theon of Alexandria; astronomer and mathematician.
Irregular crater (41 × 28 km).

Hypatia, Rimae [1°S, 23°E]
System of rilles, length 180 km.

Ibn Rushd (Cyrillus B) [11.7°S, 21.7°E] Ibn Rushd, Averroes, 1126–1198. Arabian philosopher, physician and lawyer at the Royal Court in Moslem Spain.
Crater (33 km).

Kant [10.6°S, 20.1°E] Immanuel Kant, 1724–1804. German philosopher, author of the nebular hypothesis of the formation of the solar system.
Crater with central peak (32 km/3120 m)

Moltke [0.6°S, 24.2°E] Helmuth Karl von Moltke, 1800–1891. Prussian Field Marshall. Secured publication of Schmidt's map of the Moon.
Circular crater with bright halo (6.5 km/1310 m).

Penck, Mons [10°S, 22°E] Albrecht Penck, 1858–1945. German geographer.
Mountain massif, height 4000 m, *diameter* 30 km.

Taylor [5.3°S, 16.7°E] Brook Taylor, 1685–1731. English mathematician and philosopher.
Elliptical crater (41 × 34 km).

Theon Junior. See map 45.

Theophilus [11.4°S, 26.4°E] St Theophilus, d. 412 AD. Bishop of Alexandria from 385 AD.
Ring mountain (100 km/4400 m); *the rim of its wall rises* 1200 m *above the surrounding terrain, its central peaks reach a height of* 1400 m.

Zöllner [8.0°S, 18.9°E] Johann K. F. Zöllner, 1834–1882. German astronomer, inventor of the polarizing astrophotometer.
Elongated crater with disintegrated wall (47 × 36 km).

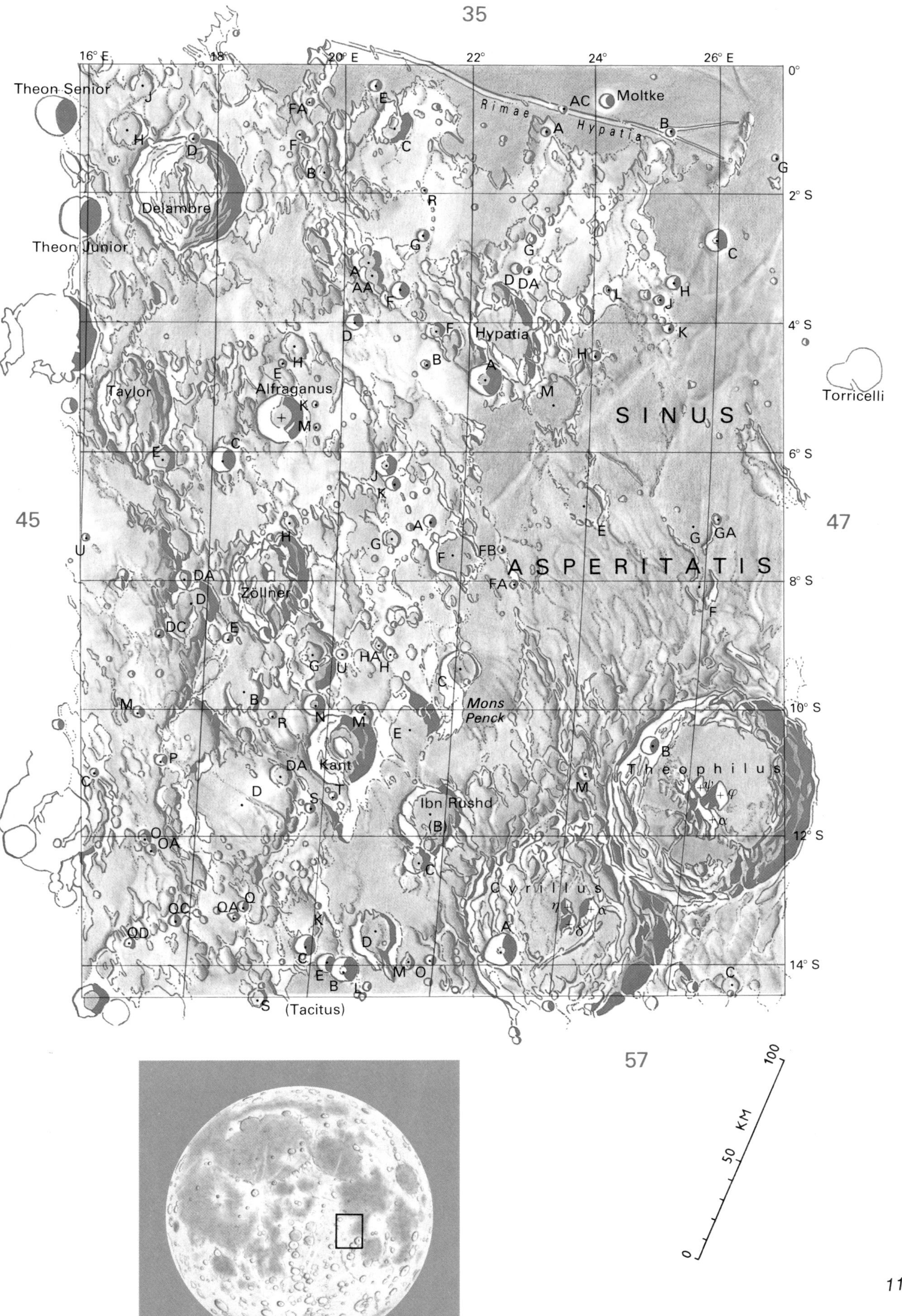
35
45
47
57
16° E
18°
20° E
22°
24°
26° E
0°
2° S
4° S
6° S
8° S
10° S
12° S
14° S
Theon Senior
Theon Junior
Delambre
Moltke
Rimae Hypatia
Hypatia
Taylor
Alfraganus
Torricelli
SINUS
ASPERITATIS
Zöllner
Mons Penck
Kant
Ibn Rushd
(B)
Theophilus
Cyrillus
(Tacitus)
KM
0
50
100

47. CAPELLA

This 'continental' promontory separates Mare Tranquillitatis from Mare Nectaris. The 'continental' area is traversed by the rilles Rimae Gutenberg. The edge of M. Nectaris is dominated by a prominent pair of craters, Capella and Isidorus. The double, pear-shaped crater Torricelli and the small crater Censorinus, which is one of the brightest objects on the Moon, also contribute to the character of this region.

Asperitatis, Sinus (Bay of Asperity). See map 46.

Capella [7.6°S, 34.9°E] Martianus Capella, fifth century AD. Carthaginian lawyer; Copernicus refers to his theory that Mercury and Venus orbit the Sun, and that the Sun, together with the rest of the planets, revolves around the Earth.
Crater (49 km).

Capella, Vallis [7°S, 35°E]
Valley about 110 km *long, through the crater Capella.*

Censorinus [0.4°S, 32.7°E] Censorinus, *c.* 238 AD. Roman astronomer. His letter *De Die Natali* deals with the influence of stars and chronology.
Small circular crater with exceptionally bright halo (3.8 km).

Daguerre [11.9°S, 33.6°E] Louis Daguerre, 1789–1851. French landscape painter, inventor of 'daguerreotype' photographic process.
Horseshoe-shaped circular depression, diameter 46 km.

Gaudibert [10.9°S, 37.8°E] Casimir M. Gaudibert, 1823–1901. French amateur astronomer and selenographer.
Inconspicuous crater divided by central peaks, mountain ranges and ridges (33 km).

Gutenberg, Rimae [5°S, 38°E]. Named after the crater Gutenberg (map 48).
System of wide rilles (length 330 km*) visible through even quite a small telescope.*

Isidorus [8.0°S, 33.5°E] St Isidore of Seville, *c.* 560–636 AD. Bishop of Seville, interested in astronomy, believed the Earth to be a sphere.
Crater (42 km/1580 m).

Leakey (Censorinus F) [3.2°S, 37.4°E] Louis S. B. Leakey, 1903–1972. British archaeologist and palaeoanthropologist.
Crater (13 km).

Mädler [11.0°S, 29.8°E] Johann H. Mädler, 1794–1874. German selenographer, author of the monograph *Der Mond,* which includes a map of the Moon.
Regular crater (28 km/2670 m).

Nectaris, Mare (Sea of Nectar). See map 58.

Torricelli [4.6°S, 28.5°E] Evangelista Torricelli, 1608–1647. Italian physicist (contemporary of Galileo), inventor of the mercury barometer.
Crater 23 km *in diameter; its western wall is open and linked with a smaller crater, so that the whole formation appears pear-shaped.*

Tranquillitatis, Mare (Sea of Tranquillity). See maps 35 and 36.

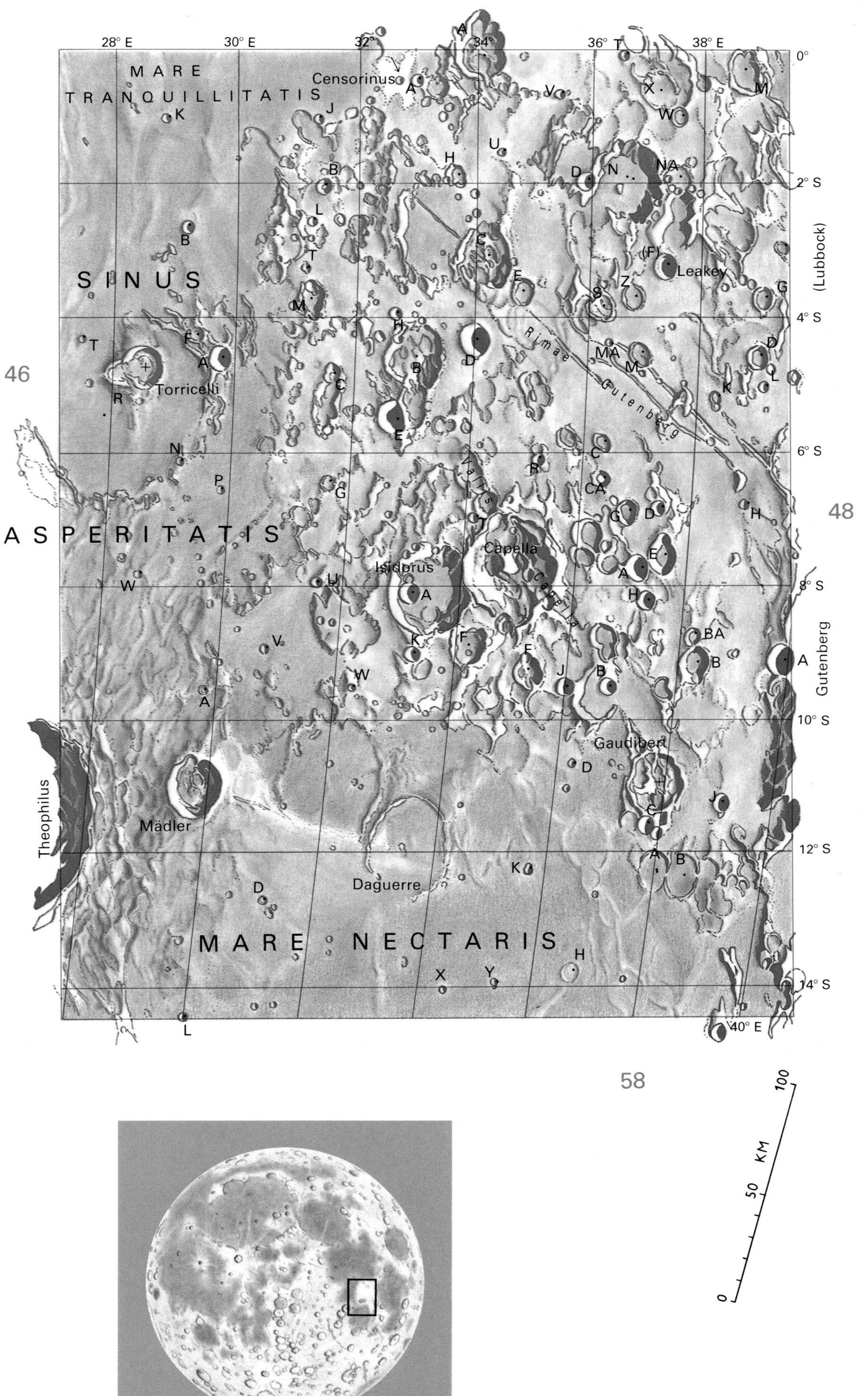
36
46
48
58
28° E
30° E
32°
34°
36°
38° E
40° E
0°
2° S
4° S
6° S
8° S
10° S
12° S
14° S
MARE
TRANQUILLITATIS
SINUS
ASPERITATIS
MARE NECTARIS
Censorinus
Torricelli
Isidorus
Capella
Vallis Capella
Rimae Gutenberg
Leakey
(F)
(Lubbock)
Gutenberg
Gaudibert
Mädler
Theophilus
Daguerre
100
50
KM
0

48. MESSIER

The vast area of Mare Fecunditatis, which is crossed by numerous wrinkle ridges, is dominated by the well-known pair of craters Messier and Messier A, the latter of which is the source of two bright rays radiating to the west. In oblique illumination 'ghost craters' can be seen on the mare surface and also a system of rilles around the crater Gutenberg.

Al-Marrakushi (Langrenus D) [10.4°S, 55.8°E] Al-Marrakushi, *c.* 1262 AD. Arabian astronomer.
Crater (8 km).

Amontons [5.3°S, 46.8°E] Guillaume Amontons, 1663–1705. French physicist.
Crater (3 km).

Bellot [12.4°S, 48.2°E] Joseph R. Bellot, 1826–1853. French seaman, participated in two Antarctic expeditions, died while attempting to rescue Franklin in the Arctic.
Ring crater (17 km).

Crozier [13.5°S, 50.8°E] Francis R. M. Crozier, 1796–1848. English naval captain, participated in the Arctic expedition of Parry, and accompanied Ross to the Antarctic. Died in the Arctic with Franklin.
Flooded crater (22 km).

Fecunditatis, Mare (Sea of Fertility).
Irregularly shaped mare, surface area 326 000 sq km. (See also maps 37, 49 and 59.)

Geikie, Dorsa [3°S, 53°E] Sir Archibald Geikie, 1835–1924. Scottish geologist.
Large system of wrinkle ridges, length 240 km.

Goclenius [10.0°S, 45.0°E] Rudolf Gockel, 1572–1621. German physician, physicist, and mathematician.
Irregular crater (54 × 72 km) *with clefts on its floor.*

Goclenius, Rimae [8°S, 43°E]
System of wide rilles, length 240 km.

Gutenberg [8.6°S, 41.2°E] Johann Gutenberg, *c.* 1398–1468. German goldsmith, invented movable-type printing and developed printing presses.
Crater 74 km *in diameter; its eastern wall is broken by the flooded crater* **Gutenberg E** *and is connected to the south with the crater* **Gutenberg C;** *its floor has a number of peaks and clefts; the crater* **Gutenberg A** (15 km/3430 m) *lies on the south-western wall.*

Ibn Battuta (Goclenius A) [6.9°S, 50.4°E] Abu Abd Allah Mohammed Ibn Abd Allah, 1304–1377. Arabian geographer.
Crater (12 km).

Lindbergh (Messier G) [5.4°S, 52.9°E] Charles A. Lindbergh, 1902–1974. American pilot – first solo flight across the Atlantic.
Crater (13 km).

Lubbock [3.9°S, 41.8°E] Sir John W. Lubbock, 1803–1865. English mathematician and astronomer.
Crater (14.5 km).

Magelhaens [11.9°S, 44.1°E] Fernão de Magalhães (Magellan) 1480–1521. Famous Portuguese navigator, the first to sail around Cape Horn; his fleet completed the first circumnavigation of the world.
Flooded crater with a dark floor (41 km).

Mawson, Dorsa [7°S, 53°E] Sir Douglas Mawson, 1882–1958. Australian antarctic explorer.
System of wrinkle ridges, length 180 km.

Messier [1.9°S, 47.6°E] Charles Messier, 1730–1817. French astronomer, discovered 14 comets, author of the catalogue of star clusters, nebulae, etc. known as Messier's Catalogue.
Oval crater (9 × 11 km).

Messier A (previously W. H. Pickering)
Double crater (13 × 11 km), *which is the source of two bright rays radiating to the west.*

Messier, Rima [1°S, 45°E]
Narrow, barely visible rille, length 100 km.

Pyrenaeus, Montes (Pyrenees) [14°S, 41°E] Mädler's name for a mountain range situated south of crater Gutenberg.
Length 250 km. (See also maps 47, 58 and 59.)

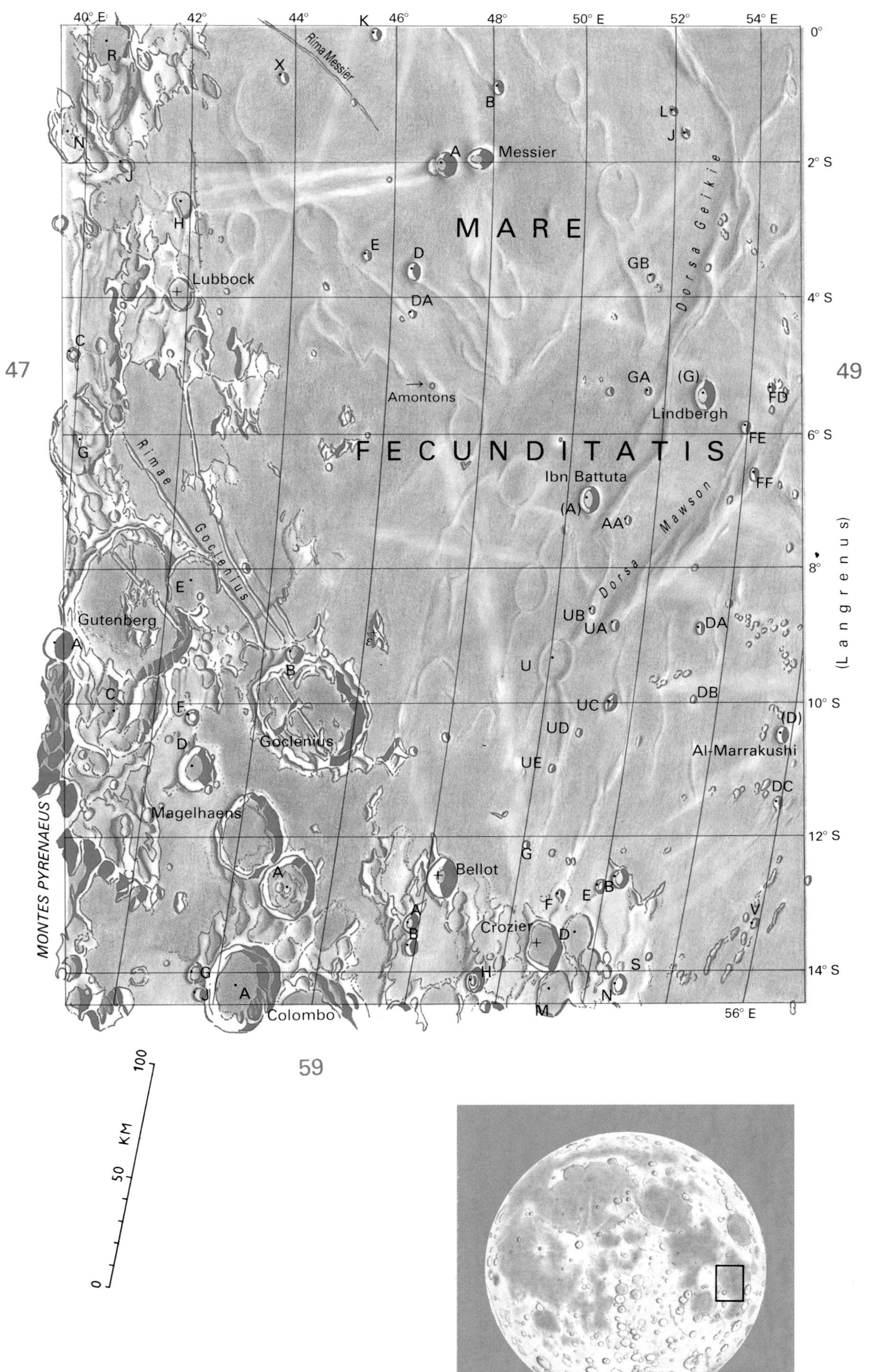
37
47
49
59
40° E
42°
44°
46°
48°
50° E
52°
54° E
56° E
0°
2° S
4° S
6° S
8°
10° S
12° S
14° S
MARE
FECUNDITATIS
Rima Messier
Messier
Lubbock
Amontons
Lindbergh
Ibn Battuta
Dorsa Geikie
Dorsa Mawson
Rimae Goclenius
Gutenberg
Goclenius
Magelhaens
Bellot
Crozier
Colombo
Al-Marrakushi
(Langrenus)
MONTES PYRENAEUS
0
50
100
KM

49. LANGRENUS

The eastern edge of Mare Fecunditatis and the eastern limb of the Moon are intersected by a wide and dense field of craters, in which orientation is not easy. The crater Langrenus is a beautiful sight through a telescope. During favourable librations the dark surface of Mare Smythii is visible close to the limb of the Moon.

Acosta (Langrenus C) [5.6°S, 60.1°E] Cristobal Acosta, 1515–1580. Portuguese physician and historian.
Crater (13 km).

Andrusov, Dorsa [1°S, 57°E] Nikolai I. Andrusov, 1861–1924. Soviet geologist.
System of wrinkle ridges, length 160 km.

Ansgarius [12.7°S, 79.7°E] St Ansgar, 801–864 AD, German theologian.
Prominent crater with terraced walls (94 km).

Atwood (Langrenus K) [5.8°S, 57.7°E] G. Atwood, 1745–1807. British mathematician and physicist.
Crater (29 km).

Avery (Gilbert U) [1.4°S, 81.4°E] Oswald T. Avery, 1877–1955. Canadian physicist.
Crater (9 km).

Barkla (Langrenus A) [10.7°S, 67.2°E] C. G. Barkla, 1877–1944. British physicist (Nobel Laureate).
Crater (43 km).

Bilharz (Langrenus F) [5.8°S, 56.3°E] T. Bilharz, 1825–1862. German physician.
Crater (43 km).

Black (Kästner F) [9.2°S, 80.4°E] Joseph Black, 1728–1799. French chemist. *Crater* (18 km).

Born (Maclaurin Y) [6.0°S, 66.8°E] Max Born, 1882–1970. German physicist and optician.
Crater (15 km).

Carrillo [2.2°S, 80.9°E] Flores N. Carrillo, 1911–1967. Mexican soil engineer. *Crater* (16 km).

Dale [9.6°S, 82.9°E] Sir Henry H. Dale, 1875–1968. British physiologist, Nobel Laureate.
Crater (22 km).

Elmer [10.1°S, 84.1°E] Charles W. Elmer, 1872–1954. American astronomer. *Crater* (17 km).

Fecunditatis, Mare (Sea of Fertility). See also maps 37, 48 and 59.

Geissler (Gilbert D) [2.6°S, 76.5°E] Heinrich Geissler, 1814–1879. German physicist.
Crater (16 km).

Gilbert [3.2°S, 76.0°E] Grove K. Gilbert, 1843–1918. American geologist.
Walled plain (107 km).

Haldane [1.7°S, 84.1°E] John B. S. Haldane, 1892–1964. British biologist, geneticist and popularizer of science.
Crater (38 km).

Hargreaves (Maclaurin S) [2.2°S, 64.0°E] Frederick J. Hargreaves, 1891–1970. British astronomer and optician.
Crater (16 km).

Houtermans [9.4°S, 87.2°E] Friedrich G. Houtermans, 1903–1966. German physicist.
Crater (30 km).

Kapteyn [10.8°S, 70.6°E] Jacobus C. Kapteyn, 1851–1922. Dutch astronomer.
Prominent crater (49 km).

Kästner [7.0°S, 79.1°E] Abraham G. Kästner, 1719–1800. German mathematician and physicist.
Walled plain (105 km).

Kiess [6.4°S, 84.0°E] Carl C. Kiess, 1887–1967. American astrophysicist.
Crater (63 km).

Kreiken [9.0°S, 84.6°E] E. A. Kreiken, 1896–1964. Dutch astronomer.
Crater (23 km).

Lamé. See map 60.

Langrenus [8.9°S, 60.9°E] Michel Florent van Langren, *c.* 1600–1675. Belgian engineer and mathematician. Drew the first map of the Moon with the names of formations.
Very prominent crater with terraced walls and hills and central peaks on its floor (132 km).

la Pérouse [10.7°S, 76.3°E] Jean François de Galoup, Comte de la Pérouse, 1741–1788. French navigator.
Crater (78 km).

Lohse [13.7°S, 60.2°E] Oswald Lohse, 1845–1915. German astronomer, photographed the planets, mapped Mars.
Crater (42 km) *by the northern edge of the crater Vendelinus* (map 60).

Maclaurin [1.9°S, 68.0°E] Colin Maclaurin, 1698–1746. Scottish professor of mathematics at Aberdeen and Edinburgh.
Inconspicuous crater with central peak (50 km).

Morley (Maclaurin R) [2.8°S, 64.6°E] Edward W. Morley, 1838–1923. American chemist.
Crater (14 km).

Naonobu (Langrenus B) [4.6°S, 57.8°E] Ajima Naonobu, 1732–1798. Japanese mathematician.
Crater (35 km).

Rankine [3.9°S, 71.5°E] William J. M. Rankine, 1820–1872. Scottish physicist.
Crater (9 km).

Smythii, Mare (Smyth's Sea). See map 38.

Somerville (Langrenus J) [8.3°S, 64.9°E] Mary F. Somerville, 1780–1872. Scottish physicist and mathematician.
Crater (15 km).

Van Vleck (Gilbert M) [1.9°S, 78.3°E] John M. Van Vleck, 1833–1912. American astronomer and mathematician.
Crater (31 km).

von Behring (Maclaurin F) [7.8°S, 71.8°E] Emil A. von Behring, 1854–1917. German bacteriologist.
Crater (39 km).

Webb [0.9°S, 60.0°E] Thomas W. Webb, 1806–1885. English astronomer, author of *Celestial Objects for Common Telescopes.*
Crater (22 km).

Weierstrass (Gilbert N) [1.3°S, 77.2°E] Karl Weierstrass, 1815–1897. German mathematician.
Crater (33 km).

Widmanstätten [6.1°S, 85.5°E] Alois B. Widmanstätten, 1754–1849. Austrian scientist. 'Widmanstätten patterns' can be seen on the etched and polished surfaces of iron meteorites, and reveal their crystalline structure.
Crater (46 km).

Note: Nine further named craters in Mare Smythii are marked on the libration zone map, on page 185: Helmert, Kao, Lebesgue, Runge, Slocum, Swasey, Talbot, Tucker, Warner.

38

48

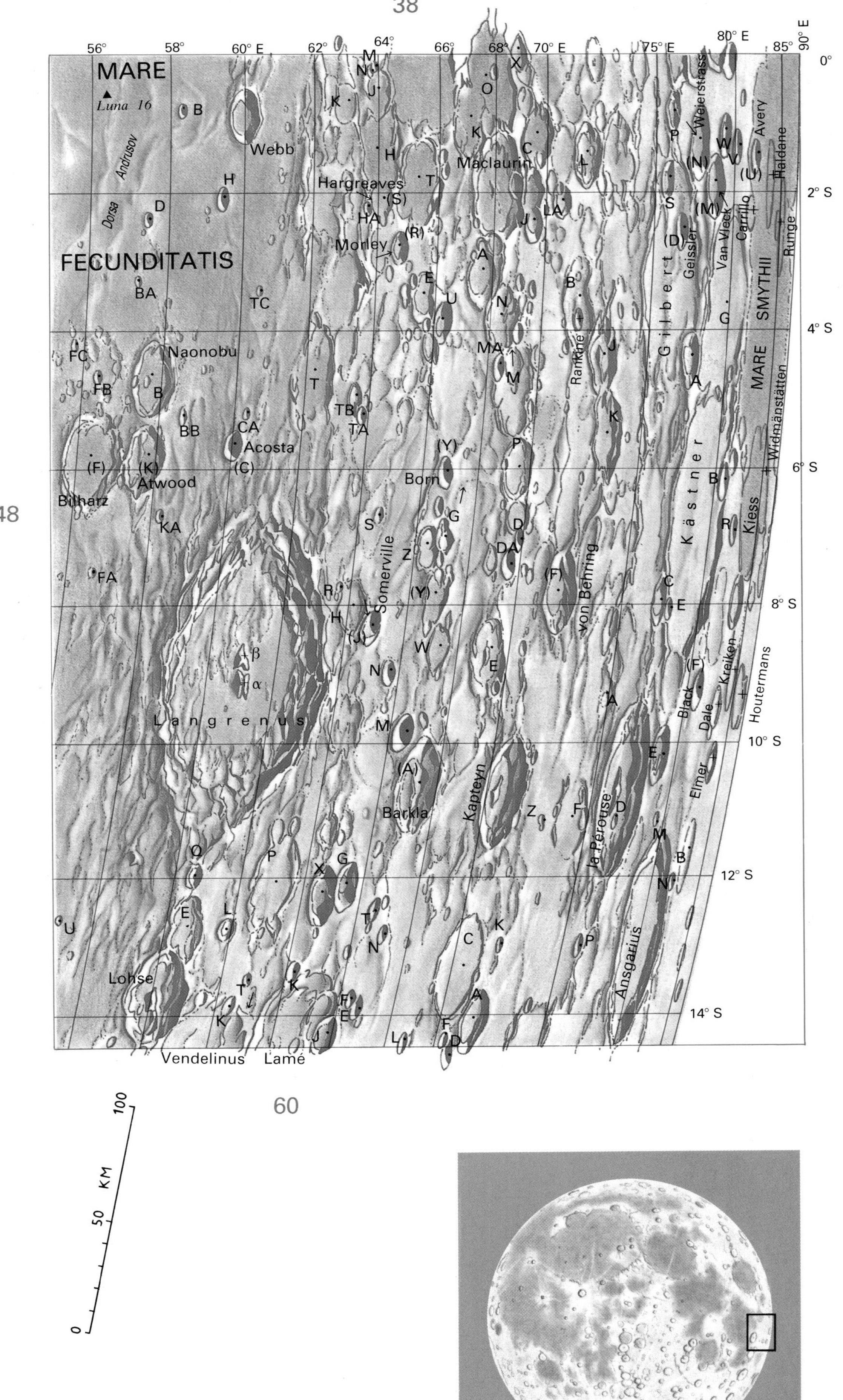

56°
58°
60° E
62°
64°
66°
68°
70° E
75° E
80° E
85°
90° E
0°
2° S
4° S
6° S
8° S
10° S
12° S
14° S
MARE
Luna 16
FECUNDITATIS
Dorsa Andrusov
Webb
Hargreaves
Morley
Maclaurin
Weierstrass
Avery
Haldane
Van Vleck
Carrillo
Runge
Geissler
Gilbert
MARE SMYTHII
Widmanstätten
Rankine
Naonobu
Acosta
Atwood
Bilharz
Born
Somerville
von Behring
Kästner
Kiess
Langrenus
Kreiken
Black
Dale
Houtermans
Elmer
Barkla
Kapteyn
la Pérouse
Ansgarius
Lohse
Vendelinus
Lamé
0
50
KM
100

60

50. DARWIN

The western limb of the Moon, on which part of the remarkable basin of Mare Orientalis is situated. The observer's attention will be attracted by the wide Rima Sirsalis, which continues on map 39.

Aestatis, Lacus (Summer Lake). See map 39.

Autumni, Lacus (Autumn Lake). See map 39.

Byrgius [24.7°S, 65.3°W] Joost Bürgi, 1552–1632. Swiss clockmaker, outstanding mechanic; made astrometric instruments, including Tycho Brahe's sextant.
Crater (87 km); **Byrgius A** *is the centre of a bright ray system.*

Cordillera, Montes (Cordillera Mountains) [20°S, 80°W]
Eastern part of a circular mountain chain 900 km *in diameter* (see map 39).

Crüger [16.7°S, 66.8°W] Peter Crüger, 1580–1639. German mathematician, teacher of Hevelius.
Crater with a very dark floor (46 km).

Darwin [19.8°S, 69.1°W] Charles R. Darwin, 1809–1882. English naturalist, author of the theory of evolution through natural selection.
Disintegrated walled plain (130 km).

Darwin, Rimae [20°S, 67°W]
System of rilles, length 280 km.

Eichstadt [22.6°S, 78.3°W] Lorenz Eichstadt, 1596–1660. German physician, mathematician and astronomer.
Prominent crater on the edge of the Cordillera Mountains (49 km).

Kopff [17.4°S, 89.6°W] August Kopff, 1882–1960. German astronomer.
Crater (42 km).

Krasnov [29.9°S, 79.6°W] Alexander V. Krasnov, 1866–1907. Russian astronomer. Measured lunar librations with a heliometer.
Crater (41 km).

Lamarck [22.9°S, 69.8°W] Jean Baptiste P. A. de M. Lamarck, 1744–1829. French naturalist, founder of the study of the zoology of invertebrates.
Considerably disintegrated crater (115 km).

Nicholson [26.2°S, 85.1°W] Seth B. Nicholson, 1891–1963. American solar astronomer. With Pettit he invented a thermocouple for measurement of the surface temperatures of the planets.
Crater in the Rook Mountains, (38 km).

Orientale, Mare (Eastern Sea) [20°S, 95°W]
Flooded centre, one of the youngest lunar basins. The whole of M. Orientale is situated on the far-side of the Moon and is visible only during very favourable librations. Diameter 300 km.

Pettit [27.5°S, 86.6°W] Edison Pettit, 1889–1962. American astronomer. Research on solar prominences.
Crater (35 km), *one of a pair with Nicholson.*

Rook, Montes (Rook Mountains) [20°S, 83°W] Lawrence Rooke, 1622–1666. English astronomer, observer of Jupiter's satellites.
One of the inner circular mountain chains that surround the basin of Mare Orientale, length about 900 km.

Sirsalis, Rima [17°S, 62°W]
Sizeable, wide rille, visible through even a small telescope.

Veris, Lacus [13°S, 87°W] (Spring Lake)
Narrow 'mare' on the inner edge of the Rook Mountains; it is formed by separated dark areas. The total span is 540 km, *the total surface area is* 12 000 sq km.

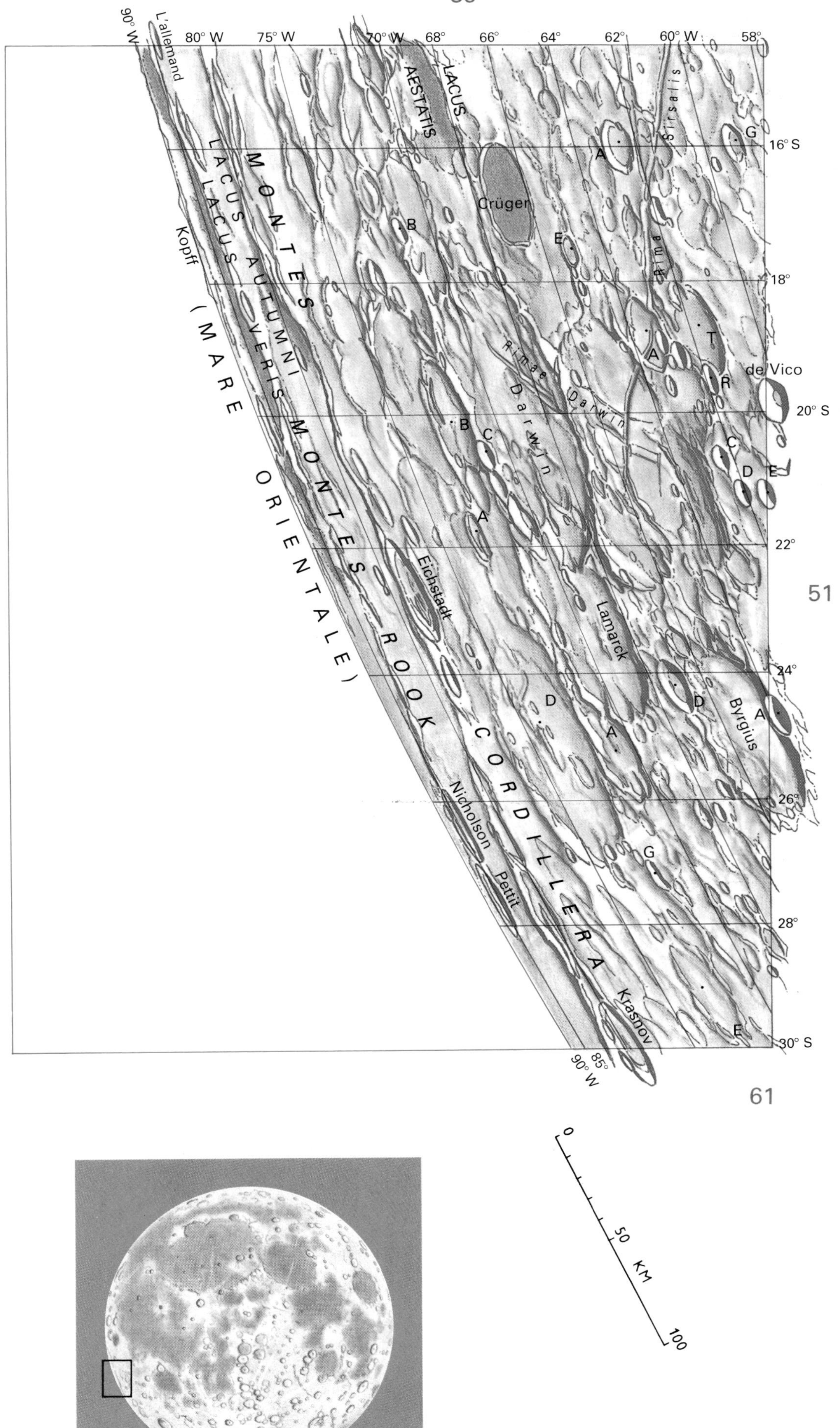
39
51
61
90° W
80° W
75° W
70° W
68°
66°
64°
62°
60° W
58°
16° S
18°
20° S
22°
24°
26°
28°
30° S
90° W
85°
L'allemand
LACUS
AESTATIS
Rima Sirsalis
Crüger
MONTES
LACUS AUTUMNI
LACUS VERIS
Kopff
(MARE ORIENTALE)
MONTES ROOK
CORDILLERA
Rimae Darwin
Darwin
de Vico
Eichstadt
Lamarck
Byrgius
Nicholson
Pettit
Krasnov
0
50
KM
100

51. MERSENIUS

A hilly region between the western edge of Mare Humorum and the western limb of the Moon. Clefts and rilles run parallel with the edge of Mare Humorum.

Billy, Rima [15°S, 48°W]. Named after the crater Billy (map 40).
Rille, length 70 km.

Cavendish [24.5°S, 53.7°W] Henry Cavendish, 1731–1810. English chemist and physicist, discovered hydrogen. 'Cavendish's experiment' with a torsion balance to determine the mass of the Earth.
Crater (56 km) *whose wall is interrupted by the crater* **Cavendish E.**

de Gasparis [25.9°S, 50.7°W] Annibale de Gasparis, 1819–1892. Italian astronomer, discovered 9 asteroids.
Flooded crater with clefts on its floor (30 km).

de Gasparis, Rimae [25°S, 50°W]
System of rilles covering an area 130 km *in diameter.*

de Vico [19.7°S, 60.2°W] Francesco de Vico, 1805–1848. Italian astronomer, observer of Venus. Discovered six comets.
Crater (20 km).

Fontana [16.1°S, 56.6°W] Francesco Fontana, *c.* 1585–1656. Italian lawyer and amateur astronomer, observer of the planets.
Crater (31 km).

Henry [24.0°S, 56.8°W] Joseph Henry, 1792–1878. American physicist. Invented electric motor and electric relay.
Crater (41 km).

Henry Frères [23.5°S, 58.9°W] Henry brothers; Paul Henry, 1848–1905; Prosper Henry, 1849–1903. French astronomers; pioneered astrophotography and designed astrographic telescopes used to produce a photographic chart of the sky (*Carte du Ciel*), by international co-operation. Constructed large refracting telescopes.
Crater (42 km).

Humorum, Mare (Sea of Moisture). See p. 130.

Liebig [24.3°S, 48.2°W] Justus von Liebig, 1803–1873. German chemist. Invented a process for silvering glass, used for astronomical telescope mirrors.
Crater (37 km).

Liebig, Rupes [25°S, 46°W]
Fault at the western edge of Mare Humorum, length 180 km.

Mersenius [21.5°S, 49.2°W] Marin Mersenne, 1588–1648. French theologian, mathematician and physicist.
Flooded crater with convex floor (84 km).

Mersenius, Rimae [20°S, 45°W]
System of wide, clearly visible rilles, length 230 km.

Palmieri [28.6°S, 47.7°W] Luigi Palmieri, 1807–1896. Italian mathematician and geophysicist.
Flooded crater (41 km).

Palmieri, Rimae [28°S, 47°W]
System of narrow rilles, length 150 km.

Vieta [29.2°S, 56.3°W] François Viète, 1540–1603. French lawyer and mathematician.
Crater (87 km).

Zupus [17.2°S, 52.3°W] Giovanni B. Zupi, *c.* 1590–1650. Italian Jesuit astronomer.
Remains of a flooded crater (38 km).

Zupus, Rimae [15°S, 53°W]
Indistinct rilles, difficult to observe, length 120 km.

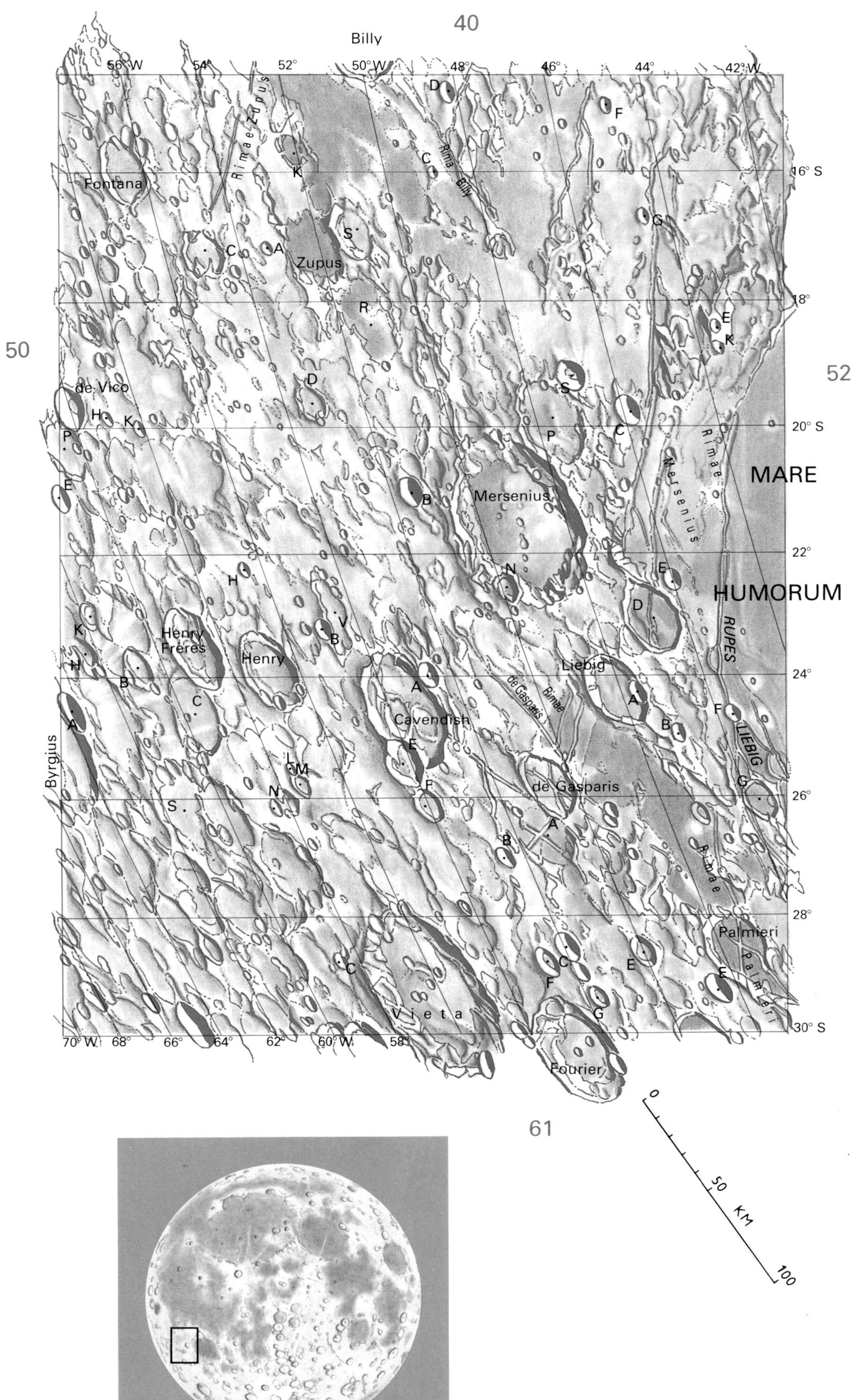

40
Billy
50
52
61
Rimae Zupus
Rima Billy
Fontana
Zupus
de Vico
MARE
HUMORUM
Mersenius
Rimae Mersenius
RUPES LIEBIG
Henry Frères
Henry
Liebig
Rimae de Gasparis
Cavendish
de Gasparis
Byrgius
Rimae Palmieri
Palmieri
Vieta
Fourier
16° S
18°
20° S
22°
24°
26°
28°
30° S
56° W
54°
52°
50° W
48°
46°
44°
42° W
70° W
68°
66°
64°
62°
60° W
58°
0
50
100
KM

52. GASSENDI

The Mare Humorum basin and an interesting walled plain Gassendi, which resembles a diamond ring, are among the most prominent features of the south-western quadrant of the nearside of the Moon. The system of concentric wrinkle ridges in M. Humorum and the neighbouring rilles Rimae Hippalus are good objects for telescopic observation.

Agatharchides [19.8°S, 30.9°W] Agatharchides, *c.* second century BC. Greek geographer and historian.
Flooded crater (49 km/1180 m).

Doppelmayer [28.5°S, 41.4°W] Johann G. Doppelmayer, 1671–1750. German mathematician and astronomer, author of a map of the Moon.
Greatly eroded crater (64 km).

Doppelmayer, Rimae [26°S, 45°W]
System of narrow rilles, total length about 130 km.

Gassendi [17.5°S, 39.9°W] Pierre Gassendi, 1592–1655. French theologian, mathematician and astronomer. Supported Copernicus' theories, exchanged letters with Kepler and Galileo. The first to observe a transit of Mercury across the Sun in 1631, which was forecast by Kepler.
Prominent walled plain (110 km/1860 m), *with numerous clefts, hills and central mountains on its floor; the wall is interrupted by the crater* **Gassendi A** (33 km/3600 m).

Gassendi, Rimae
Complex system of clefts inside Gassendi.

Hippalus [24.8°S, 30.2°W] Hippalus, *c.* 120 AD. Greek navigator, sailed the open sea from Arabia to India, discovered the importance of the monsoons in navigation.
Remains of a crater (58 km/1230 m).

Hippalus, Rimae [25°S, 29°W]
Sizeable system of wide rilles, length 240 km, *visible through even a small telescope.*

Humorum, Mare (Sea of Moisture) [24°S, 39°W]
Circular lunar mare with a surface area of 113 000 sq km *and a diameter of* 380 km.

Kelvin, Promontorium (Cape Kelvin) [27°S, 33°W]. William Thomson, Lord Kelvin, 1824–1907. British physicist. Worked in the fields of thermodynamics and electricity, made over 60 inventions, constructed submarine cables.

Kelvin, Rupes [28°S, 33°W]
Fault at the edge of Mare Humorum, length about 150 km.

Loewy [22.7°S, 32.8°W] Moritz Loewy, 1833–1907. French astronomer, Director of Paris Observatory. Designed the 'equatorial coudé' type of telescope, worked in the field of astrometry. Photographic *Atlas of the Moon* (with Puiseux).
Flooded crater (22 × 26 km/1090 m).

Puiseux [27.8°S, 39.0°W] Pierre Puiseux, 1855–1928. French astronomer, made over 6000 photographs of the Moon using the 'equatorial coudé' (see Loewy), co-author with Loewy of the famous Paris *Atlas of the Moon.*
Flooded crater (25 km/400 m).

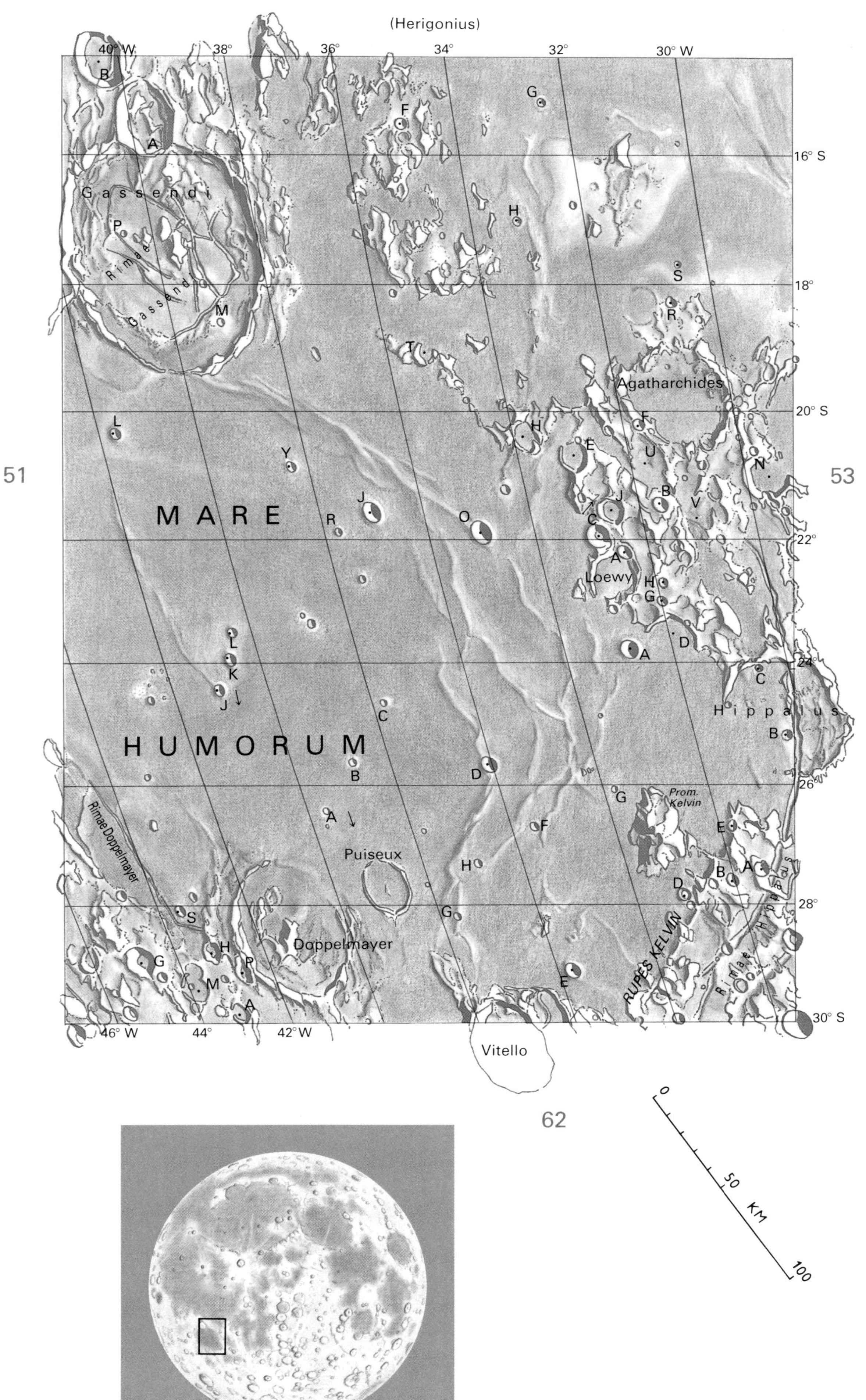

41
(Herigonius)
51
53
62
MARE
HUMORUM
Gassendi
Rimae Gassendi
Agatharchides
Loewy
Hippalus
Prom. Kelvin
Puiseux
Doppelmayer
Rimae Doppelmayer
RUPES KELVIN
Rimae Hippalus
Vitello
40° W
38°
36°
34°
32°
30° W
46° W
44°
42° W
16° S
18°
20° S
22°
24°
26°
28°
30° S
0
50
KM
100

53. BULLIALDUS

The western part of Mare Nubium with the prominent Bullialdus, which is one of the most attractive lunar craters. Other interesting objects include an elevated 'causeway' across the valley Bullialdus W close to the crater Agatharchides O, a typical lunar dome Kies Pi, and the prominent rilles Rimae Hippalus and Rima Hesiodus.

Agatharchides, Rima [20°S, 28°W]
Rille, length about 50 km, *named after the crater Agatharchides* (map 52).

Bullialdus [20.7°S, 22.2°W] Ismaël Boulliau, 1605–1694. French astronomer, historian and theologian.
Very prominent crater with terraced walls and central peaks (61 km/3510 m), *interesting radial structure on the outside of the crater.*

Campanus [28.0°S, 27.8°W] Giovanni Campano, thirteenth century. Italian theologian, astronomer and astrologer.
Crater (48 km/2080 m).

Darney: see map 42.

Epidemiarum, Palus (Marsh of Epidemics) [32°S, 27°W]
Diameter 300 km, *surface area* 27 000 sq km.

Gould [19.2°S, 17.2°W] Benjamin A. Gould, 1824–1896. American astronomer, founder of the *Astronomical Journal,* used the transatlantic cable to determine longitude differences between Europe and America.
Remains of a crater (34 km).

Hippalus: see map 52.

Hippalus, Rimae: see map 52.

Kies [26.3°S, 22.5°W] Johann Kies, 1713–1781. German mathematician and astronomer.
Flooded crater (44 km/380 m).

König [24.1°S, 24.6°W] Rudolf König, 1865–1927. Austrian selenographer, musician and merchant. Built his own observatory, made 47 000 measurements of lunar formations; König's telescope, made by Zeiss, is still in use at the Prague Observatory in Czechoslovakia.
Crater (23 km/2440 m).

Lubiniezky [17.8°S, 23.8°W] Stanislaus Lubiniezky, 1623–1675. Polish astronomer, studied and published details of the movements of 415 comets.
Flooded crater (44 km/770 m).

Mercator [29.3°S, 26.1°W] Gerard de Kremer (Gerhardus Mercator), 1512–1594. Belgian cartographer, originated 'Mercator's projection', frequently used in terrestrial and astronomical maps.
Crater with flooded floor (47 km/1760 m).

Mercator, Rupes [30°S, 23°W]
Fault on the south-western edge of Mare Nubium, length about 180 km.

Nubium, Mare (Sea of Clouds). See also map 54.

Opelt [16.3°S, 17.5°W] Friedrich W. Opelt, 1794–1863. German financier, patron of selenographers Lohrmann and Schmidt.
Remains of a crater (49 km).

Small craters: **Kies E** (6.5 km/1120 m)
Opelt E (8.0 km/1370 m).

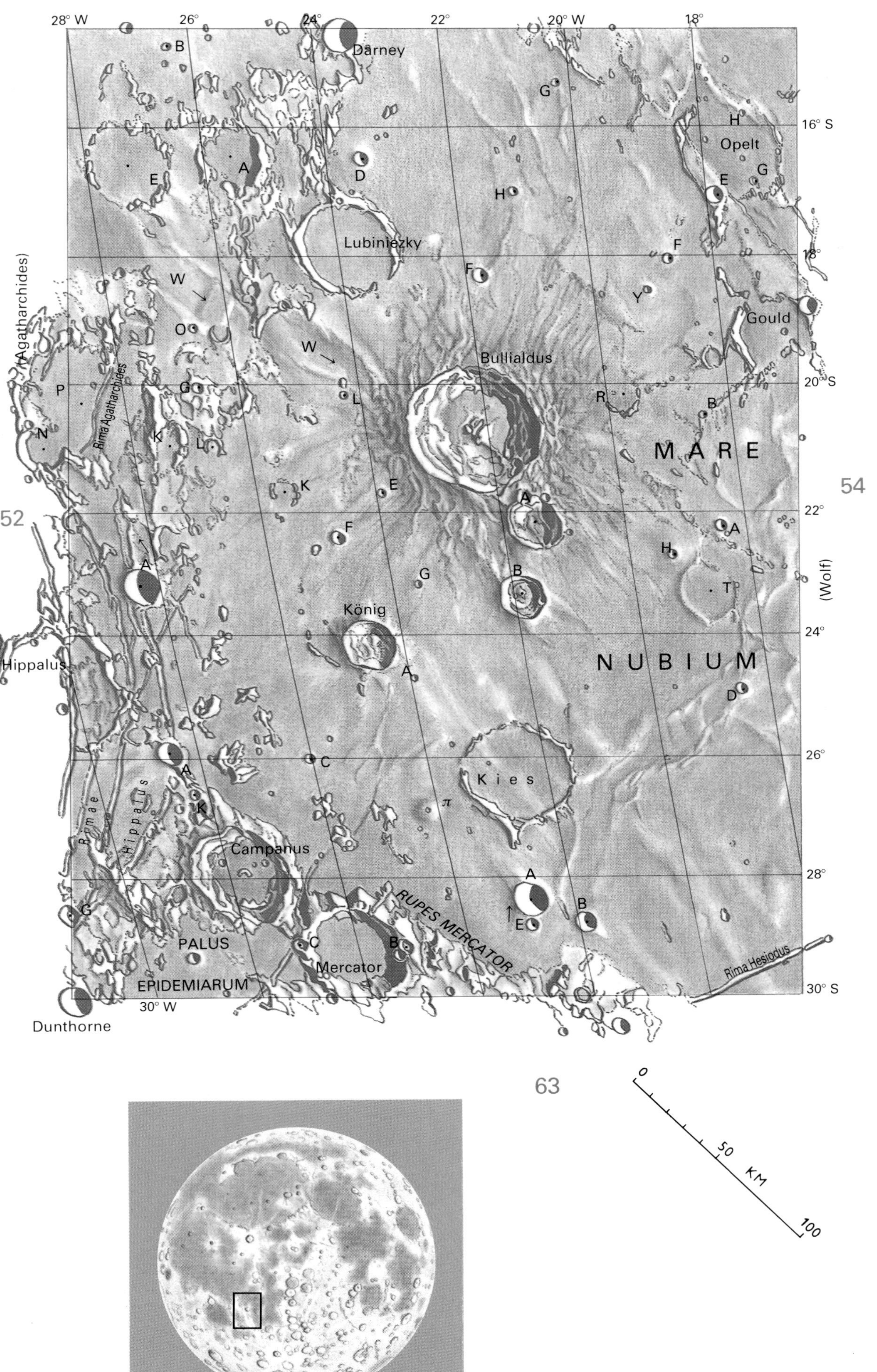
28° W
26°
24°
22°
20° W
18°
16° S
18°
20° S
22°
24°
26°
28°
30° S
30° W
Darney
Opelt
Lubiniezky
Gould
Bullialdus
MARE
NUBIUM
König
Kies
Campanus
Mercator
RUPES MERCATOR
Rima Hesiodus
PALUS
EPIDEMIARUM
Hippalus
Rimae Hippalus
Rima Agatharchides
(Agatharchides)
(Wolf)
Dunthorne
52
54
0
50
100
KM

63

54. BIRT

The eastern part of Mare Nubium with many wrinkle ridges. Not far from the crater Birt is Rupes Recta, the Straight Wall', the most remarkable fault on the Moon. When illuminated from the east it casts a wide shadow which is clearly visible through even a small telescope; in the setting Sun it resembles a fine white line. Larger telescopes reveal a cleft Rima Birt between the craters Birt E (4.9 × 2.9 km/600 m) and Birt F (3.1 km/470 m).

Birt [22.4°S, 8.5°W] William R. Birt, 1804–1881. English astronomer and selenographer.
Crater (17 km/3470 m); **Birt A** *is situated at the edge of its wall* (6.8 km/1040 m).

Birt, Rima [21°S, 9°W]
A rille, about 50 km *long, which connects the small craters* **Birt E** *and* **Birt F.**

Hesiodus [29.4°S, 16.3°W] Hesiod, *c.* 700 BC. Greek poet.
Flooded crater (43 km); **Hesiodus A** *has double concentric walls.*

Lassell [15.5°S, 7.9°W] William Lassell, 1799–1880. English amateur astronomer. With his home-made telescope he discovered four planetary satellites: one of Neptune, one of Saturn and two of Uranus. He also discovered 600 nebulae in the course of two years.
Crater (23 km/910 m); **Lassell D** (1.7 km/400 m) *looks like a bright spot.*

Lippershey [25.9°S, 10.3°W] Hans Lippershey (Jan Lapprey), d. 1619. Dutch spectacle-maker, reputed inventor of the telescope.
Crater (6.8 km/1350 m).

Nicollet [21.9°S, 12.5°W] Jean N. Nicollet, 1788–1843. French selenographer.
Regular crater (15.2 km/2030 m).

Nubium, Mare (Sea of Clouds) [20°S, 15°W]
Surface area 254 000 sq km, *approximately circular shape; its northern border is not clearly defined.*

Pitatus [29.8°S, 13.5° W] Pietro Pitati, sixteenth century. Italian mathematician and astronomer.
Flooded walled plain (97 km).

Pitatus, Rimae
System of clefts inside Pitatus, length about 100 km.

Recta, Rupes (Straight Fault) [22°S, 7°W]. (Previously called the 'Straight Wall', 'Wall Beta'.)
Length 110 km, *height* 240–300 m, *apparent width about* 2.5 km; *it is not a steep slope, as was believed in the past, but a rather moderate slope of gradient about* 7° (1:9).

Taenarium, Promontorium (Cape Taenarium) [19°S, 8°W]. Name given by Hevelius to Cape Matapan (Tainaron) on Peloponnesus.

Wolf [22.7°S, 16.6°W] Maximilian F. J. C. Wolf, 1863–1932. German astronomer. Developed a photographic method for discovering asteroids and with his assistants discovered over 300 of them.
Remains of a flooded crater (25 km); *its wall is linked to that of the crater* **Wolf B.**

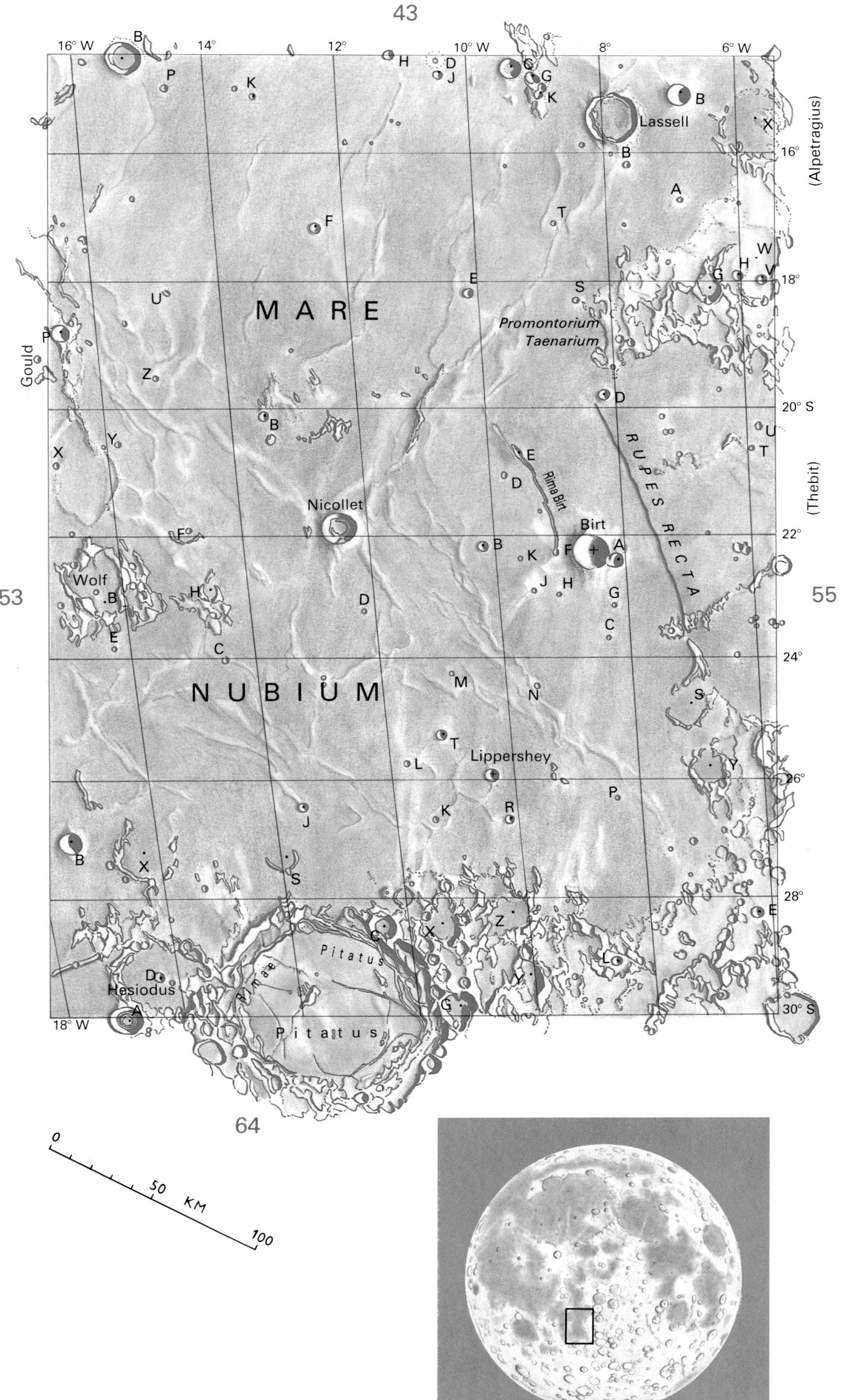
43
16° W
14°
12°
10° W
8°
6° W
Lassell
16°
(Alpetragius)
18°
MARE
Promontorium
Taenarium
Gould
20° S
RUPES RECTA
Rima Birt
Nicollet
Birt
(Thebit)
22°
Wolf
53
55
24°
NUBIUM
Lippershey
26°
28°
Pitatus
Rimae Pitatus
Hesiodus
30° S
18° W
64
0
50
KM
100

55. ARZACHEL

The area surrounding the prime meridian, south of the walled plains Ptolemaeus and Alphonsus, is a region of large craters. These two, together with Arzachel, form an impressive trio. On the western edge of the map the 'continental' structures give way to the darker Mare Nubium (map 54).

Aliacensis. See map 65; forms a pair with the crater Werner.

Alpetragius [16.0°S, 4.5°W] Nur ed-din al Betrugi, twelfth century. Arabian astronomer. Attempted to improve the Ptolemaic system.
Crater with central massif (40 km/3900 m).

Alphonsus. See map 44.

Arzachel [18.2°S, 1.9°W] Al Zarkala, *c.* 1028–1087 AD. Arabian astronomer from Muslem Spain, author of *Toledo Tables.*
Very prominent crater with terraced clefts in its floor (97 km/3610 m); *formations E and F are valleys parallel to and within the elevated southern walls.*

Arzachel, Rimae
Rilles inside Arzachel, length 50 km.

Blanchinus [25.4°S, 2.5°E] Giovanni Bianchini (Johannes Blanchinus), *c.* 1458. Italian teacher of astronomy in Ferrara.
Crater (58 × 68 km).

Delaunay [22.2°S, 2.5°E] Charles E. Delaunay, 1816–1872. French astronomer.
Heart-shaped formation 46 km *in diameter, divided by a central mountain range.*

Donati [20.7°S, 5.2°E] Giovanni B. Donati, 1826–1873. Italian astronomer, discovered seven comets – notably 'Donati's comet' in 1858.
Crater with central peak (36 km).

Faye [21.4°S, 3.9°E] Hervé Faye, 1814–1902. French astronomer, discovered 'Faye's comet' in 1843.
Considerably disintegrated crater with central peak (37 km).

Krusenstern [26.2°S, 5.9°E] Adam J. von Krusenstern, 1770–1846. Russian naval officer. Circumnavigated the world in 1803–1806.
Crater with a flat floor (47 km).

la Caille [23.8°S, 1.1°E] Nicolas Louis de La Caille, 1713–1762. French astronomer. Mapped the sky, gave names to several southern constellations.
Flooded crater (68 km).

Parrot [14.5°S, 3.3°E] Johann J. F. W. Parrot, 1792–1840. German surgeon and physicist, traveller and explorer.
Remains of a walled plain (70 km).

Purbach [25.5°S, 1.9°W] Georg von Peuerbach, 1423–1461. Austrian astronomer.
Walled plain (118 km/2980 m).

Regiomontanus [28.4°S, 1.0°W] Johann Müller, 1436–1476. Prominent German astronomer. Critically assessed Ptolemy's *Almagest.*
Irregular walled plain (126 × 110 km/1730 m); *central peak with crater* **Regiomontanus A** (5.6 km/1200 m).

Thebit [22.0°S, 4.0°W] Thebit ben Korra, 826–901 AD. Arabian astronomer, translated Ptolemy's *Almagest* into Arabic.
Crater (57 km/3270 m); **Thebit A** (20 km/2720 m) *overlaps the main crater; a third crater,* **Thebit L** (10 km) *overlaps that, making a very well-known triple formation.*

Werner [28.0°S, 3.3°E] Johannes Werner, 1468–1528. German astronomer.
Prominent crater with terraced walls (70 km/4220 m).

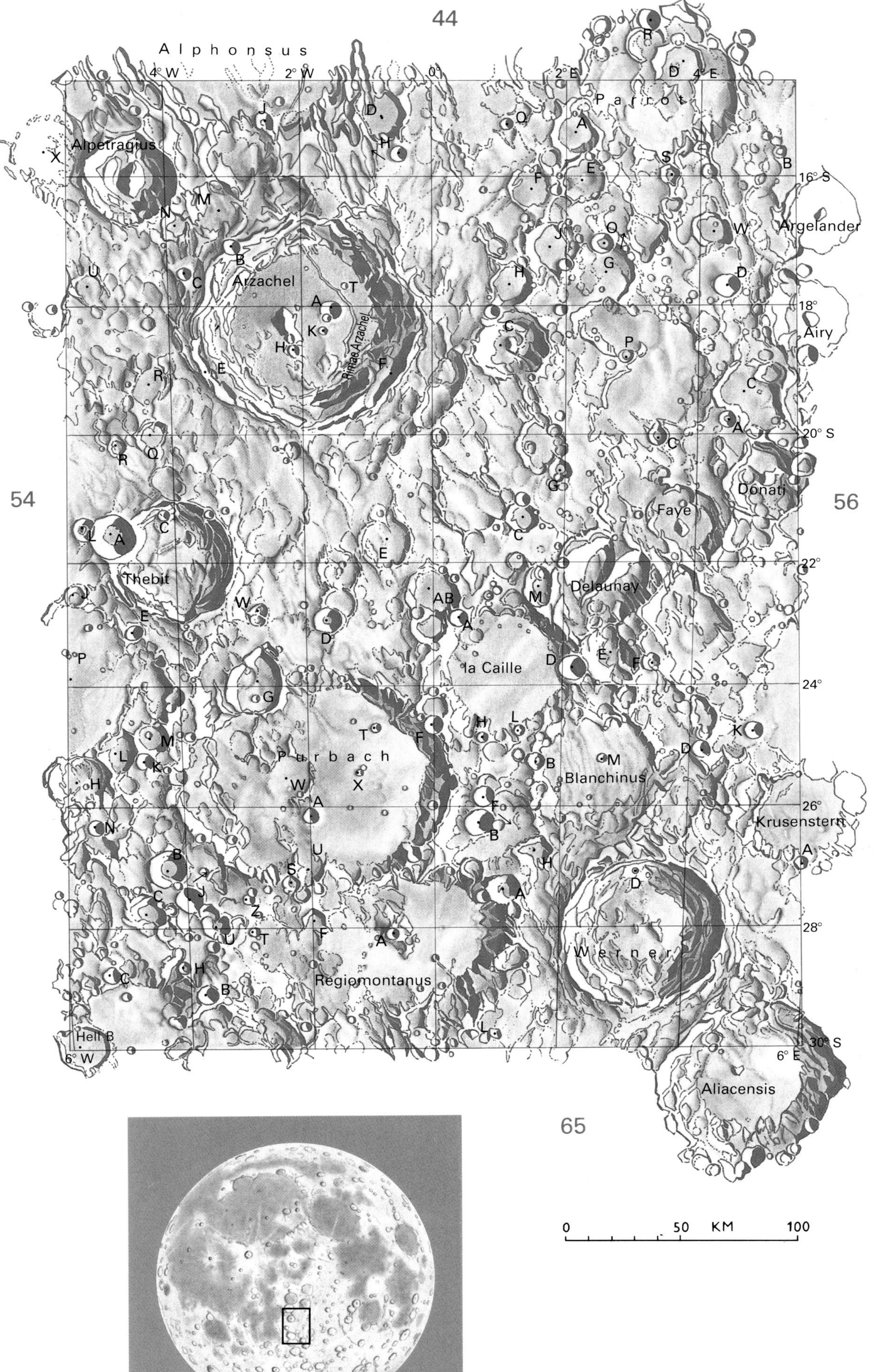
44
Alphonsus
4° W
2° W
0°
2° E
4° E
Parrot
Alpetragius
16° S
Argelander
Arzachel
Rimae Arzachel
18°
Airy
20° S
54
56
Donati
Faye
Thebit
22°
Delaunay
la Caille
24°
Purbach
Blanchinus
26°
Krusenstern
28°
Werner
Regiomontanus
Hell B
6° W
30° S
6° E
Aliacensis
65
0
50
KM
100

56. AZOPHI

A 'continental' region with a number of craters rather difficult to identify. However, the pair Azophi and Abenezra and the prominent craters Almanon, Geber, Playfair and Apianus are useful points of orientation. Also conspicuous is the row of craters Airy, Argelander and Vogel.

Abenezra [21.0°S, 11.9°E] Abraham bar Rabbi ben Ezra, *c.* 1092–1167. Jewish scholar from Toledo, Spain. Theologian, philosopher, mathematician and astronomer.
Polygonal crater (42 km/3730 m).

Abulfeda, Catena [17°S, 17°E] Named after the crater Abulfeda (map 45).
Crater chain, length 210 km (continued on map 57).

Airy [18.1°S, 5.7°E] George B. Airy, 1801–1892. English, Seventh Astronomer Royal.
Crater with angular perimeter (37 km).

Almanon [16.8°S, 15.2°E] Abdalla Al Mamun, 786–833 AD. Caliph of Baghdad, son of Harun al-Raschid, patron of sciences.
Crater (49 km/2480 m).

Apianus [26.9°S, 7.9°E] Peter Bienewitz, 1495–1552. German mathematician and astronomer, author of *Astronomicum Caesareum.*
Crater (63 km/2080 m).

Argelander [16.5°S, 5.8°E] Friedrich W. A. Argelander, 1799–1875. German astronomer, author of *Bonner Durchmusterung* – positional catalogue of 300 000 stars in the northern sky.
Crater (34 km/2980 m).

Azophi [22.1°S, 12.7°E] Abderrahman Al-Sufi, 903–986 AD. Persian astronomer, compiled a star catalogue.
Crater (48 km/3730 m).

Geber [19.4°S, 13.9°E] Gabir ben Aflah, d. *c.* 1145. Spanish–Arabian astronomer.
Crater (45 km/3510 m).

Krusenstern. See map 55.

Playfair [23.5°S, 8.4°E] John Playfair, 1748–1819. Scottish mathematician and geologist.
Crater (48 km/2910 m).

Pontanus [28.4°S, 14.4°E] Giovanni G. Pontano, 1427–1503. Italian poet and astronomer.
Crater (58 km).

Sacrobosco [23.7°S, 16.7°E] John Holywood (Johannes Sacrobuschus), 1200–1256. English (Yorkshire-born) teacher of mathematics, later of Oxford.
Crater (98 km).

Vogel [15.1°S, 5.9°E] Hermann K. Vogel, 1841–1907. German astrophysicist. Applied spectroscopy, spectral classification of the stars.
Crater (27 km/2780 m).

Small craters: **Sacrobosco A** (17.7 km/1830 m)
Sacrobosco B (14.4 km/1210 m)
Sacrobosco C (13.4 km/2630 m)

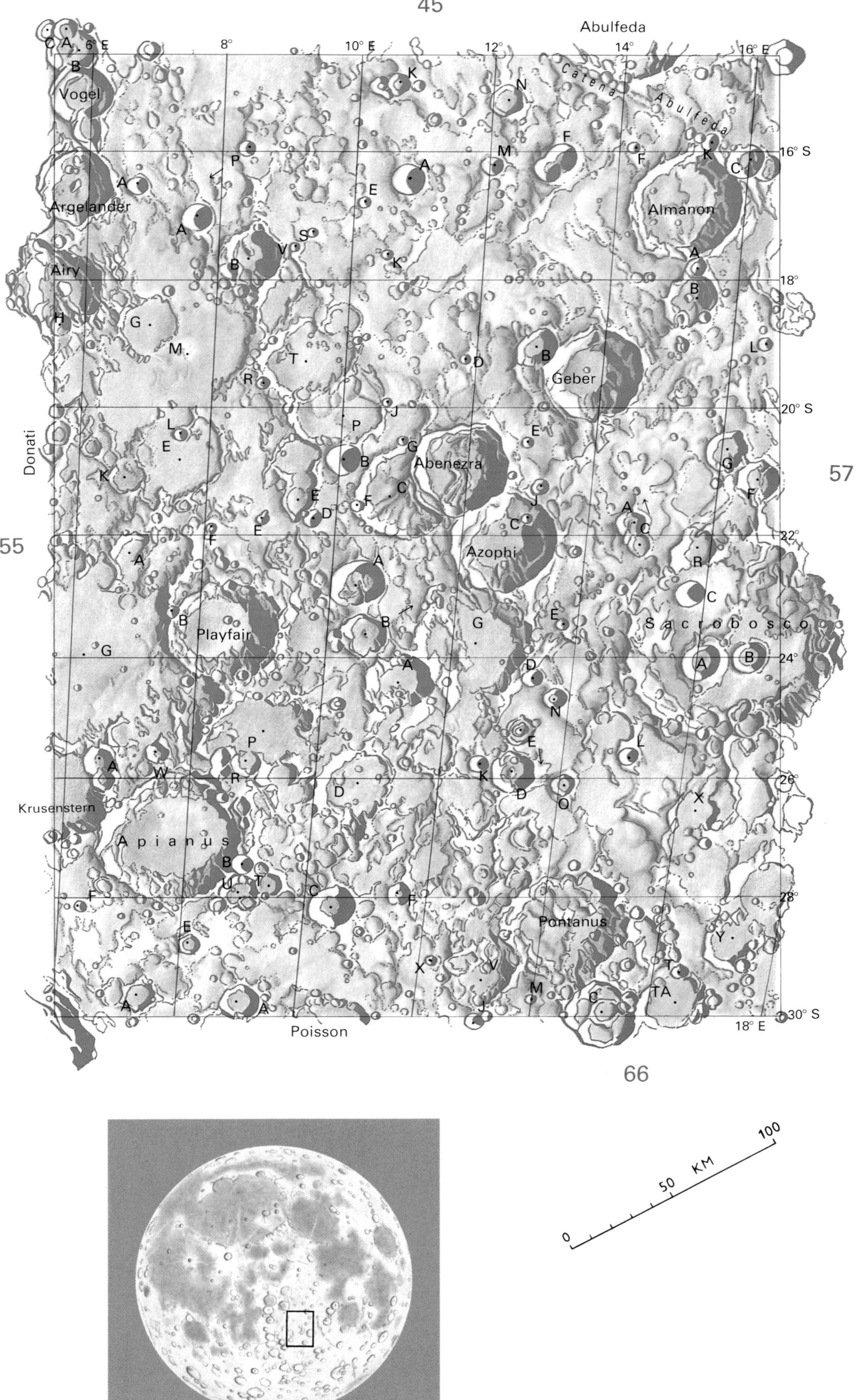
45
Abulfeda
Catena Abulfeda
Vogel
Argelander
Airy
Almanon
Geber
Abenezra
Azophi
Donati
Sacrobosco
Playfair
Apianus
Krusenstern
Pontanus
Poisson
55
57
66
6° E
8°
10° E
12°
14°
16° E
18° E
16° S
18°
20° S
22°
24°
26°
28°
30° S
0
50
KM
100

57. CATHARINA

The Rupes Altai, in this map illuminated by light coming from the east, commences near the crater Tacitus and cross the region to the south and south-east. A bright, wide ray from the crater Tycho (map 64) runs across the craters Polybius A and B. The large crater Catharina forms a conspicuous trio with Cyrillus and Theophilus (map 46).

Abulfeda, Catena. See map 56.

Altai, Rupes [24°S, 23°E] 'Altai Scarp' (Altai Range).
A mountain range on the perimeter of the basin of Mare Nectaris, named by Mädler; the range resembles a fault, sloping down to the basin; length 480 km.

Catharina [18.0°S, 23.6°E] St Catharina of Alexandria, d. 307 AD. Patron of Christian philosophers.
Considerably damaged ring mountain (100 km/3130 m).

Fermat [22.6°S, 19.8°E] Pierre de Fermat, 1601–1665. French scholar and mathematician; made discoveries in the theory of numbers.
Crater (39 km).

Polybius [22.4°S, 25.6°E] Polybius, 200–120 BC. Greek historian and statesman.
Crater (41 km/2050 m).

Pons [25.3°S, 21.5°E] Jean L. Pons, 1761–1831. French astronomer, discovered 36 comets, Director of the Museum Observatory, Florence.
Irregular crater (44 × 31 km).

Tacitus (16.2°S, 19.0°E] Cornelius Tacitus, *c.* 55–120 AD. Roman historian, author of the *Life of Agricola, Germania,* etc.
Prominent crater (40 km/2840 m).

Wilkins [29.4°S, 19.6°E] H. Percival Wilkins, 1896–1960. British selenographer, author of very detailed maps of the Moon.
Considerably disintegrated flooded crater (57 km).

Small craters: **Polybius A** (16.8 km/3720 m)
Polybius B (12.8 km/2630 m)
Pons B (13.9 km/3050 m)
Tacitus N (7.1 km/1050 m).

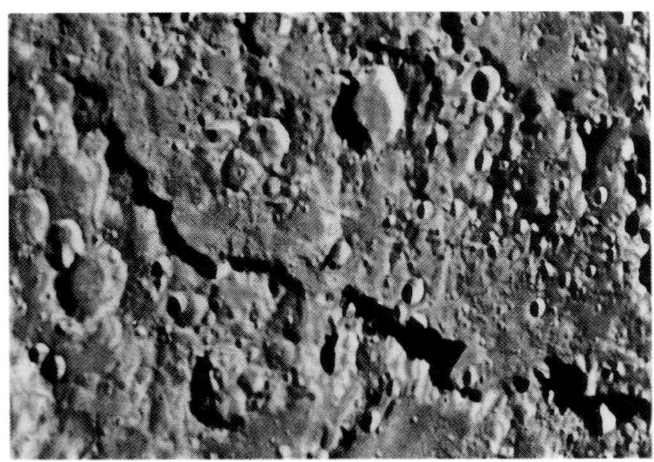

At sunset, the mountain range ***Rupes Altai*** *is bordered by conspicuous shadows which reveal the drop of more than 1000 metres to the lower-lying area.*

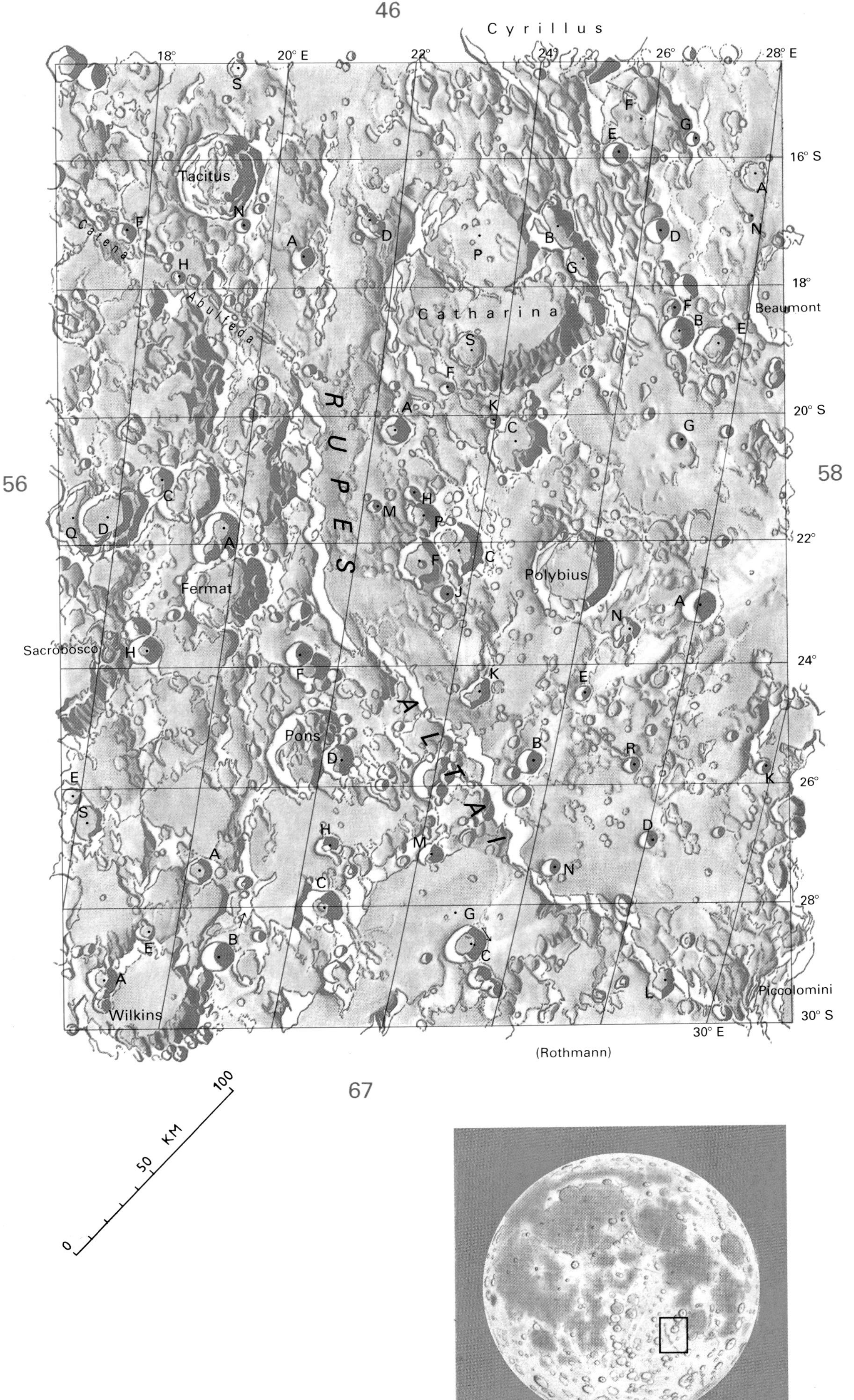
46
Cyrillus
18°
20° E
22°
24°
26°
28° E
16° S
18°
20° S
22°
24°
26°
28°
30° S
30° E
56
58
67
Tacitus
Catena Abulfeda
Catharina
Beaumont
RUPES ALTAI
Polybius
Fermat
Sacrobosco
Pons
Wilkins
Piccolomini
(Rothmann)
A
B
C
D
E
F
G
H
J
K
L
M
N
P
Q
R
S
0
50
100
KM

58. FRACASTORIUS

The dark surface of Mare Nectaris, the flooded walled plain Fracastorius and the beautiful crater Piccolomini are the most prominent formations in this part of the Moon. A low mountain range runs north of the crater Beaumont.

Beaumont [18.0°S, 28.8°E] Léonce Élie de Beaumont, 1798–1874. French geologist. Developed von Buch's theory of orology and demonstrated the method of determining the relative ages of individual rock layers.
Crater with interrupted wall to the east (53 km).

Bohnenberger [16.2°S, 40.0°E] Johann G. F. von Bohnenberger, 1765–1831. German mathematician and astronomer.
Crater with uneven hilly floor (33 km/1060 m).

Fracastorius [21.2°S, 33.0°E] Girolamo Fracastoro, 1483–1553. Italian physician, astronomer and poet. In his *Homocentrica* he attempted to replace Ptolemy's system with an inflexible system of homocentric spheres.
Walled plain (124 km), *whose wall to the north is missing, its floor continuing into M. Nectaris.*

Nectaris, Mare (Sea of Nectar) [15°S, 35°E].
Circular mare (diameter about 350 km, *surface area* 100 000 sq km); *it is the central part of a lunar basin, flooded with lava. The outer wall of the basin follows the Rupes Altai* (map 57).

Piccolomini [29.7°S, 32.2°E] Alessandro Piccolomini, 1508–1578. Italian archbishop and astronomer. Made star maps in which the stars were identified for the first time by letters of the Latin alphabet; Bayer's system using Greek letters was adopted at a later date.
Very prominent crater with central mountain massif (88 km).

Pyrenaeus, Montes (Pyrenees Mountains). See also maps 47, 48 and 59.

Rosse [17.9°S, 35.0°E] William Parsons, Third Earl of Rosse, 1800–1867. Irish nobleman, astronomer. Erected giant reflecting telescope of 72 inches (183 cm) aperture at Parsonstown, Ireland; studied nebulae and discovered the spiral structure of some galaxies; named various nebulae: the Owl Nebula, Crab Nebula, Dumb-bell Nebula, and others.
Circular crater (12 km/2420 m).

Weinek [27.5°S, 37.0°E] Ladislaus Weinek, 1848–1913. Austrian astronomer, who in 1883 became Director of the Prague Observatory at Klementinum; prepared an atlas of the Moon.
Crater (32 km/3370 m).

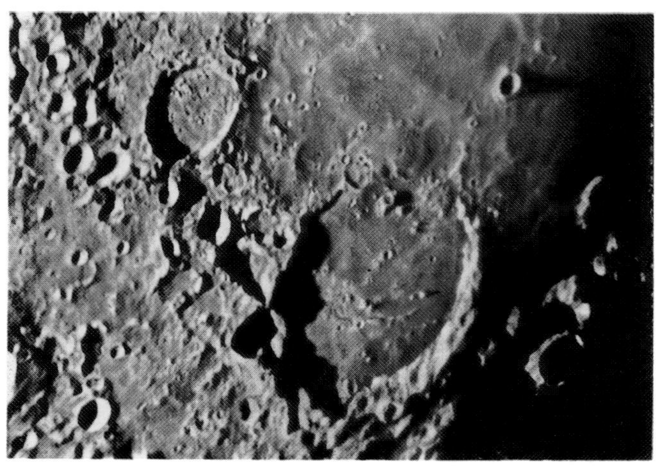

*The northern part of the wall of the walled plain **Fracastorius** is disturbed and flooded by lava flows which once filled the central part of the lunar basin Mare Nectaris.*

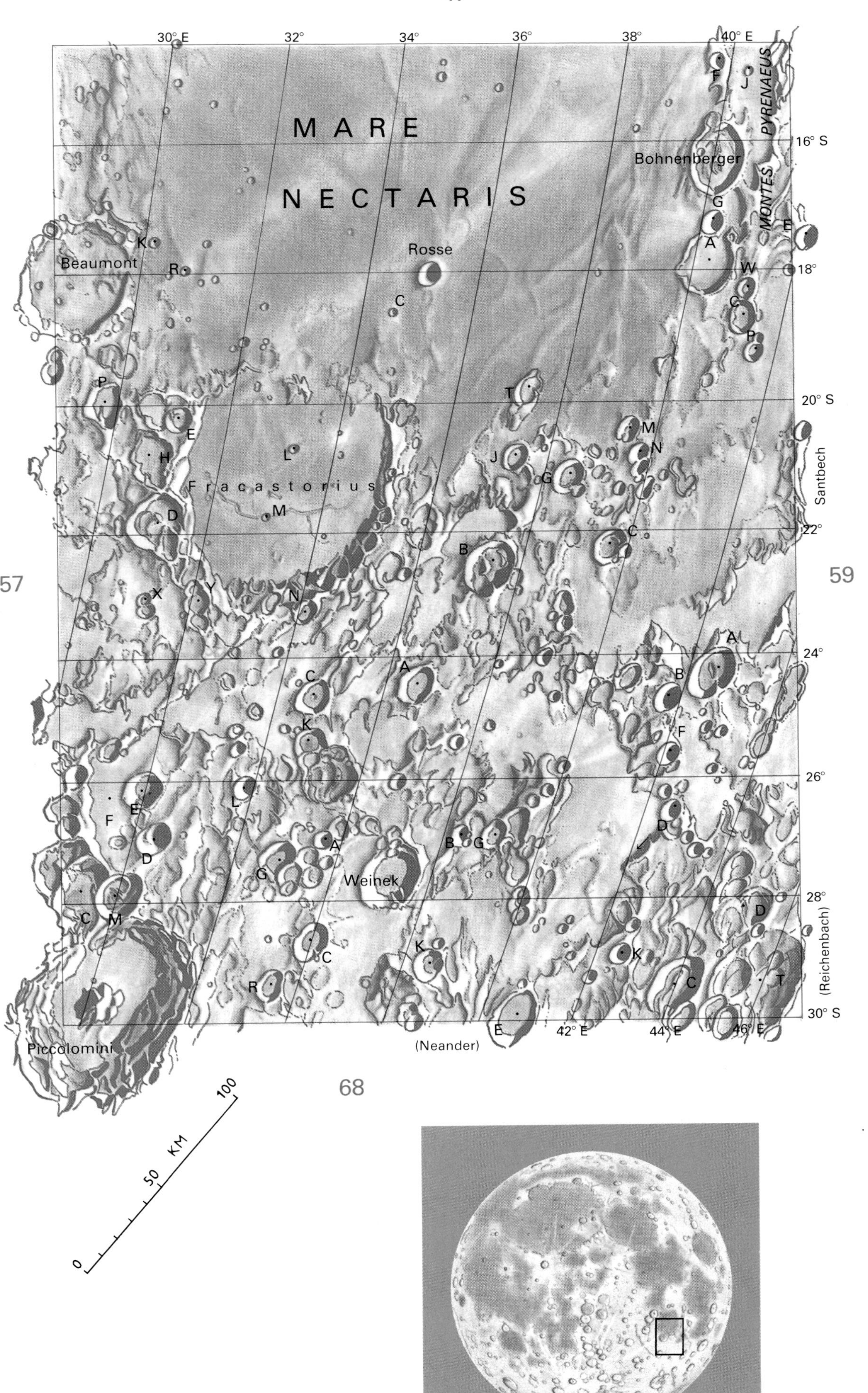
47
30° E
32°
34°
36°
38°
40° E
MARE
NECTARIS
PYRENAEUS
MONTES
Bohnenberger
16° S
Rosse
Beaumont
18°
20° S
Fracastorius
Santbech
22°
57
59
24°
26°
Weinek
28°
(Reichenbach)
Piccolomini
(Neander)
42° E
44° E
46° E
30° S
68
0
50
100
KM

59. PETAVIUS

The southern extension of Mare Fecunditatis, with bright rays from the craters Petavius B and Snellius A, includes the giant crater Petavius (which forms a magnificent chain with Langrenus, Vendelinus and Furnerius) and a group of craters named after famous navigators.

Biot [22.6°S, 51.1°E] Jean-Baptiste Biot, 1774–1862. French astronomer, geodesist and historian of astronomy.
Crater (13 km).

Borda [25.1°S, 46.6°E] Jean C. Borda, 1733–1799. French naval officer and astronomer.
Crater with disintegrated wall and central peak (44 km).

Colombo [15.1°S, 45.8°E] Cristoforo Colombo (Columbus), 1451–1506. Italian-born Spanish navigator, discovered America in 1492.
Prominent crater with central peaks (76 km).

Cook [17.5°S, 48.9°E] James Cook, 1728–1779. English naval captain and explorer, twice circumnavigated the world.
Flooded crater with a low wall (47 km).

Fecunditatis, Mare (Sea of Fertility). See also map 37, 48 and 49.

Hase [29.4°S, 62.5°E] Johann M. Hase, 1684–1742. German mathematician and cartographer.
Disintegrated crater (83 km).

McClure [15.3°S, 50.3°E] Robert le M. McClure, 1807–1873. British naval officer. Discovered the North-West Passage.
Crater (24 km).

Monge [19.2°S, 47.6°E] Gaspard Monge, 1746–1818. French mathematician. One of the founders of descriptive geometry.
Crater (37 km).

Palitzsch [28.0°S, 64.5°E] Johann G. Palitzsch, 1723–1788. German amateur astronomer. First to find Halley's comet on its return in 1758.
Inconspicuous crater (41 km).

Palitzsch, Vallis [25°S, 65°E]
A valley 110 km *long that follows the eastern wall of the crater Petavius.*

Petavius [25.3°S, 60.4°E] Denis Petau, 1583–1652. French theologian and historian; studies in chronology.
Ring mountain with central mountain chain, clefts and dark patches on the floor (177 km).

Petavius, Rimae
System of large rilles inside Petavius, length about 80 km.

Santbech [20.9°S, 44.0°E] Daniel Santbech Noviomagus, *c.* 1561. Dutch mathematician and astronomer.
Crater (64 km).

Snellius [29.3°S, 55.7°E] Willibrord van Roijen Snell (Snellius), 1591–1626. Dutch astronomer and geodesist. 'Snell's Law' in optics.
Crater (83 km).

Snellius, Vallis [31°S, 59°E]
One of the longest valleys on the Moon (500 km), *continues on map* 69; *the valley is directed towards the centre of the Mare Nectaris basin and is obviously connected with its origin.*

Wrottesley [23.9°S, 56.8°E] John, First Baron Wrottesley, 1798–1867. English astronomer. Worked in the field of astrometry; catalogue of double stars.
Prominent crater (57 km).

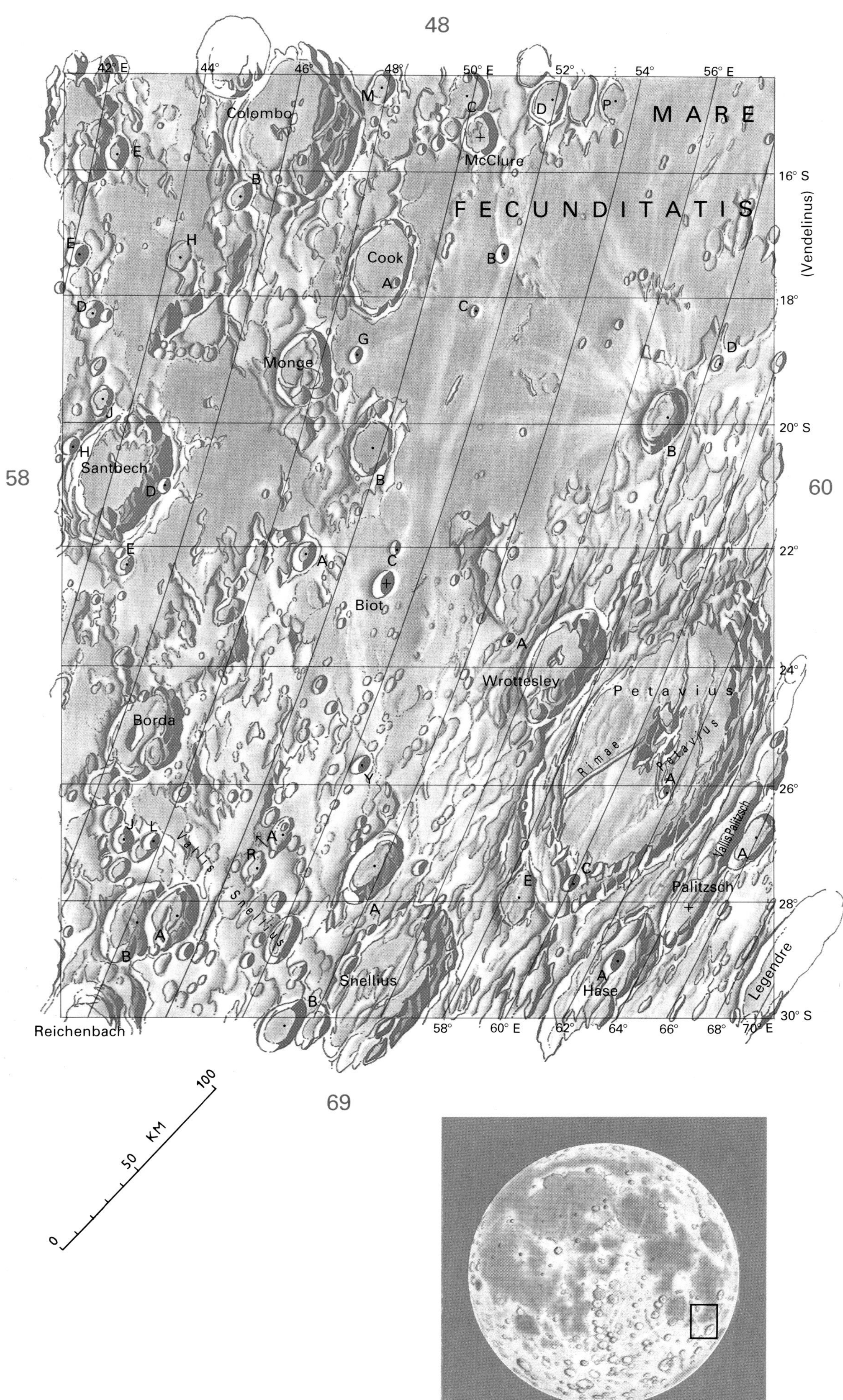
48
58
60
69
MARE
FECUNDITATIS
(Vendelinus)
Colombo
McClure
Cook
Monge
Santbech
Biot
Wrottesley
Petavius
Rimae Petavius
Borda
Vallis Snellius
Vallis Palitzsch
Palitzsch
Snellius
Hase
Legendre
Reichenbach
42° E
44°
46°
48°
50° E
52°
54°
56° E
16° S
18°
20° S
22°
24°
26°
28°
30° S
58°
60° E
62°
64°
66°
68°
70° E
0
50
KM
100

60. VENDELINUS

This region of the eastern limb of the near-side of the Moon contains the large walled plain Humboldt, which is best seen shortly after Full Moon. The walled plain Vendelinus belongs to the conspicuous line of craters Langrenus, Vendelinus, Petavius and Furnerius.

Balmer [20.1°S, 70.6°E] Johann J. Balmer, 1825–1898. Swiss mathematician and physicist. 'Balmer series' of lines in the spectrum of hydrogen.
Remains of a flooded walled plain (112 km).

Behaim [16.5°S, 79.4°E] Martin Behaim (Behem), 1459–1506. German navigator and cartographer.
Regular crater with terraced walls and a central peak (55 km).

Gibbs [18.4°S, 84.3°E] Josiah W. Gibbs, 1839–1903. American mathematician and physicist.
Crater (77 km).

Hecataeus [21.8°S, 79.6°E] Hecataeus, 550–480 BC. Greek geographer of Miletus, author of a description of the world, with map.
Walled plain (127 km).

Holden [19.1°S, 62.5°E] Edward S. Holden, 1846–1914. American astronomer, first Director of Lick Observatory.
Crater (47 km).

Humboldt [27.2°S, 80.9°E] Wilhelm von Humboldt, 1767–1835. German statesman and philologist. Brother of Alexander von Humboldt.
Typical walled plain, whose floor has a central mountain range, a network of concentric and radial clefts and dark patches close to the wall (207 km).

Humboldt, Catena [22°S, 85°E]
Chain of small craters about 160 km *long, directed towards the centre of Humboldt; the chain, which resembles a wide rille, is visible from the Earth only at extreme librations.*

Lamé [14.7°S, 64.5°E] Gabriel Lamé, 1795–1870. French mathematician.
Crater (84 km).

Legendre [28.9°S, 70.2°E] Adrien M. Legendre, 1752–1833. French mathematician. Elliptic functions, theory of numbers.
Walled plain (79 km).

Phillips [26.6°S, 76.0°E] John Phillips, 1800–1874. British geologist, popularizer of sciences, observed Mars and the Moon.
Walled plain (124 km).

Schorr [19.5°S, 89.7°E] Richard Schorr, 1867–1951. German astronomer.
Crater (53 km).

Vendelinus [16.3°S, 61.8°E] Godefroid Wendelin, 1580–1667. Belgian astronomer.
Walled plain (147 km).

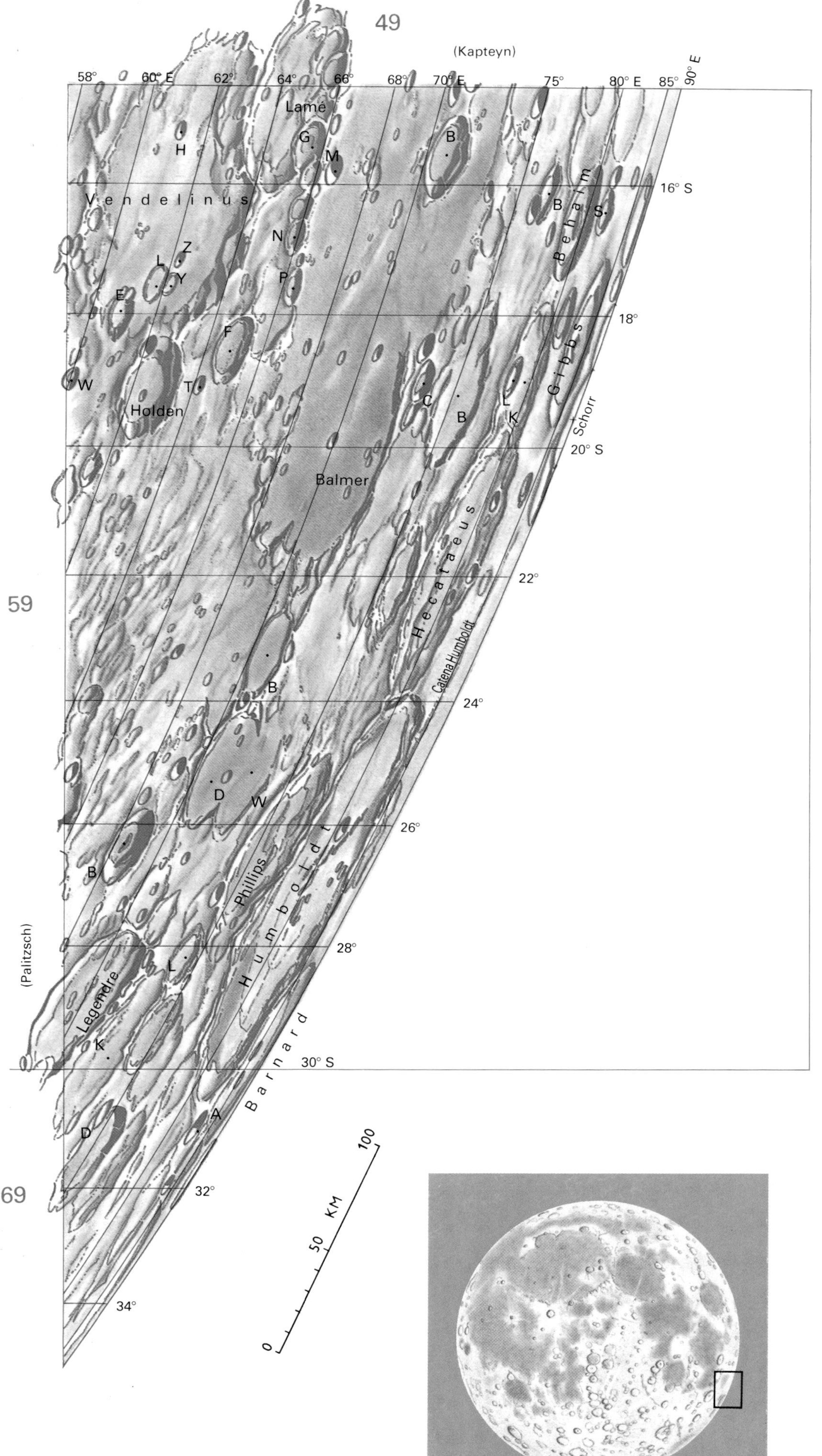

49
(Kapteyn)
58°
60° E
62°
64°
66°
68°
70° E
75°
80° E
85°
90° E
Lamé
G
M
H
B
16° S
Vendelinus
B
S
Behaim
N
Z
L
Y
P
E
18°
F
Gibbs
W
T
C
B
L
K
Holden
Schorr
20° S
Balmer
Hecataeus
22°
59
Catena Humboldt
B
24°
D
W
26°
B
Phillips
Humboldt
(Palitzsch)
L
28°
Legendre
K
Barnard
30° S
A
D
69
32°
34°
0
50
KM
100

61. PIAZZI

The south-western limb of the Moon. This area displays a rich system of radial mountain ranges and valleys, which are directed towards the centre of the basin of Mare Orientale, e.g. Vallis Bouvard, Vallis Inghirami and Vallis Baade. However, the first of these is difficult to observe from the Earth.

Baade [44.8°S, 81.8°W] Walter Baade, 1893–1960. German-born American astronomer, who made significant contributions to understanding our own Galaxy and other galaxies.
Crater (55 km).

Baade, Vallis [46°S, 76°W]
Valley, length 160 km.

Bouvard, Vallis (Bouvard's Valley) [39°S, 83°W]. Alexis Bouvard, 1767–1843. French mathematician and astronomer, discovered several comets.
Valley, length about 280 km, *width* 40 km.

Catalán [45.7°S, 87.3°W] Miguel A. Catalán, 1894–1957. Spanish physicist and mathematician. Research in the field of spectroscopy.
Crater (25 km).

Fourier [30.3°S, 53.0°W] Jean-B. J. Fourier, 1768–1830. French physicist and mathematician. 'Fourier series'.
Crater (52 km).

Graff [42.4°S, 88.6°W] Kasimir R. Graff, 1878–1950. Polish-born Viennese astronomer.
Crater (36 km).

Inghirami, Vallis [44°S, 73°W] *Named after the* crater Inghirami (map 62).
Valley, length 140 km.

Lacroix [37.9°S, 59.0°W] Sylvestre F. de Lacroix, 1765–1843. French mathematician and teacher.
Crater (38 km).

Lagrange [33.2°S, 72.0°W] Joseph L. Lagrange, 1736–1813. Outstanding French mathematician, author of *Mécanique Analytique*.
Damaged walled plain (160 km).

Lehmann. See also map 62.

Piazzi [36.2°S, 67.9°W] Giuseppe Piazzi, 1746–1826. Italian astronomer, discovered the first asteroid (Ceres).
Eroded walled plain (101 km).

Shaler [32.9°S, 85.2°W] Nathaniel S. Shaler, 1841–1906. American geologist and palaeontologist; geological interpretation of photographs of the Moon.
Crater (48 km).

Wright [31.6°S, 86.6°W] (1) Frederick E. Wright, 1878–1953; American astronomer and selenographer. (2) Thomas Wright, 1711–1786; British natural philosopher. (3) William H. Wright, 1871–1959; American astronomer: photographs of Mars.
Crater (40 km).

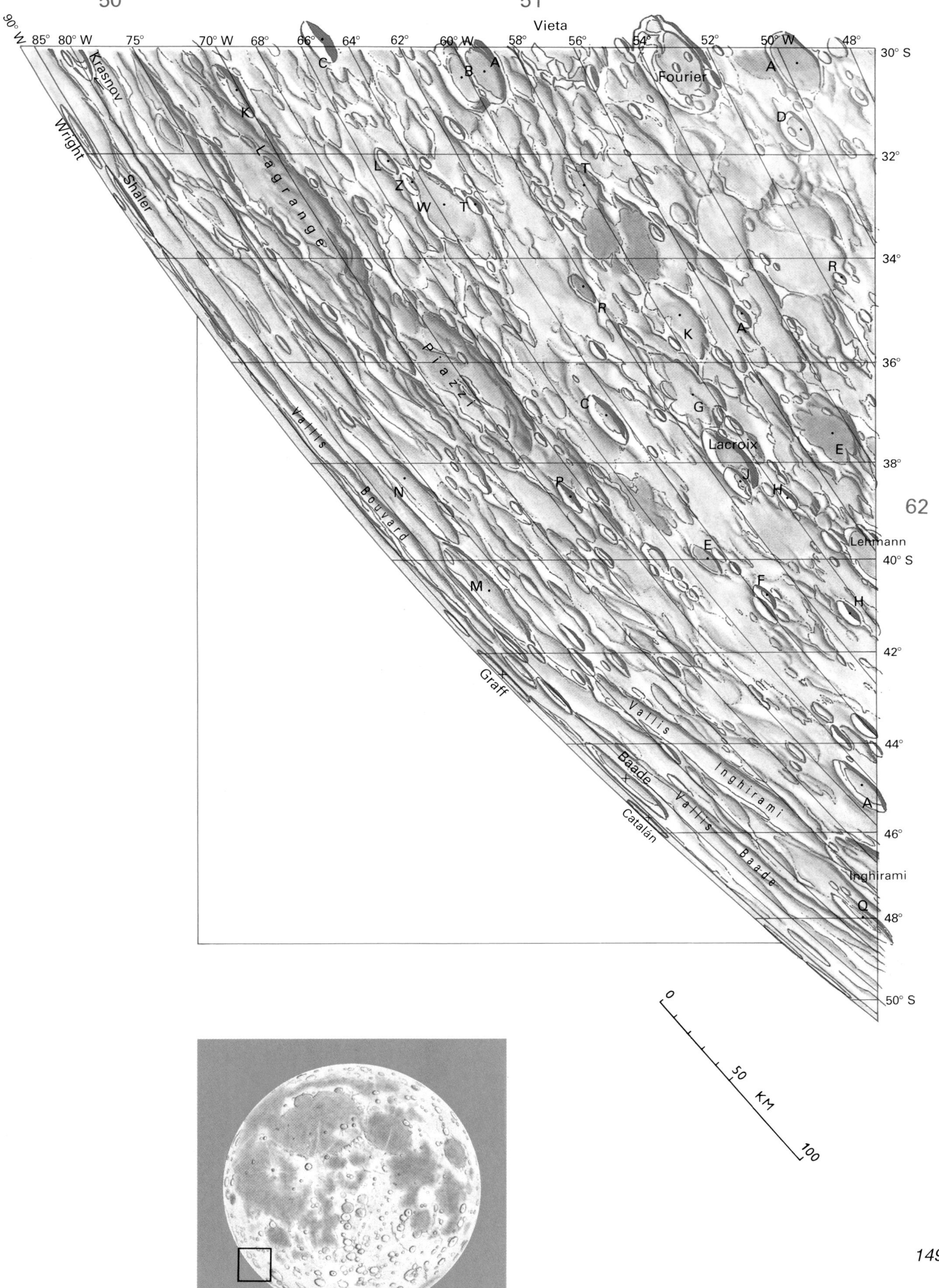
50
51
62
Vieta
90° W
85°
80° W
75°
70° W
68°
66°
64°
62°
60° W
58°
56°
54°
52°
50° W
48°
30° S
32°
34°
36°
38°
40° S
42°
44°
46°
48°
50° S
Krasnov
Wright
Shaler
Lagrange
Piazzi
Vallis Bouvard
Graff
Vallis Inghirami
Baade
Catalán
Vallis Baade
Fourier
Lacroix
Lehmann
Inghirami
A
B
C
D
K
L
Z
W
T
R
G
E
J
H
P
N
M
F
Q
0
50
KM
100

62. SCHICKARD

Mountainous region along the south-western limb of the Moon, which in the north adjoins the edge of Mare Humorum. One of the largest walled plains, Schickard, with characteristic dark patches on its floor, is in this area.

Clausius [36.9°S, 43.8°W] Rudolf J. E. Clausius, 1822–1888. German physicist. Thermodynamics, kinetic theory of gases.
Crater with a flooded floor (25 km).

Drebbel [40.9°S, 49.0°W] Cornelius Drebbel, 1572–1634. Dutch physicist. Claimant to the invention of the telescope and microscope.
Crater (30 km).

Excellentiae, Lacus (Lake of Excellence) [36°S, 43°W]. Ill-defined area of the mare close to the crater Clausius. The name was accepted by the IAU in 1976.
Maximum length about 150 km.

Inghirami [47.5°S, 68.8°W] Giovanni Inghirami, 1779–1851. Italian astronomer.
Prominent crater (91 km) *crossed by radial mountain ranges and valleys running from the centre of Mare Orientale.*

Lee [30.7°S, 40.7°W] John Lee, 1783–1866. English selenographer and collector of antiquities.
Remains of a flooded crater (41 km/1340 m).

Lehmann [40.0°S, 56.0°W] Jacob H. W. Lehmann, 1800–1863. German theologian and astronomer. Worked in the field of celestial mechanics.
Considerably eroded crater (53 km).

Lepaute [33.3°S, 33.6°W] Madame Lepaute, *née* Nicole Reine Etable de la Brière, 1723–1788. French mathematician, co-operated with Clairaut and Lalande.
Crater (16 km/2070 m).

Schickard [44.4°S, 54.6°W] Wilhelm Schickard, 1592–1635. German mathematician and astronomer. First to attempt determination of the path of a meteor by simultaneous observations from different places.
Vast walled plain with a partially flooded floor (227 km).

Vitello [30.4°S, 37.5°W] Erazmus C. Witelo, 1225–1290 AD. Polish mathematician and physicist, worked at Padua in Italy.
Crater (42 km/1730 m).

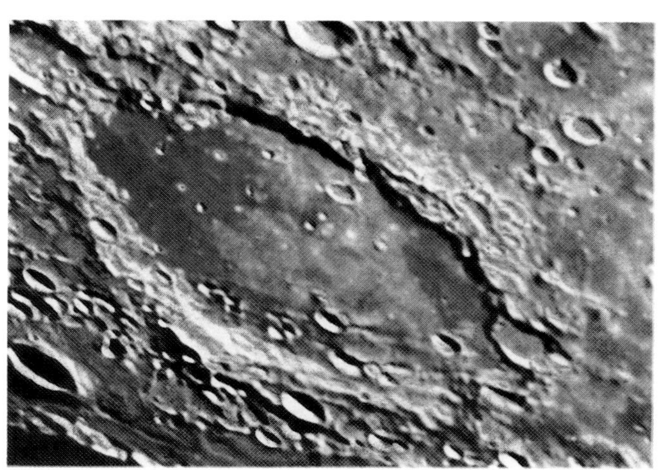

The large walled plain ***Schickard,*** *one of the most prominent lunar formations of its kind, lies on the southwestern limb of the Moon.*

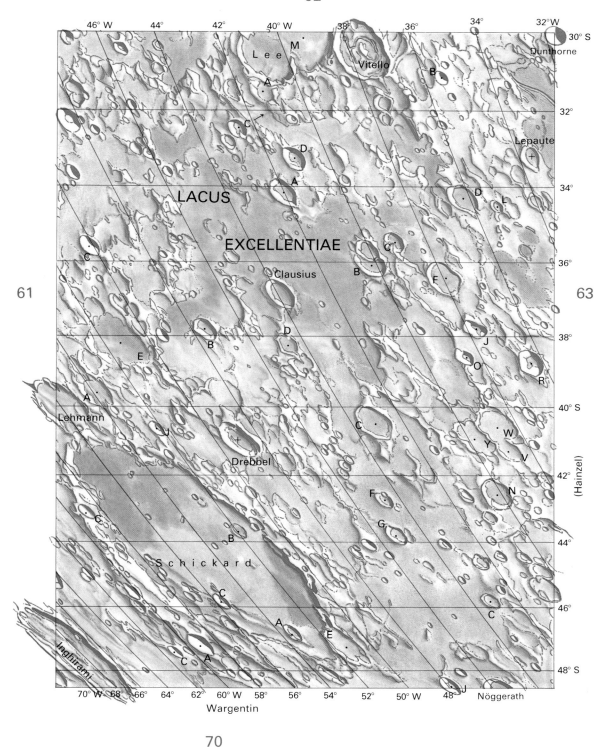

46° W 44° 42° 40° W 38° 36° 34° 32° W

30° S

Lee M A Vitello B Dunthorne

32°

C D Lepaute

A

34°

LACUS D L

EXCELLENTIAE

C C

36°

Clausius B F

61 63

D

38°

B J

E O R

A

Lehmann

40° S

J C W

Drebbel Y V

42° (Hainzel)

F N

C G

B

44°

Schickard

C

46°

C

A E

A

C

Inghirami

48° S

J

70° W 68° 66° 64° 62° 60° W 58° 56° 54° 52° 50° W 48° Nöggerath

Wargentin

70

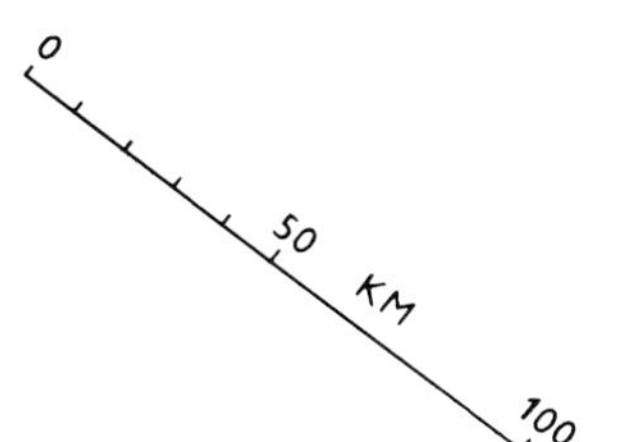

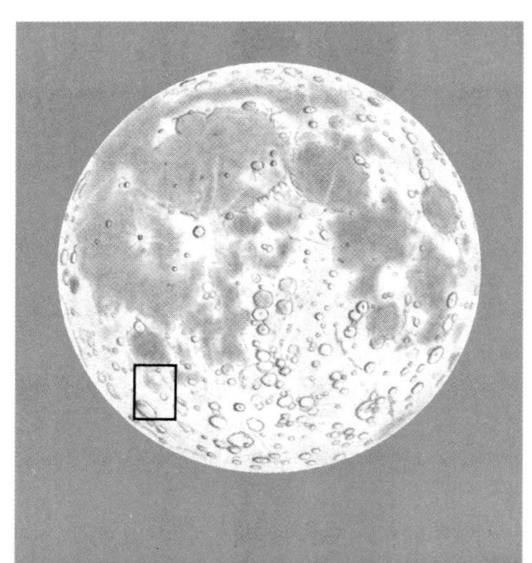

63. CAPUANUS

Area in the south-western sector of the Moon. On the surface of Palus Epidemiarum there is a system of rilles close to the crater Ramsden. Notice also the wide rille Rima Hesiodus, which continues on maps 53 and 54. The floor of the crater Capuanus contains a group of domes.

Capuanus [34.1°S, 26.7°W] Francesco Capuano di Manfredonia, fifteenth century. Italian theologian and astronomer.
Flooded crater (60 km).

Cichus [33.3°S, 21.1°W] Francesco degli Stabili, also known as Cecco d'Ascoli, 1257–1327. Italian astronomer and astrologer; accused of heresy and burnt in Florence.
Crater (41 km/2760 m); *on its western wall is* **Cichus C** (11.1 km/1250 m).

Dunthorne [30.1°S, 31.6°W] Richard Dunthorne, 1711–1775. British geodesist and astronomer.
Crater (16 km/2780 m).

Elger [35.3°S, 29.8°W] T. Gwyn Elger, 1838–1897. British selenographer, produced a map of the Moon (1895).
Crater (21 km/1250 m).

Epidemiarum, Palus (Marsh of Epidemics). See map 53.

Epimenides [40.9°S, 30.2°W] Epimenides, late seventh century BC. Cretan poet and prophet.
Crater (27 km).

Haidinger [39.2°S, 25.0°W] Wilhelm Karl von Haidinger, 1795–1871. Austrian geologist and physicist.
Crater (22 km/2330 m); *on its south-eastern wall is* **Haidinger B** (10.3 km/1500 m).

Hainzel [41.3°S, 33.5°W] Paul Hainzel, *c.* 1570. German astronomer, co-operated with Tycho Brahe.
Complex formation consisting of three overlapping craters, the largest of which (Hainzel) is 70 km *in diameter; the two smaller ones are Hainzel* **A** *and Hainzel* **C**.

Hesiodus, Rima [30°S, 21°W] Named after the crater Hesiodus (map 54).
Wide rille, length 300 km.

Lagalla [44.6°S, 22.5°W] Giulio Cesare Lagalla, 1571–1624. Italian philosopher, one of the earliest telescopic observers of the Moon.
Remains of a crater (85 km).

Marth [31.1°S, 29.3°W] Albert Marth, 1828–1897. German astronomer.
Interesting crater with a double wall.

Mee [43.7°S, 35.0°W] Arthur B. P. Mee, 1860–1926. Scottish astronomer and writer. Observer of the Moon and Mars.
Eroded crater (132 km).

Ramsden [32.9°S, 31.8°W] Jesse Ramsden, 1735–1800. British mechanic, maker of astronomical instruments.
Flooded crater (25 km/1990 m).

Ramsden, Rimae [33°S, 31°W]
System of jointed rilles which occupy a surface area about 130 km *in diameter.*

Timoris, Lacus (Lake of Fear) [39°S, 28°W]
Elongated region of mare material, surrounded by mountain massifs; its length is about 130 km.

Weiss [31.8°S, 19.5°W] Edmund Weiss, 1837–1917. Austrian astronomer, co-founder and Director of the new Vienna University Observatory.
Remains of a flooded crater (66 km).

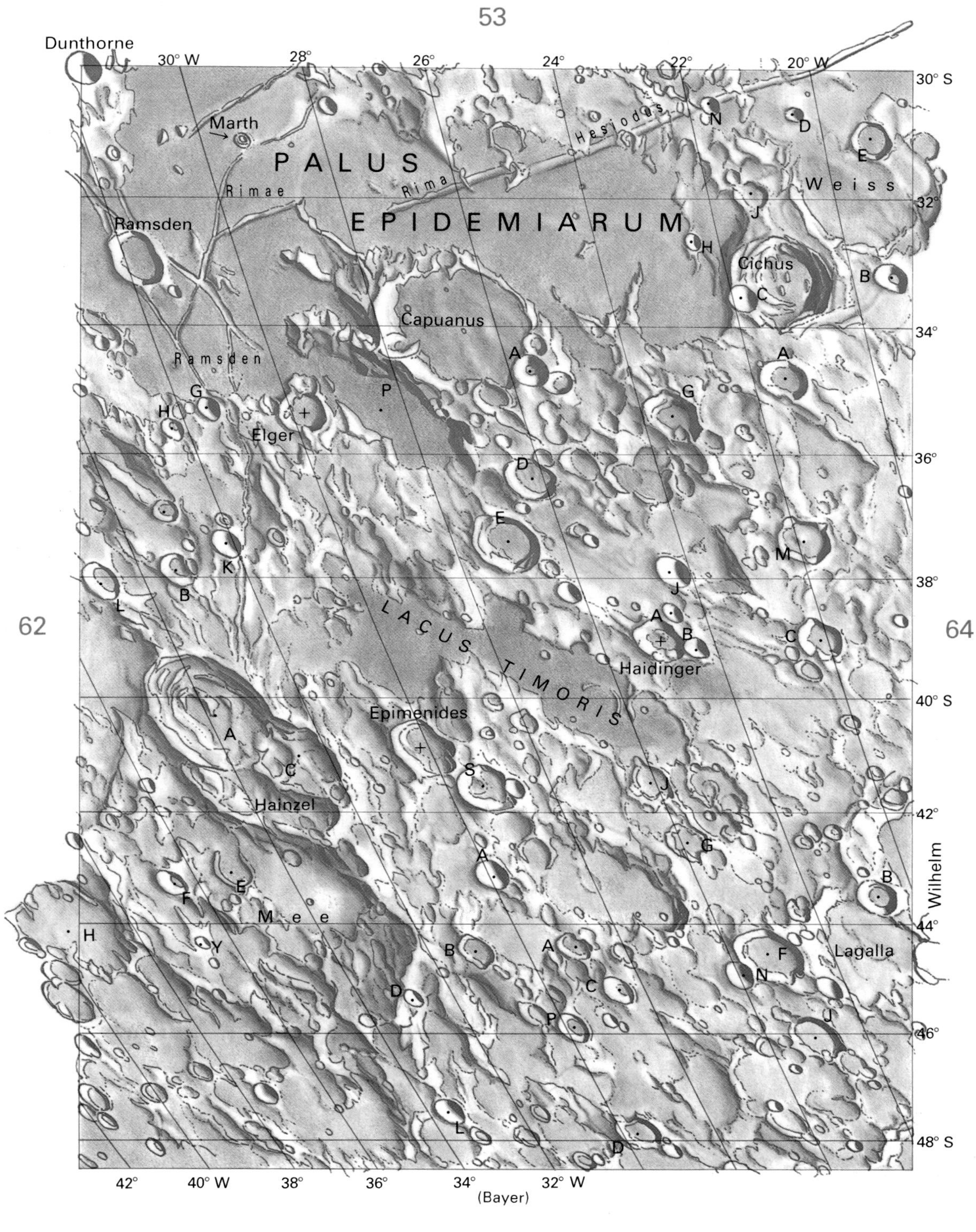

53

Dunthorne

30° W 28° 26° 24° 22° 20° W

30° S

Marth

PALUS

Rimae

Rima Hesiodus

EPIDEMIARUM

N D E Weiss

32°

Ramsden

J H Cichus B C

Capuanus

34°

Ramsden

A A

G P G

H Elger

36°

D

E M

K J

B L

38°

62

LACUS TIMORIS

A B C

Haidinger

64

40° S

Epimenides

A

C S J

Hainzel

42°

G

A B

F E Wilhelm

Mee

H Y

44°

B A F Lagalla

N C D

P J

46°

L D

48° S

42° 40° W 38° 36° 34° 32° W

(Bayer)

71

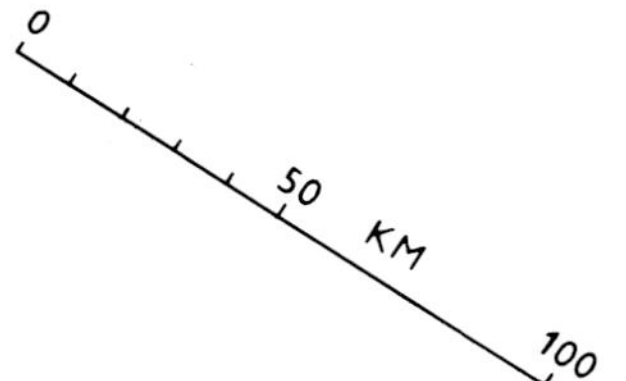

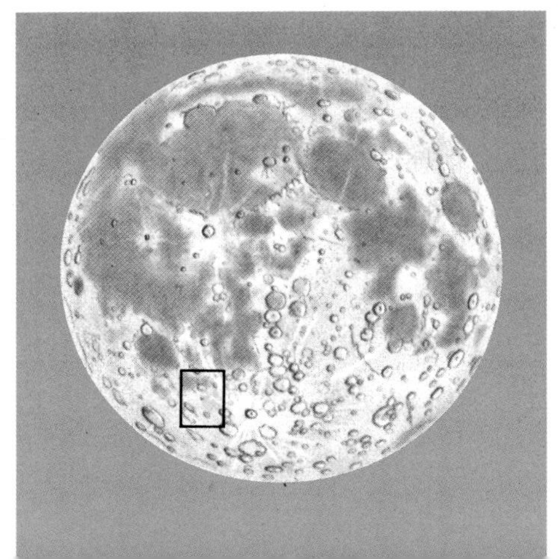

64. TYCHO

Crater field south of Mare Nubium. This area is dominated by the crater Tycho, which is one of the most prominent craters on the Moon and the centre of the most extensive system of bright rays. *Surveyor 7* soft-landed close to the northern rim of Tycho.

Ball [35.9°S, 8.4°W] William Ball, d. 1690. English amateur astronomer. Confirmed the Huygens' observations of Saturn's rings.
Crater (41 km/2810 m).

Brown [46.4°S, 17.9°W] Ernest W. Brown, 1866–1938. English-born American astronomer, author of theory of the motion of the Moon.
Crater (34 km); *its wall is penetrated by the crater* **Brown E.**

Deslandres. See also map 65.

Gauricus [33.8°S, 12.6°W] Luca Gaurico, 1476–1558. Italian theologian, astronomer and astrologer, translator of Ptolemy's *Almagest.*
Considerably eroded crater (79 km).

Heinsius [39.5°S, 17.7°W] Gottfried Heinsius, 1709–1769. German mathematician and astronomer.
Crater (64km/2650 m), *merges with* **Heinsius A** (20 km/3270 m), **Heinsius B** *and* **Heinsius C**.

Hell [32.4°S, 7.8°W] Maximilian Hell, 1720–1792. Hungarian astronomer, founder of the original Vienna Observatory; observed transit of Venus in 1769.
Crater (33 km/2200 m).

Montanari [45.8°S, 20.6°W] Geminiano Montanari, 1633–1687. Italian astronomer. First micrometric measurements for the mapping of the Moon.
Distorted crater (77 km).

Pictet [43.6°S, 7.4°W] Marc A. Pictet, 1752–1825. Swiss astronomer and naturalist. Director of Geneva Observatory.
Crater (62 km).

Pitatus. See map 54.

Sasserides [39.1°S, 9.3°W] Gellio Sasceride, 1562–1612. Danish physician and astronomer, assistant to Tycho Brahe.
Ruined crater (90 km).

Street [46.5°S, 10.5°W] Thomas Streete (Street), 1621–1689. English astronomer, author of *Astronomia Carolina.*
Crater (58 km).

Tycho [43.3°S, 11.2°W] Tycho Brahe, 1546–1601. Danish astronomer, prominent observer and organizer of scientific research. His accurate observations enabled Kepler to discover the laws of planetary motion.
Crater (85 km/4850 m) *with the most extensive ray system.*

Wilhelm [43.1°S, 20.8°W] Wilhelm IV, Landgrave of Hesse, 'The Wise', 1532–1592. German prince and astronomer.
Walled plain (107 km).

Wurzelbauer [33.9°S, 15.9°W] Johann P. von Wurzelbauer, 1651–1725. German astronomer, observer of the Sun.
Greatly eroded crater (88 km).

54

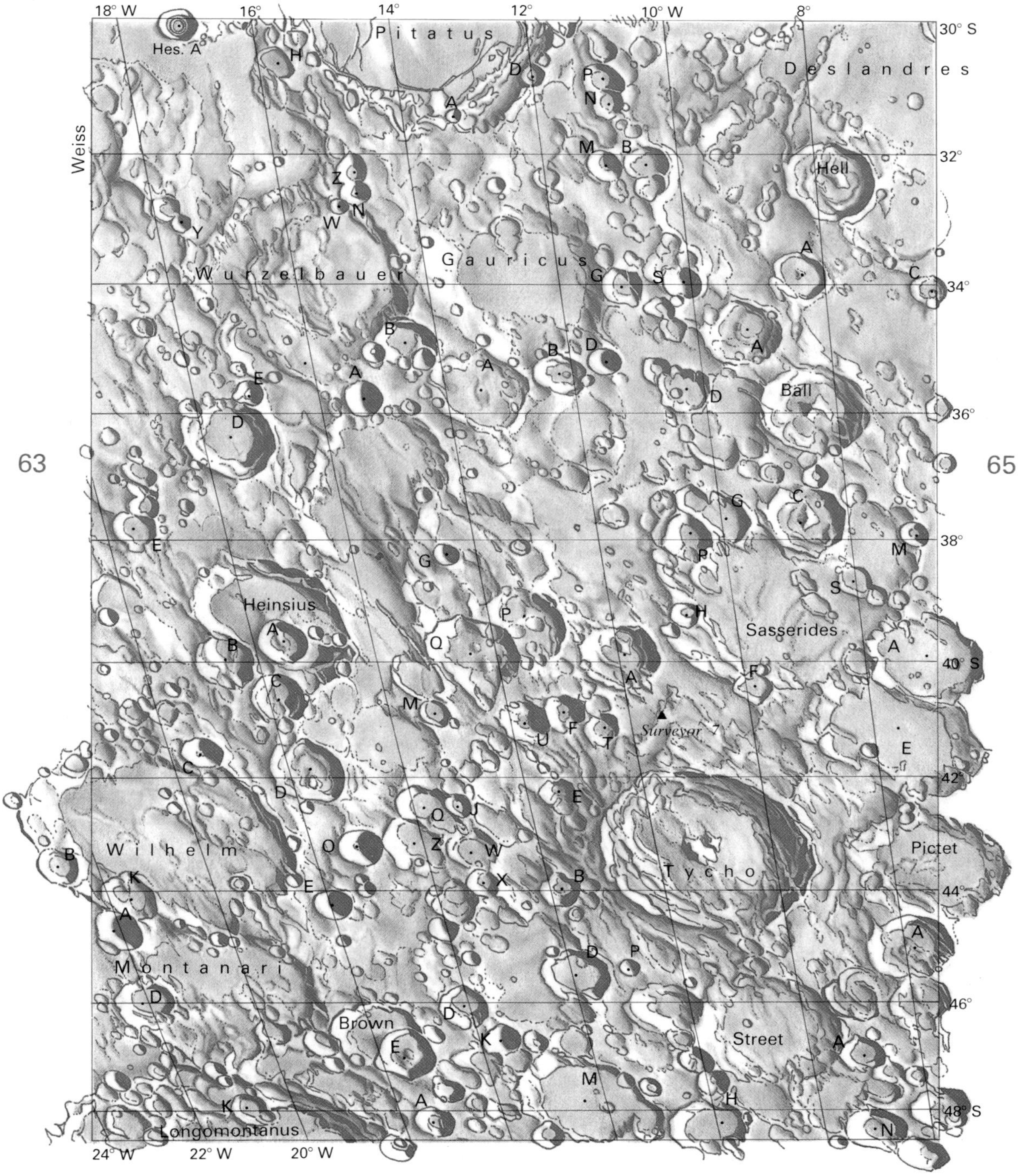

18° W 16° 14° 12° 10° W 8°

30° S 32° 34° 36° 38° 40° S 42° 44° 46° 48° S

24° W 22° W 20° W

Pitatus
Hes. A
Deslandres
Weiss
Hell
Gauricus
Wurzelbauer
Ball
Heinsius
Sasserides
Surveyor 7
Wilhelm
Tycho
Pictet
Montanari
Brown
Street
Longomontanus

63

65

72

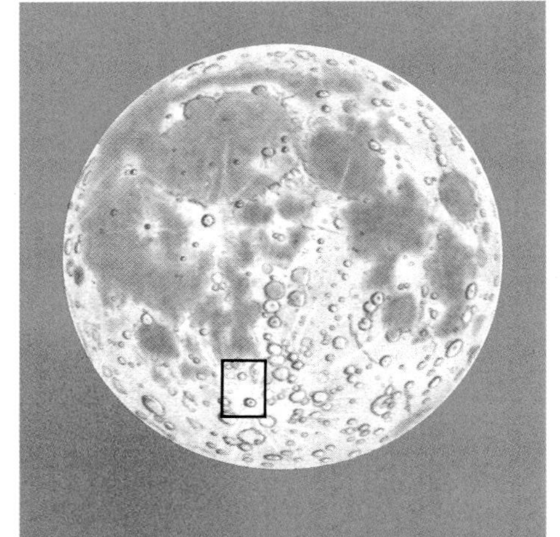

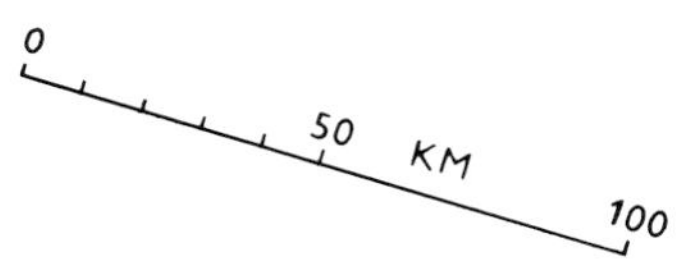

65. WALTER

This is a dense crater field in the vicinity of the prime meridian on the southern hemisphere of the Moon. Identification is facilitated by the presence of the walled plain Stöfler and the crater Aliacensis to the north-east. Note also the group of craters bordering Huggins.

Aliacensis [30.6°S, 5.2°E] Pierre d'Ailly, 1350–1420. French theologian and geographer.
Crater (80 km/3680 m).

Deslandres [32.5°S, 5.2°W] Henri A. Deslandres, 1853–1948. French astronomer; observer of the Sun, invented spectroheliograph, Director of Paris Observatory at Meudon.
Ruined walled plain (234 km).

Fernelius [38.1°S, 4.9°E] Jean Fernel, 1497–1558. French physicist.
Crater with a flooded floor (65 km).

Huggins [41.1°S, 1.4°W] Sir William Huggins, 1824–1910. English astronomer, pioneered astronomical spectroscopy.
Crater (65 km), *merges with the crater Nasireddin.*

Lexell [35.8°S, 4.2°W] Anders J. Lexell, 1740–1784. Swedish mathematician and astronomer. Worked in the field of celestial mechanics.
Crater (63 km).

Licetus [47.1°S, 6.7°E] Fortunio Liceti, 1577–1657. Italian physicist and philosopher.
Crater (75 km).

Miller [39.3°S, 0.8°E] William A. Miller, 1817–1870. English chemist.
Crater (75 km).

Nasireddin [41.0°S, 0.2°E] Nasir-al-Din, Mohammed Ibn Hassan, 1201–1274 AD. Persian astronomer.
Crater (52 km).

Nonius [34.8°S, 3.8°E] Pedro Nuñez, 1492–1577. Portuguese mathematician. Devised an early type of vernier for reading divided circles and scales.
Disintegrated polygonal crater (70 km/2990 m).

Orontius [40.3°S, 4.0°W] Orontius Finaeus, 1494–1555. French mathematician and cartographer.
Walled plain (122 km).

Pictet. See map 64.

Proctor [46.4°S, 5.1°W] Mary Proctor, 1862–1957. Daughter of the astronomer R. A. Proctor; astronomer and popularizer of astronomy.
Crater (52 km).

Saussure [43.4°S, 3.8°W] Horace B. de Saussure, 1740–1799. Swiss philosopher and natural historian.
Crater (54 km/1880 m).

Stöfler [41.1°S, 6.0°E] Johann Stöfler, 1452–1534. German mathematician, astronomer and astrologer.
Walled plain with a flooded floor (126 km/2760 m).

Walter [33.0°S, 0.7°E] Bernard Walter, 1430–1504. German astronomer.
Walled plain (132 × 140 km/4130 m).

55
64
66
73
6° W
4° W
2° W
0°
2° E
4° E
6° E
30° S
32°
34°
36°
38°
40° S
42°
44°
46°
48° S
Deslandres
Aliacensis
Walter
Nonius
Lexell
Kaiser
Fernelius
Miller
Huggins
Orontius
Nasireddin
Stöfler
Faraday
Pictet
Saussure
Proctor
Maginus
Licetus
0
50
KM
100

66. MAUROLYCUS

Crater field in the south-eastern sector of the Moon. The prominent crater Maurolycus forms a characteristic trio with the neighbouring craters Faraday and Stöfler (map 65). The crater Gemma Frisius is remarkable for its exceptionally high wall – over 5000 m.

Barocius [44.9°S, 16.8°E] Francesco Barozzi, *c.* 1570. Italian mathematician.
Crater (82 km), *whose wall is ruined in the north and intersected by the craters Barocius* **B** *and Barocius* **C**.

Breislak [48.2°S, 18.3°E] Scipione Breislak, 1748–1826. Italian geologist, chemist and mathematician.
Crater (50 km).

Buch [38.8°S, 17.7°E] Christian L. von Buch, 1774–1853. German geologist.
Crater (54 km/1440 m).

Büsching [38.0°S, 20.0°E] Anton F. Büsching, 1724–1793. German geographer and philosopher.
Crater (52 km).

Clairaut [47.7°S, 13.9°E] Alexis C. Clairaut, 1713–1765. Eminent French mathematician, geodesist and astronomer.
Crater (75 km), *the southern part of whose wall is interrupted by the craters Clairaut* **A** *and Clairaut* **B**.

Faraday [42.4°S, 8.7°E] Michael Faraday, 1791–1867. English chemist and physicist, known for his discoveries in electricity, magnetism, etc.
Crater (70 km/4090 m).

Gemma Frisius [34.2°S, 13.3°E] Reinier Jemma, 1508–1555. Dutch physician (born in Friesland), cartographer and astronomer.
Crater with damaged wall (88 km/5160 m).

Goodacre [32.7°S, 14.1°E] Walter Goodacre, 1856–1938. English selenographer, made a map of the Moon.
Crater (46 km/3190 m).

Kaiser [36.5°S, 6.5°E] Frederick Kaiser, 1808–1872. Dutch astronomer, observer of double stars and Mars.
Crater (52 km); *on its eastern wall is* **Kaiser A** (21 × 14 km/2330 m).

Maurolycus [41.8°S, 14.0°E] Francesco Maurolico, 1494–1575. Italian mathematician, opponent of the Copernican theory.
Vast walled plain with central peaks (114 km/4730 m).

Poisson [30.4°S, 10.6°E] Siméon D. Poisson, 1781–1840. French mathematician. Worked in the field of celestial mechanics. Friend of Lagrange and Laplace.
Remains of a crater (42 km), *the floor of which is connected to that of* **Poisson T** *to the south-west.*

The walled plain ***Maurolycus*** *with terraced walls and central peaks stands out in the southeastern sector of the near-side of the Moon, which is densely covered with craters.*

56

8° 10° E 12° 14° 16° 18° E

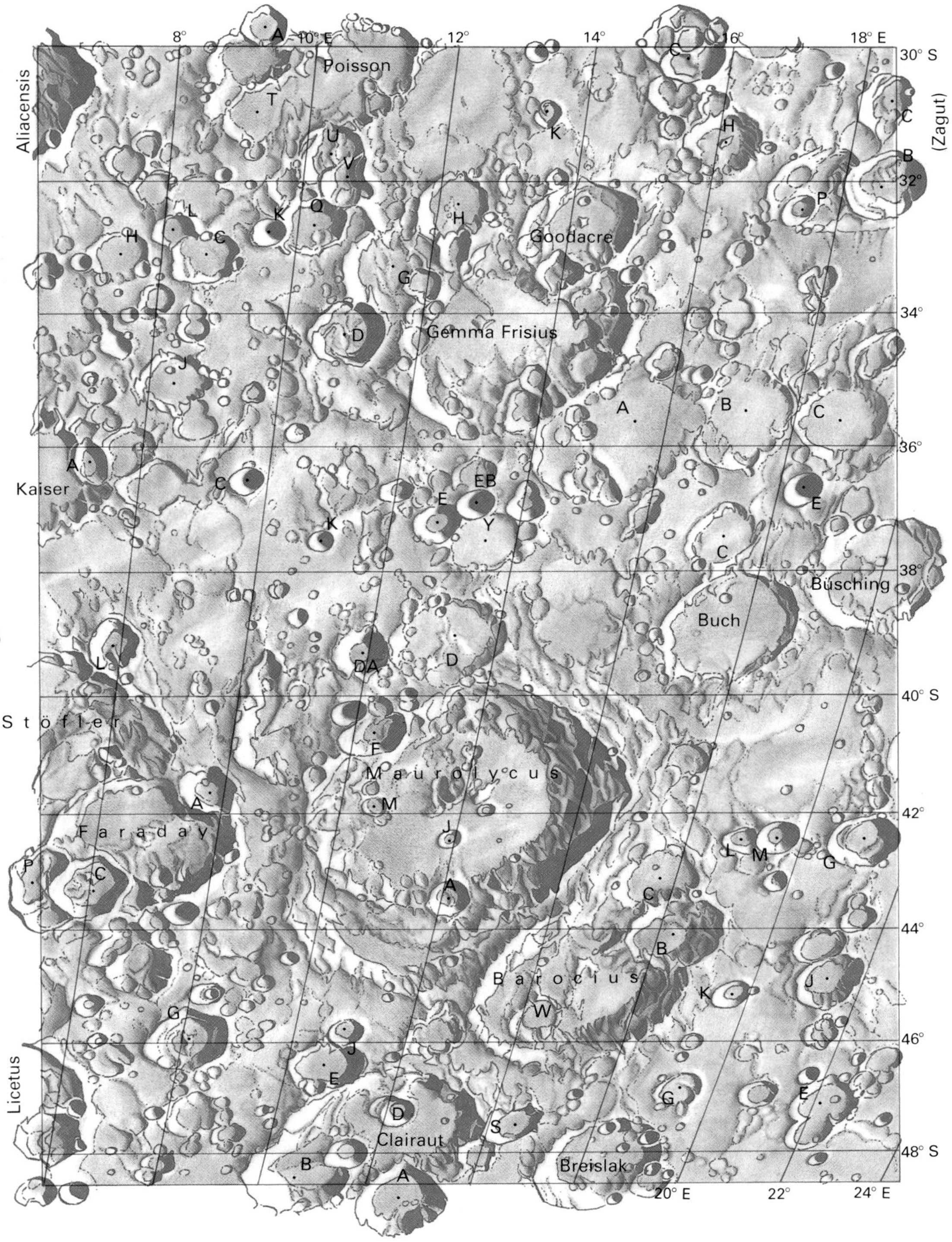

30° S

Aliacensis

Poisson

(Zagut)

A T U V K Q L H C G H K C H B P

32°

Goodacre

34°

D

Gemma Frisius

J

A B C

36°

Kaiser

A C K E EB Y E C

38°

Büsching

Buch

65

67

L DA D

40° S

Stöfler

F

Maurolycus

A M J

42°

Faraday

P C L M G

A C

44°

B

Barocius

J K W G

46°

J E

Licetus

D G E S

Clairaut

48° S

B A

Breislak

20° E 22° 24° E

74

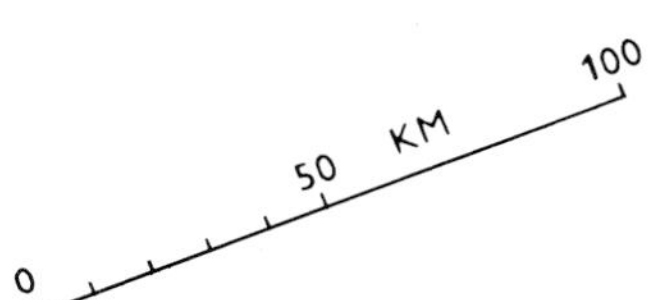

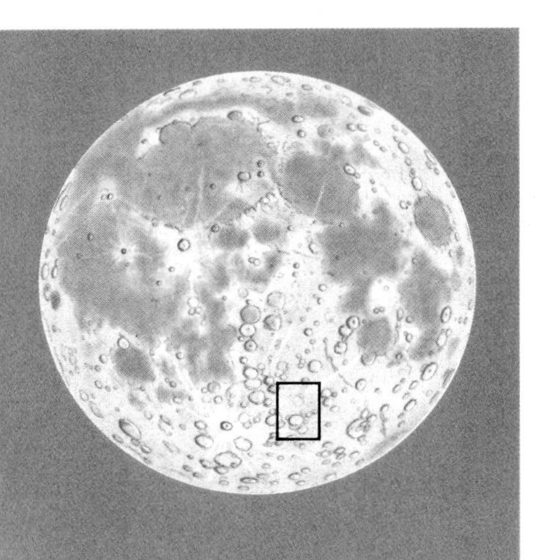

67. RABBI LEVI

Very dense and chaotic crater field in the south-eastern sector of the Moon. An important point of orientation is the crater Rabbi Levi, which contains two pairs of small craters. At the bottom right of the map is a part of the walled plain Janssen with a wide rille on its floor.

Celsius [34.1°S, 20.1°E]. Anders Celsius, 1701–1744. Swedish physicist and astronomer, inventor of the centigrade temperature scale.
Crater (36 km).

Dove [46.7°S, 31.5°E] Heinrich W. Dove, 1803–1879. German physicist. Research in the field of meteorology and electricity.
Crater (30 km).

Janssen. See map 68.

Janssen, Rimae
System of rilles on the floor of Janssen, length 140 km; *suitable for observation with small telescopes.*

Lindenau [32.3°S, 24.9°E] Bernhard von Lindenau, 1780–1854. German astronomer, soldier and politician.
Prominent crater with terraced walls and central peaks (53 km/2930 m).

Lockyer [46.2°S, 36.7°E] Sir Norman Lockyer, 1836–1920. English astrophysicist, discovered helium in the Sun.
Crater (34 km) *on the wall of Janssen.*

Nicolai [42.4°S, 25.9°E] Friedrich B. G. Nicolai, 1793–1846. German astronomer. Worked in the field of celestial mechanics, computed the orbits of comets.
Crater (42 km).

Rabbi Levi [34.7°S, 23.6°E] Levi ben Gershom, 1288–1344 AD. French Jewish philosopher, mathematician and astronomer.
Crater (81 km); *craters* **Rabbi Levi L** (12.6 km/2410 m), **Rabbi Levi A** (12.1 km/1350 m) *and others are situated on its floor.*

Riccius [36.9°S, 26.5°E] Matteo Ricci, 1552–1610. Italian missionary in China, teacher of mathematics and astronomy, geographer.
Ruined crater (71 km); *south of it is* **Riccius E** (22 km/3520 m).

Rothmann [30.8°S, 27.7°E] Christopher Rothmann, d. c. 1600. German astronomer; theoretician.
Crater (42 km/4220 m).

Spallanzani [46.3°S, 24.7°E] Lazzaro Spallanzani, 1729–1799. Italian scientist, physiologist and traveller.
Crater (32 km).

Stiborius [34.4°S, 32.0°E] Andreas Stoberl, 1465–1515. Austrian philosopher, theologian and astronomer.
Crater (44 km/3750 m).

Wöhler [38.2°S, 31.4°E] Friedrich Wöhler, 1800–1882. German chemist. Discovered beryllium and yttrium.
Crater (27 km/2050 m).

Zagut [32.0°S, 22.1°E] Abraham ben S. Zaguth, late fifteenth century. Spanish Jewish astronomer and astrologer.
Crater (84 km); *west of it is* **Zagut B** (32 km/3410 m).

57

67

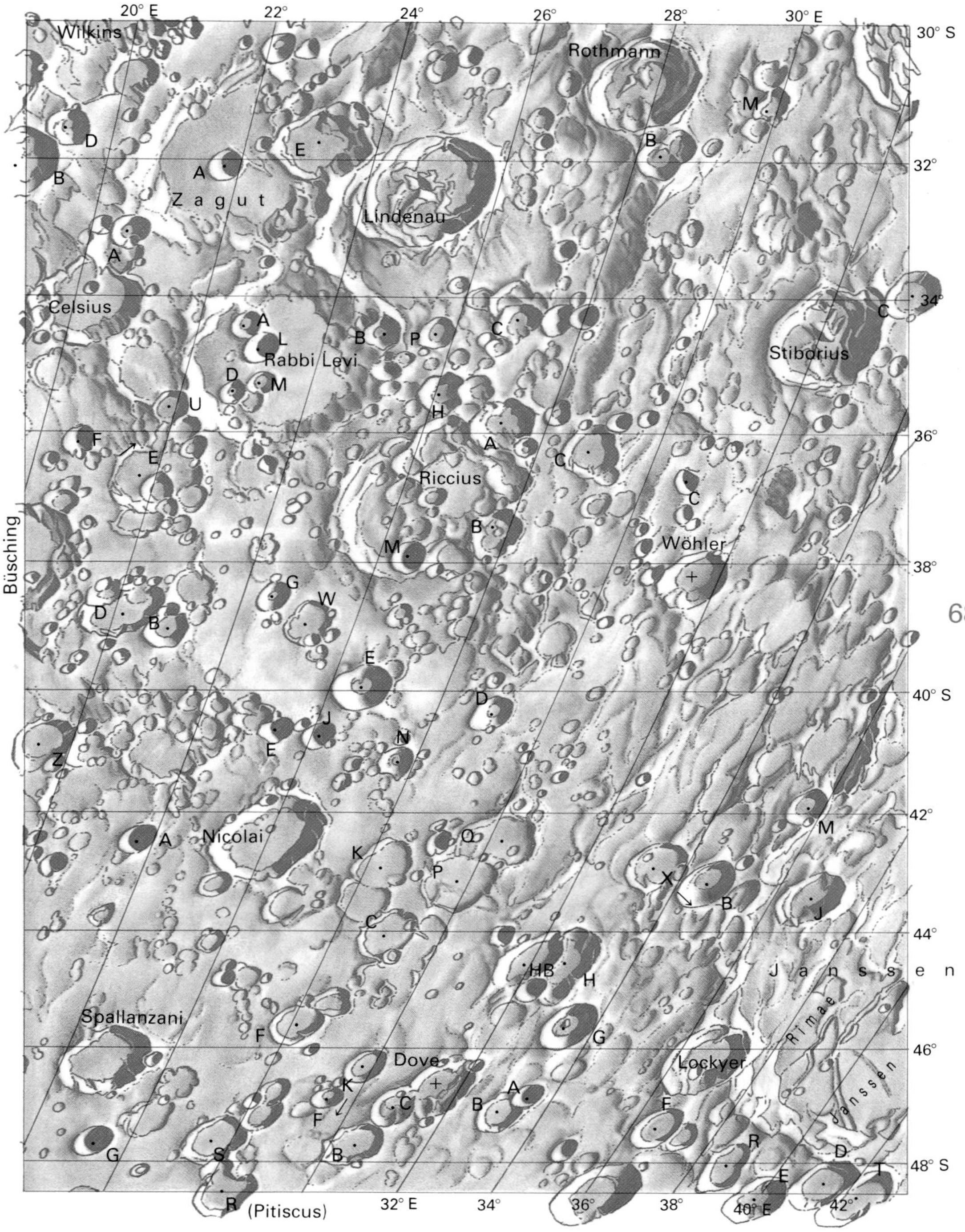

20° E
22°
24°
26°
28°
30° E
30° S
32°
34°
36°
38°
40° S
42°
44°
46°
48° S

Wilkins
Rothmann
Zagut
Lindenau
Celsius
Rabbi Levi
Stiborius
Riccius
Wöhler
Büsching
Nicolai
Janssen
Rimae Janssen
Spallanzani
Dove
Lockyer
(Pitiscus)

32° E
34°
36°
38°
40° E
42°

66
68
75

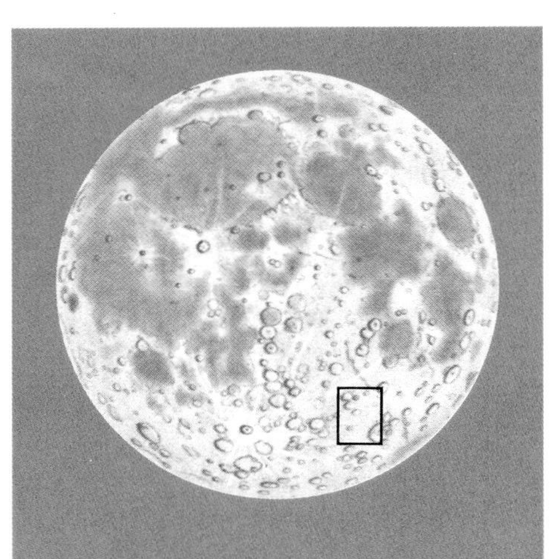

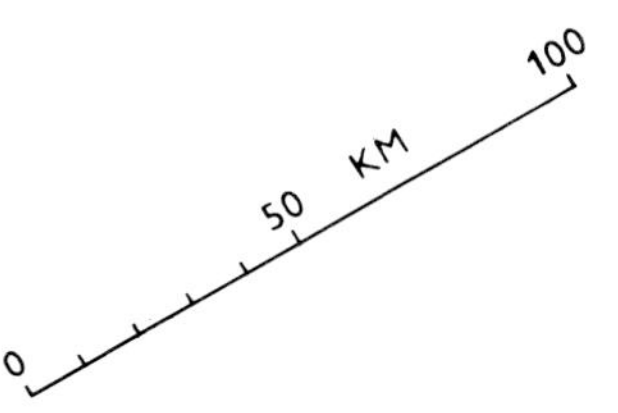

68. RHEITA

The crater field situated along the south-eastern limb of the Moon. It is intersected by a system of faults and valleys, which are directed towards the centre of the Mare Nectaris basin. The long Rheita Valley especially deserves attention.

Brenner [39.0°S, 39.3°E] Leo Brenner, 1855–1928. Austrian amateur astronomer, observer of the Moon and the planets.
Greatly eroded crater (97 km).

Fabricius [42.9°S, 42.0°E] David Goldschmidt (known as Fabricius), 1564–1617. Amateur astronomer from East Friesland.
Prominent crater (78 km).

Janssen [44.9°S, 41.5°E] Pierre J. C. Janssen, 1824–1907. French astronomer. Became Director of the Paris Observatory at Meudon in 1875.
Walled plain with a wide rille, mountain massif, etc., on its floor (190 km).

Mallet [45.4°S, 54.2°E] Robert Mallet, 1810–1881. Irish civil engineer and seismologist.
Crater (58 km) *adjacent to Vallis Rheita.*

Metius [40.3°S, 43.3°E] Adriaan Adriaanszoon (known as Metius), 1571–1635. Dutch mathematician and astronomer.
Crater (88 km).

Neander [31.3°S, 39.9°E] Michael Neumann, 1529–1581. German mathematician, physician and astronomer.
Crater with central peaks (50 km/3400 m).

Peirescius [46.5°S, 67.6°E] Nicolas C. Fabri de Peiresc, 1580–1637. French natural historian and astronomer. Discovered the Great Nebula in Orion in 1610.
Crater (62 km).

Reimarus [47.7°S, 60.3°E] Nicolai Reymers Bär (or Ursus), d. 1600. German mathematician. Was charged with plagiarism as a result of publishing a description of a planetary system very similar to that of Tycho Brahe. Professor of mathematics in Prague.
Crater (48 km).

Rheita [37.1°S, 47.2°E] Anton Maria Schyrleus (Šírek) of Rheita, 1597–1660. Czech optician and astronomer; constructed Kepler's telescope. Prepared a map of the Moon.
Crater on the northern edge of Vallis Rheita (70 km); *crater valley* **Rheita E** (66 × 32 km).

Rheita, Vallis (Rheita Valley) [42°S, 51°E]
The longest valley on the near-side of the Moon, length about 500 km; *character similar to that of Vallis Snellius* (maps 59 and 69), *i.e. it starts in the centre of Mare Nectaris and seems to share a common origin.*

Steinheil. See map 76.

Vega [45.4°S, 63.4°E] Georg F. von Vega, 1756–1802. German mathematician, author of accurate logarithmic tables.
Crater (76 km).

Young [41.5°S, 50.9°E] Thomas Young, 1773–1829. English physician, physicist and universal natural scientist; discoverer of the interference of light.
Crater (72 km).

58

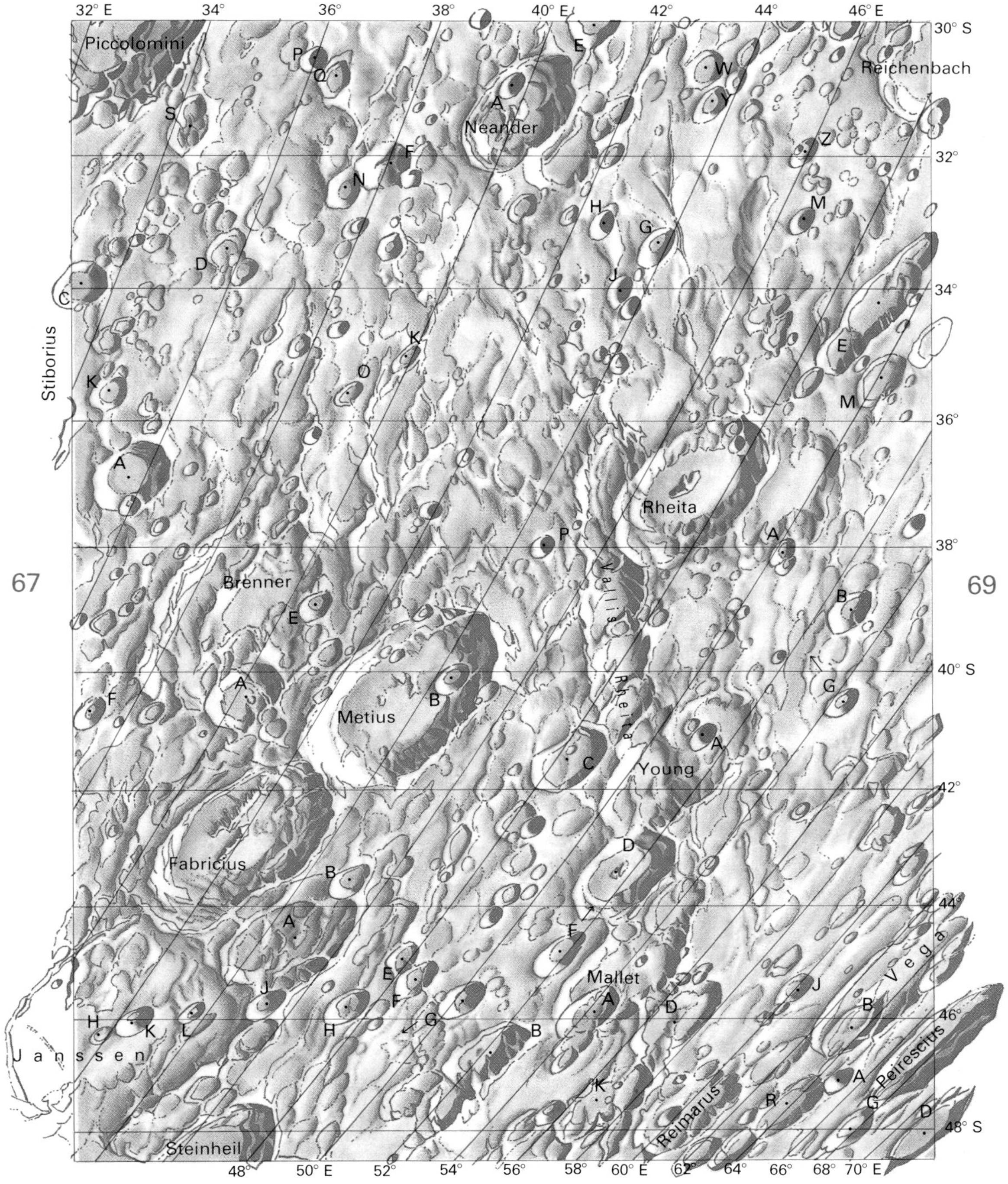

32° E 34° 36° 38° 40° E 42° 44° 46° E

30° S 32° 34° 36° 38° 40° S 42° 44° 46° 48° S

Piccolomini
P
Q
S
E
W
Y
Reichenbach
A
Neander
Z
F
N
H
M
G
D
J
C
Stiborius
K
E
K
O
M
A
Rheita
P
A
Brenner
E
B
Vallis Rheita
A
G
F
B
Metius
A
C
Young
Fabricius
D
B
A
F
E
J
Mallet
J
Vega
F
A
D
B
H
K
L
H
G
B
Janssen
A
Peirescius
K
Reimarus
R
G
D
Steinheil

48° 50° E 52° 54° 56° 58° 60° E 62° 64° 66° 68° 70° E

67

69

76

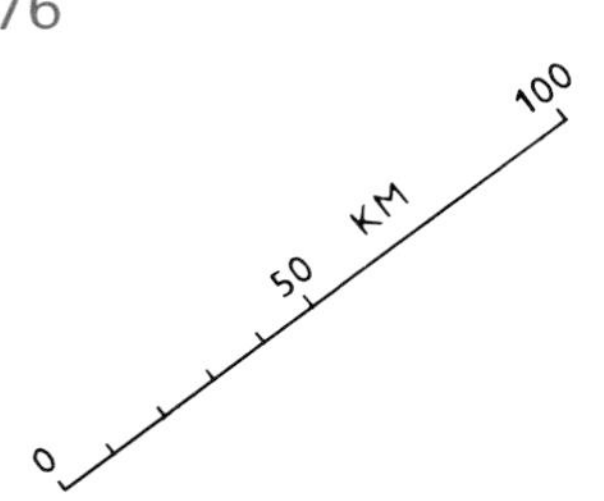

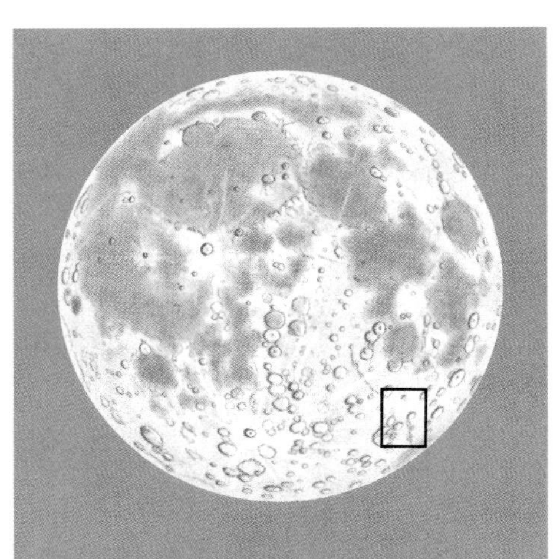

69. FURNERIUS

The south-eastern limb of the Moon. Useful points for orientation are the craters Stevinus, Furnerius and Oken (with a dark floor). From the south Mare Australe extends into this area.

Abel [34.6°S, 85.8°E] Niels H. Abel, 1802–1829. Norwegian mathematician.
Flooded walled plain (114 km).

Adams [31.9°S, 68.2°E] (1) John Couch Adams, 1819–1892; English astronomer, made calculations, independently of Le Verrier, leading to the discovery of Neptune. (2) Charles H. Adams, 1868–1951; American amateur astronomer (3) Walter S. Adams, 1876–1956. American astronomer, Director of Mt Wilson Observatory.
Crater (66 km).

Australe, Mare (Southern Sea). See map 76.

Barnard [29.6°S, 86.4°E] Edward E. Barnard, 1857–1923. American astronomer, discovered the fifth satellite of Jupiter; photographed the Galaxy. 'Barnard's star' in Ophiuchus.
Walled plain (100 km).

Fraunhofer [39.5°S, 59.1°E] Joseph von Fraunhofer, 1787–1826. German optician. Invented the diffraction grating, first to observe the 'Fraunhofer's lines' in the solar spectrum.
Crater (57 km).

Furnerius [36.3°S, 60.4°E] Georges Furner, *c.* 1643. French Jesuit, Professor of mathematics in Paris.
Prominent walled plain (125 km); **Furnerius A** *is a prominent crater with a ray system. Furnerius is part of a fine chain of craters with Langrenus, Vendelinus and Petavius.*

Furnerius, Rima
Wide rille on the floor of Furnerius, length 50 km.

Gum [40.4°S, 88.6° E] Colin S. Gum, 1924–1960. Australian astronomer.
Flooded, shallow crater (55 km).

Hamilton [42.8°S, 84.7°E] Sir William R. Hamilton, 1805–1865. Irish mathematician.
Regular, deep crater (57 km).

Hase, Rima [33°S, 66°E] Named after the crater Hase (map 59).
Wide, shallow rille, 300 km *long.*

Marinus [39.4°S, 76.5°E] Marinus of Tyre, second century AD. Eminent Greek geographer, the first to point out that Asia and Africa might be larger than Europe and that the Roman Empire did not embrace the whole world.
Crater (58 km).

Oken [43.7°S, 75.9°E] Lorenz Oken (Okenfuss), 1779–1851. German naturalist.
Flooded crater (72 km).

Reichenbach [30.3°S, 48.0°E] Georg von Reichenbach, 1772–1826. German maker of surveying and astronomical instruments.
Crater (71 km).

Snellius, Vallis
Continuation of a crater valley described on map 59.

Stevinus [32.5°S, 54.2°E] Simon Stevin, 1548–1620. Belgian-born mathematician, optician, soldier and engineer.
Prominent crater with central peak (75 km); **Stevinus A** *is a very bright crater with a ray system.*

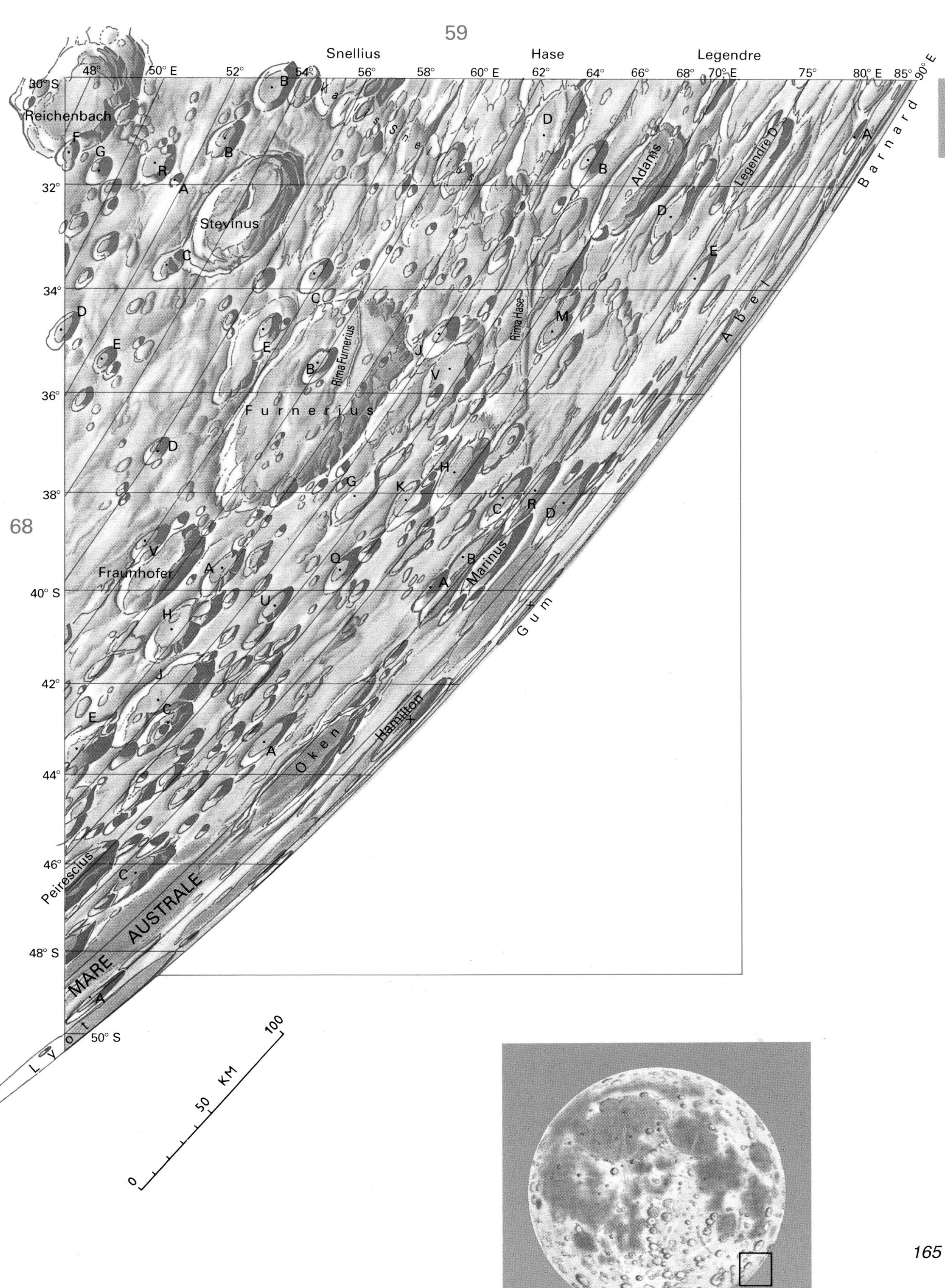
59
Snellius
Hase
Legendre
68
Reichenbach
Stevinus
Furnerius
Fraunhofer
Rima Furnerius
Rima Hase
Vallis Snellius
Adams
Legendre D
Barnard
Abel
Gum
Marinus
Hamilton
Oken
Peiresclus
MARE AUSTRALE
Lyot
30° S
32°
34°
36°
38°
40° S
42°
44°
46°
48° S
50° S
48°
50° E
52°
54°
56°
58°
60° E
62°
64°
66°
68°
70° E
75°
80° E
85°
90° E
0
50
KM
100

70. PHOCYLIDES

This area adjacent to the south-western limb of the Moon includes an exceptionally interesting crater, Wargentin, which is totally filled with lava up to its brim so that its floor forms a plateau. The crater Pingré is situated further south, on the inner edge of the wall of a lunar basin 300 km in diameter.

Nasmyth [50.5°S, 56.2°W] James Nasmyth, 1808–1890. Scottish engineer (invented the steam hammer) and selenographer; made models of the lunar surface. 'Nasmyth' telescope mounting.
Flooded crater (77 km).

Nöggerath [48.8°S, 45.7°W] Johann J. Nöggerath, 1788–1877. German geologist and mineralogist.
Crater with a flooded floor (31 km).

Phocylides [52.9°S, 57.3°W] Johannes Phocylides Holwarda (Jan Fokker), 1618–1651. Dutch astronomer. Believed that the stars had their own motions.
Prominent walled plain with a flooded floor (114 km).

Pilâtre (Hausen B) [60.3°S, 86.4°W] J. F. Pilâtre de Rozier, 1756–1785. French pioneer of aeronautics.
Crater with a considerably eroded wall (69 km).

Pingré (Pingré A) [58.7°S, 73.7°W] Alexandre G. Pingré, 1711–1796. French astronomer and theologian, author of *Cométographie* (notes on comets; includes Uranus, as it had just been discovered by W. Herschel and was still considered to be a comet).
Crater (89 km).

Wargentin [49.6°S, 60.2°W] Pehr V. Wargentin, 1717–1783. Swedish astronomer, Director of Stockholm Observatory.
The largest of a very rare type of crater, which is filled to the rim with dark marie material (lava); its diameter is 84 km *and its raised floor has numerous wrinkle ridges.*

Yakovkin (Pingré H) [54.5°S, 78.8°W] A. A. Jakovkin, 1887–1974. Soviet astronomer. Investigation of the rotation and shape of the Moon.
Crater (37 km).

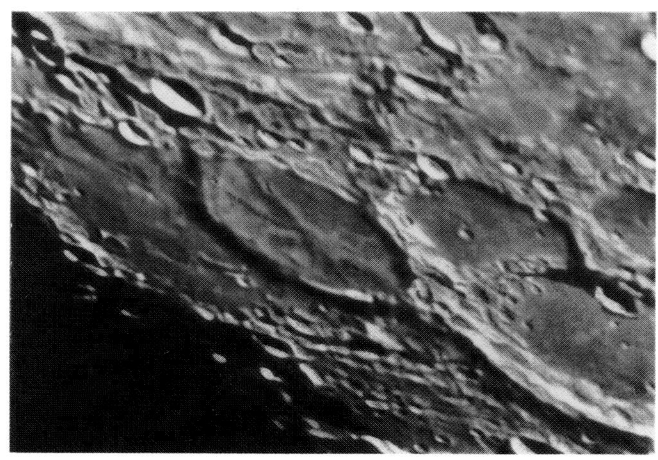

At sunrise, the western edge of the crater ***Wargentin,*** *which is filled up to the rim with lava, is bordered by a prominent shadow.*

62

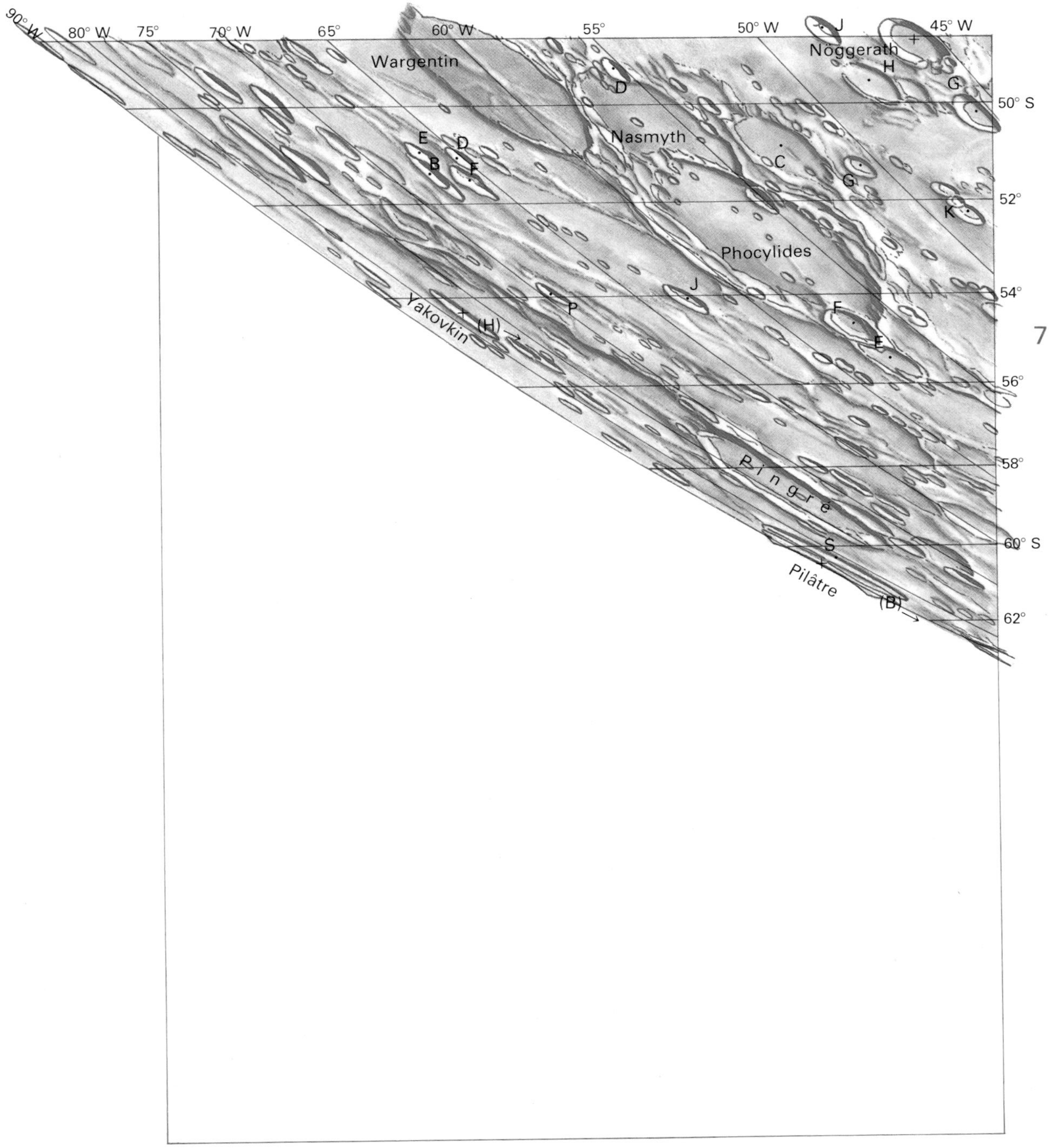

70

90° W
80° W
75°
70° W
65°
60° W
55°
50° W
45° W

Wargentin
D
Nöggerath
J
H
G
50° S
Nasmyth
E
D
B
F
C
G
52°
K
Phocylides
J
54°
P
Yakovkin
(H) →
F
E
56°
58°
Pingré
60° S
S
Pilâtre
(B) →
62°

71

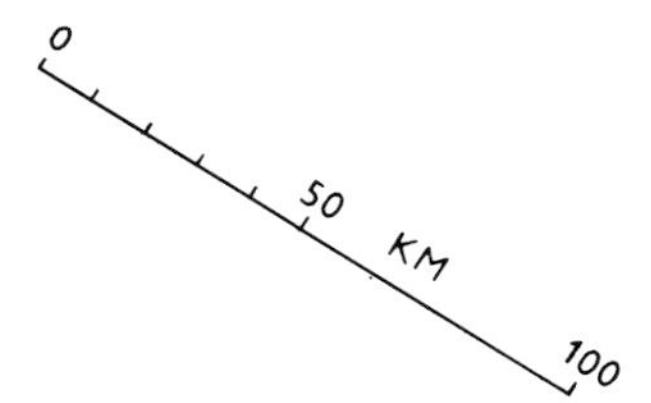

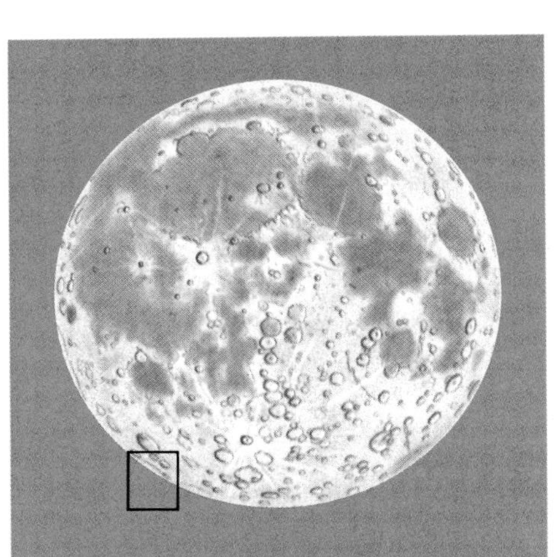

71. SCHILLER

This part of the south-western limb of the Moon contains two basins; one of them is coincident with the crater Bailly and the second, unnamed basin, 350 km in diameter, is situated between the craters Schiller, Zuchius and Phocylides (see map 70).

Bailly [66.8°S, 69.4°W] Jean Sylvain Bailly, 1736–1793. French astronomer and politician.
Vast walled plain; coincident with the outer wall of a lunar basin 303 km *in diameter.*

Bayer [51.6°S, 35.0°W] Johann Bayer, 1572–1625. German astronomer, prepared the star atlas *Uranometria,* in which he introduced Greek letters to designate the stars.
Crater (47 km).

Bettinus [63.4°S, 44.8°W] Mario Bettini, 1582–1657. Italian philosopher, mathematician and astronomer.
Crater (71 km).

Hausen [65.5°S, 88.4°W] Christian A. Hausen, 1693–1743. German astronomer, mathematician and physicist.
Ring mountain with central peaks (167 km), *visible only during favourable librations.*

Kircher [67.1°S, 45.3°W] Athanasius Kircher, 1601–1680. German mathematician and professor of Oriental languages.
Crater with a flooded floor (73 km).

Rost [56.4°S, 33.7°W] Leonhardt Rost, 1688–1727. German amateur astronomer and popularizer of astronomy.
Crater with a flooded floor (49 km).

Schiller [51.8°S, 40.0°W] Julius Schiller, d. 1627. German monk, author of a Christian atlas of the sky (*Coelum Stellarum Christianum,* Augsburg, 1627), in which the old traditional constellations were replaced by biblical characters and objects. These were not accepted.
Very elongated crater (179 × 71 km).

Segner [58.9°S, 48.3°W] Johann A. Segner, 1704–1777. German physicist. Worked on the geometry of solar and lunar eclipses.
Shallow crater with an undulating floor (67 km).

Weigel [58.2°S, 38.8°W] Erhard Weigel, 1625–1699. German mathematician and astronomer. In his *Astronomia Spherica* suggested replacing the traditional figures of the constellations by symbols of various countries ('Coelum Heraldicum').
Crater (36 km).

Zucchius [61.4°S, 50.3°W] Niccolo Zucchi, 1586–1670. Italian mathematician and astronomer, one of the first observers of the belts on Jupiter.
Crater (64 km).

63

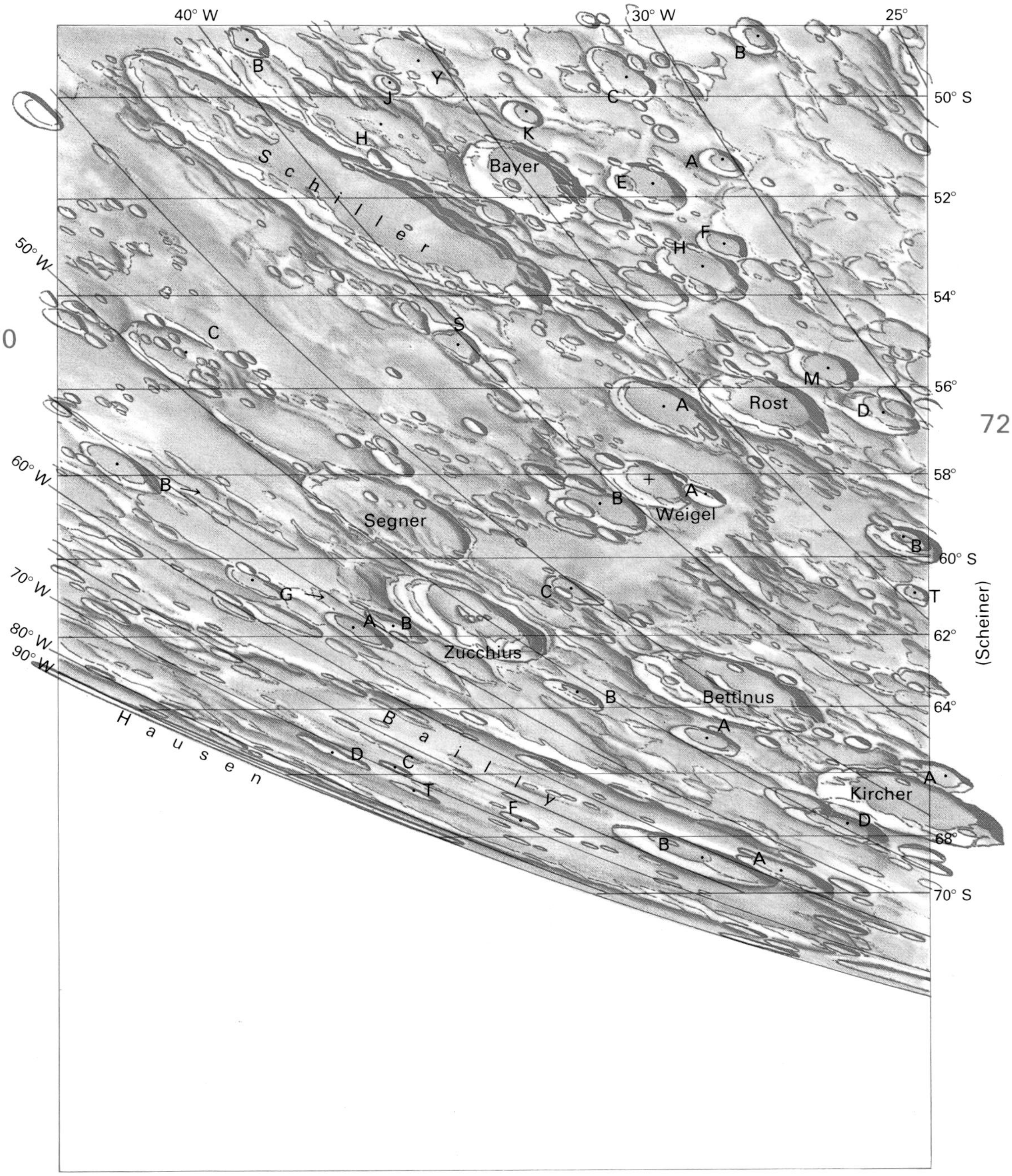

40° W
30° W
25°
50° S
52°
54°
56°
58°
60° S
62°
64°
68°
70° S
50° W
60° W
70° W
80° W
90° W

Schiller
Bayer
Rost
Weigel
Segner
Zucchius
Bettinus
Kircher
Bailly
Hausen
(Scheiner)

B
Y
J
H
K
C
A
E
F
H
S
C
M
A
D
B
B
A
B
G
C
A
B
T
B
A
A
D
C
T
F
D
B
A

70

72

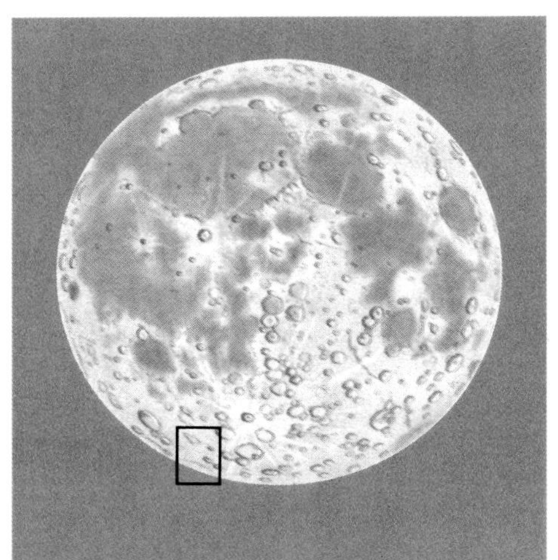

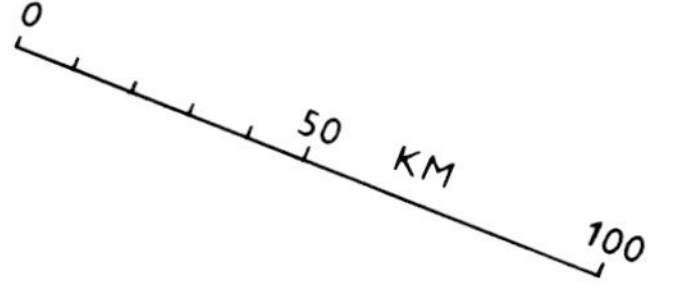

72. CLAVIUS

The limb area of the Moon adjacent to the south pole is densely covered by craters and large walled plains. The terrain is mountainous and the foreshortening close to the Moon's limb and deep shadows make observation and mapping of this area very difficult; the same applies to the limb regions on maps 73 and 74.

Blancanus [63.6°S, 21.5°W] Giuseppe Biancani, 1566–1624. Italian mathematician, geographer and astronomer.
Crater (105 km).

Casatus [72.6°S, 30.5°W] Paolo Casati, 1617–1707. Italian theologian and mathematician.
Flooded crater (111 km).

Clavius [58.4°S, 14.4°W] Christoph Klau, 1537–1612. German mathematician and astronomer, described as the Euclid of sixteenth century'.
One of the best-known walled plains (225 km); *small craters inside Clavius are suitable objects for testing the resolution of small telescopes. An interesting crescent of craters crosses the floor from Rutherfurd, decreasing in size: these are* **Clavius D, C, N, J, JA.**

Drygalski [79.7°S, 86.8°W] Erich D. von Drygalski, 1865–1949. German geographer, geophysicist and polar explorer.
Ring mountain (163 km), *only partly visible during favourable librations.*

Klaproth [69.7°S, 26.0°W] Martin H. Klaproth, 1743–1817. German chemist and mineralogist.
Flooded walled plain (119 km).

le Gentil [74.4°S, 76.5°W] Guillaume H. le Gentil, 1725–1792. French astronomer.
Considerably eroded crater (113 km).

Longomontanus [49.5°S, 21.7°W] Christian S. Longomontanus, 1562–1647. Danish astronomer, assistant to Tycho Brahe.
Walled plain (145 km).

Porter (Clavius B) [56.1°S, 10.1°W] Russell W. Porter, 1871–1949. American architect; designer of large telescopes, including the 5 m reflector at Palomar Observatory.
Crater (52 km).

Rutherfurd [60.9°S, 12.1°W] Lewis M. Rutherfurd, 1816–1892. American astronomer. Photography of the Sun and Moon.
Crater (48 × 54 km).

Scheiner [60.5°S, 27.8°W] Christoph Scheiner, 1575–1650. German mathematician and astronomer. Made the first systematic observations of the Sun.
Crater (110 km).

Wilson [69.2°S, 42.4°W] (1) Alexander Wilson, 1714–1786; Scottish astronomer, discoverer of the 'Wilson Effect' in sunspots; friend of William Herschel. (2) Charles T. R. Wilson, 1869–1959; Scottish physicist. 'Wilson's cloud chamber'. (3) Ralph E. Wilson, 1886–1960; American astronomer at Mt Wilson Observatory.
Considerably eroded crater (70 km).

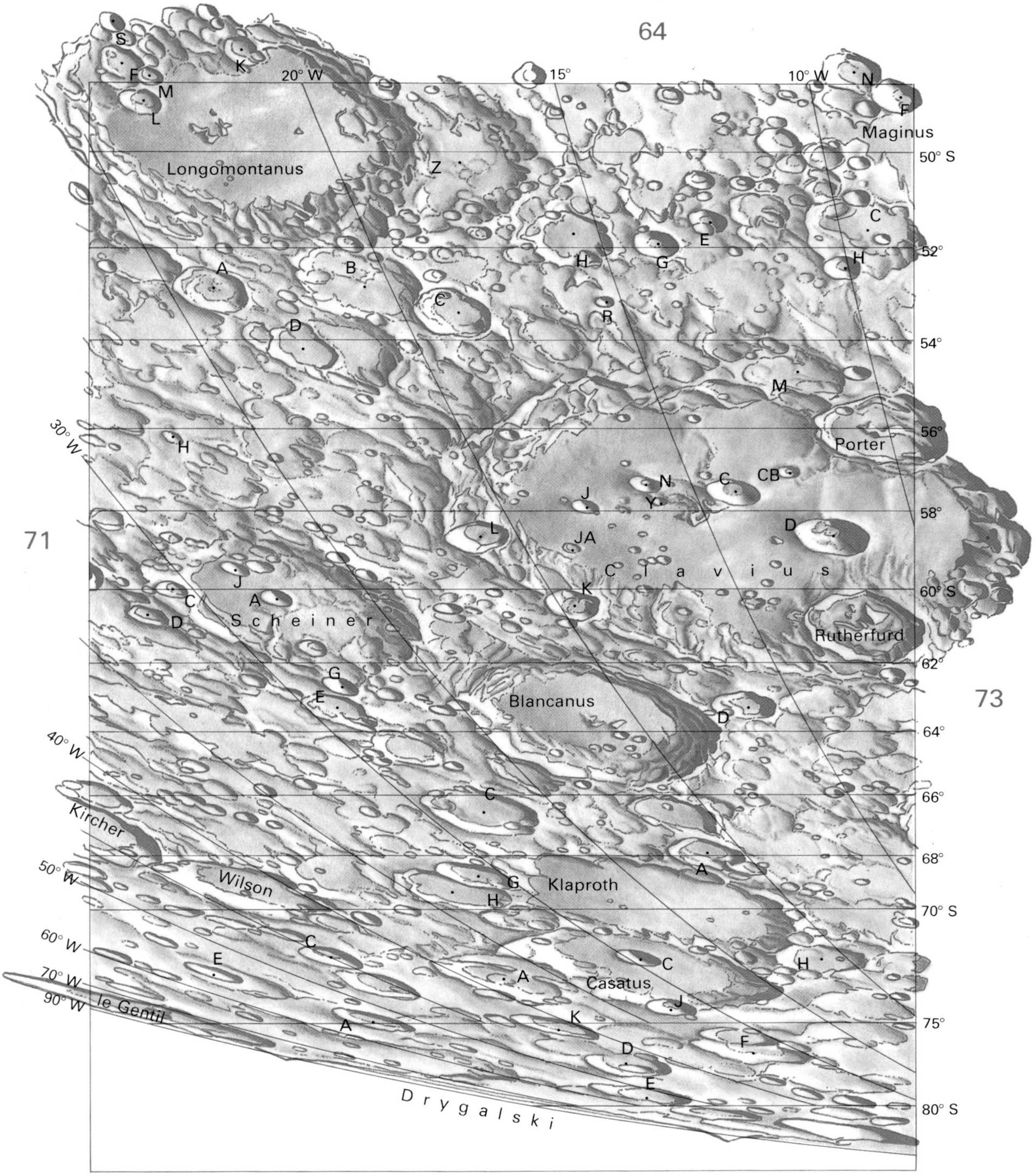

64

72

71

73

Longomontanus
Maginus
Porter
Clavius
Rutherfurd
Scheiner
Blancanus
Kircher
Wilson
Klaproth
Casatus
le Gentil
Drygalski

20° W
15°
10° W
30° W
40° W
50° W
60° W
70° W
90° W

50° S
52°
54°
56°
58°
60° S
62°
64°
66°
68°
70° S
75°
80° S

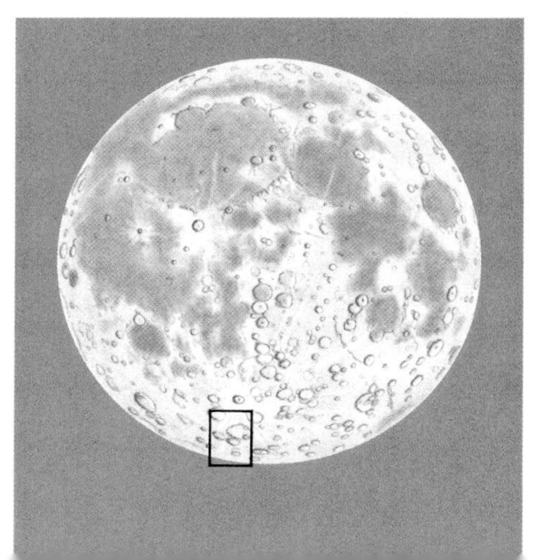

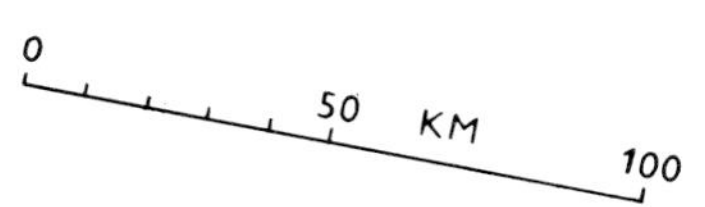

73. MORETUS

The South Pole of the Moon lies in the centre of the bottom edge of the map. Observation of the area close to the pole is very difficult and some parts are constantly hidden behind hills and crater walls. Space probes have yet to map this region in detail. (See the maps of libration zones V and VI.)

Amundsen [84.5°S, 82.8°E] Roald Amundsen, 1872–1928. Famous Norwegian polar explorer. First to reach the South Pole (1911); flew across the North Pole (1928).
Ring mountain (105 km).

Cabeus [84.9°S, 35.5°W] Niccolo Cabeo, 1586–1650. Italian mathematician, philosopher and astronomer.
Crater (about 98 km).

Clavius. See map 72.

Curtius [67.2°S, 4.4°E] Albert Curtz, 1600–1671. German astronomer; published Tycho Brahe's observations.
Crater (95 km).

Cysatus [66.2°S, 6.1°W] Jean-Baptiste Cysat, 1588–1657. Swiss mathematician and astronomer.
Crater (49 km).

Deluc [55.0°S, 2.8°W] Jean A. Deluc, 1727–1817. Swiss geologist and physicist.
Crater (47 km).

Gruemberger [66.9°S, 10.0°W] Christoph Grienberger, 1561–1636. Austrian mathematician and astronomer.
Crater (94 km).

Heraclitus [49.2°S, 6.2°E] Heraclitus, *c.* 540–480 BC. Greek philosopher of Ephesus.
Ruined crater with a central mountain range (90 km).

Lilius [54.5°S, 6.2°E] Luigi Giglio, *d.* 1576. Italian physicist and philosopher. Suggested reform of the Julian calendar.
Crater with central mountain (61 km).

Maginus [50.0°S, 6.2°W] Giovanni A. Magini, 1555–1617. Italian mathematician, astronomer and astrologer.
Vast walled plain (163 km).

Malapert [84.9°S, 12.9°E] Charles Malapert, 1581–1630. Belgian mathematician, philosopher and astronomer.
Irregular crater (*about* 69 km) *close to the south pole.*

Moretus [70.6°S, 5.5°W] Théodore Moretus, 1602–1667. Belgian mathematician.
Circular mountain range (114 km), *central peak.*

Newton [76.7°S, 16.9°W] Isaac Newton, 1643–1727. Famous English physicist, formulated the laws of gravitation and the theory of fluxions (calculus); also experimented with optics.
Crater (79 km).

Pentland [64.6°S, 11.5°E] Joseph B. Pentland, 1797–1873. Irish politician and geographer.
Crater (56 km).

Short [74.6°S, 7.3°W] James Short, 1710–1768. Scottish mathematician and optician.
Crater (71 km).

Simpelius [73.0°S, 15.2°E] Hugh Sempill (correctly Sempilius), 1596–1654. Scottish linguist and mathematician.
Crater (70 km).

Zach [60.9°S, 5.3°E] Franz X. von Zach, 1754–1832. Hungarian astronomer, born in Bratislava.
Crater (71 km).

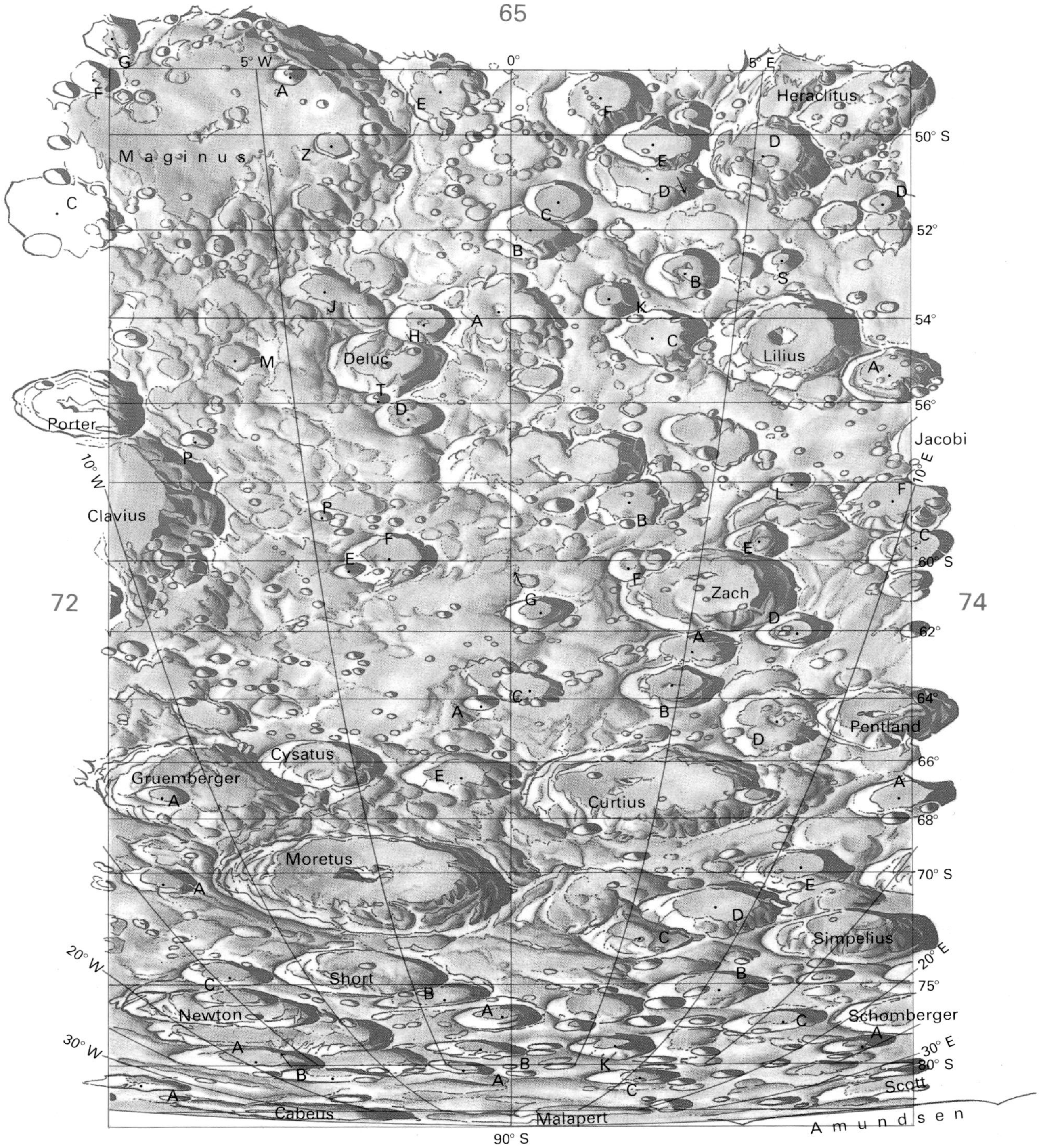

65

5° W 0° 5° E

Heraclitus

Maginus

50° S

52°

54°

Deluc

Lilius

Porter

56°

Jacobi

10° W 10° E

Clavius

60° S

72

Zach

74

62°

64°

Pentland

Cysatus

66°

Gruemberger

Curtius

68°

Moretus

70° S

Simpelius

20° W 20° E

Short

75°

Newton

Schomberger

30° W 30° E

80° S

Scott

Cabeus

Malapert

Amundsen

90° S

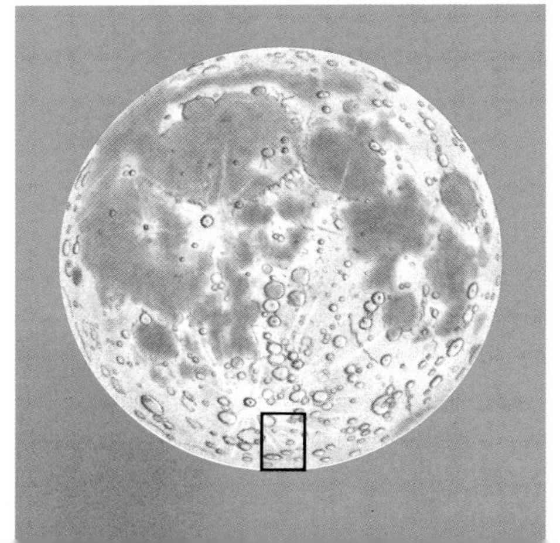

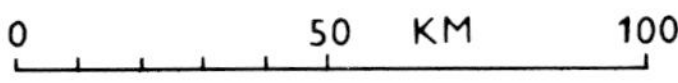

74. MANZINUS

The limb of the Moon east of the south polar region. The crater field in this area is very dense and chaotic. The pair of craters Mutus and Manzinus make orientation easier.

Asclepi [55.1°S, 25.4°E] Giuseppe Asclepi, 1706–1776. Italian Jesuit, astronomer and physicist.
Crater (43 km).

Baco [51.0°S, 19.1°E] Roger Bacon, 1214–1294. English scientist and Franciscan friar, asserted that observational and experimental methods are essential to knowledge.
Prominent crater (70 km).

Boguslawsky [72.9°S, 43.2°E] Palon H. Ludwig von Boguslawski, 1789–1851. German astronomer, Director of Breslau Observatory. Discovered a comet in April 1835.
Crater with a flooded floor (97 km).

Boussingault. See map 75.

Cuvier [50.3°S, 9.9°E] Georges Cuvier, 1769–1832. French naturalist and palaeontologist.
Crater with a flooded floor (75 km).

Demonax [78.2°S, 59.0°E] Demonax, second century BC. Cypriot-born Greek philosopher.
Crater (114 km).

Hale. See map 75.

Ideler [49.2°S, 22.3°E] Christian L. Ideler, 1766–1846. German chronologist.
Crater (39 km).

Jacobi [56.7°S, 11.4°E] Karl G. J. Jacobi, 1804–1851. German mathematician and philosopher. Invented the 'Jacobian functions'.
Crater with a flooded floor (68 km).

Kinau [60.8°S, 15.1°E] C. A. Kinau, *c.* 1850. German botanist and selenographer.
Crater with central peak (42 km).

Manzinus [67.7°S, 26.8°E] Carlo A. Manzini, 1599–1677. Italian philosopher and astronomer.
Crater with a flooded floor (98 km).

Mutus [63.6°S, 30.1°E] Vincente Mut (Muth), d. 1673. Spanish astronomer and navigator.
Crater (78 km).

Schomberger [76.7°S, 24.9°E] Georg Schoenberger, 1597–1645. Austrian mathematician and astronomer. Believed that sunspots were satellites of the Sun ('stellae solares').
Crater (85 km).

Scott [81.9°S, 45.3°E] Robert F. Scott, 1868–1912. English polar explorer. He was the second to reach the South Pole.
Crater with a ruined wall (108 km).

Tannerus [56.4°S, 22.0°E] Adam Tanner, 1572–1632. Austrian mathematician and theologian.
Crater with a sharp rim (29 km).

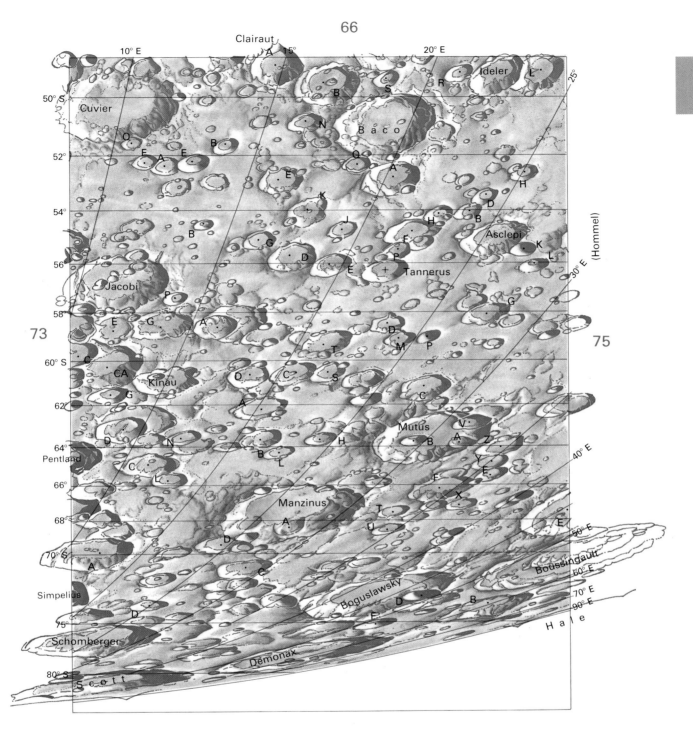

66

Clairaut

10° E 15° 20° E 25° 30° E 40° E 50° E 60° E 70° E 90° E

50° S 52° 54° 56° 58° 60° S 62° 64° 66° 68° 70° S 75° 80° S

Cuvier

Baco

Ideler

Asclepi

(Hommel)

Tannerus

Jacobi

73

75

Kinau

Mutus

Pentland

Manzinus

Simpelius

Boussingault

Boguslawsky

Schomberger

Demonax

Hale

Scott

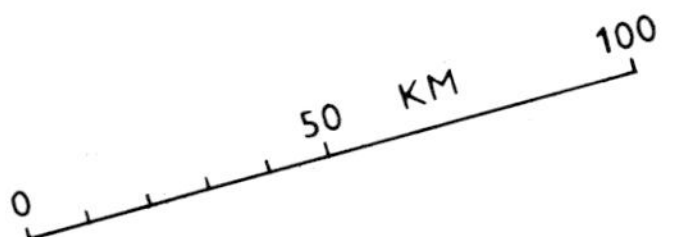

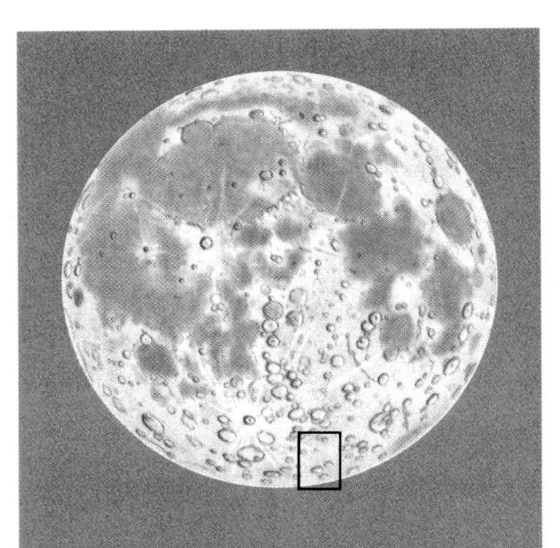

75. HAGECIUS

A part of the south-eastern limb of the Moon with a group of large craters around Hagecius. Important for orientation in this crater field is the crater Hommel, which with smaller craters on its wall forms a characteristic group.

Biela [54.9°S, 51.3°E] Wilhelm von Biela, 1782–1856. Czech-born Austrian soldier and astronomer. In 1826 discovered 'Biela's comet'.
Crater (76 km).

Boussingault [70.4°S, 54.7°E] Jean-Baptiste Boussingault, 1802–1887. French agricultural chemist and botanist.
Crater (131 km); *inside the formation is the large crater* **Boussingault A,** *so that the whole feature resembles a crater with a double wall.*

Gill [63.9°S, 75.9°E] Sir David Gill, 1843–1914. English astronomer. Compiled a catalogue of about 400 000 stars, measured the distances of 22 stars.
Crater (66 km).

Hagecius [59.8°S, 46.6°E] Tadeus Hajek of Hajek, 1525–1600. Czech naturalist, mathematician and astronomer. Tycho Brahe and Kepler were invited to Prague on his advice.
Crater (76 km).

Hale [74.2°S, 90.8°E] (1) George E. Hale, 1868–1938; American astronomer, Director of Mt Wilson Observatory. (2) William Hale, 1797–1870; English scientist in the field of rocket technology.
Crater (84 km), *stretching over to the far-side of the Moon.*

Helmholtz [68.1°S, 64.1°E] Hermann von Helmholtz, 1821–1894. German physiologist, surgeon and physicist.
Crater (95 km).

Hommel [54.6°S, 33.0°E] Johann Hommel, 1518–1562. German mathematician and astronomer, teacher of Tycho Brahe.
Crater (125 km).

Nearch [58.5°S, 39.1°E] Nearchus, *c.* 325 BC. Greek commander, friend of Alexander the Great.
Crater (76 km).

Neumayer [71.1°S, 70.7°E] Georg B. von Neumayer, 1826–1909. German meteorologist, naturalist and explorer.
Crater (76 km).

Pitiscus [50.4°S, 30.9°E] Bartholomäus Pitiscus, 1561–1613. German theologian and mathematician.
Prominent crater (82 km).

Rosenberger [55.4°S, 43.1°E] Otto A. Rosenberger, 1800–1890. German mathematician and astronomer.
Crater (96 km).

Vlacq [53.3°S, 38.8°E] Adriaan Vlacq, *c.* 1600–1667. Dutch bookseller and mathematician. In 1628 he published logarithmic tables calculated to ten places.
Prominent crater (89 km).

Wexler [69.1°S, 90.2°E] Harry Wexler, 1911–1962. American meteorologist. He elaborated the programme of weather satellites.
Crater (52 km), *stretching over to the far-side of the Moon.*

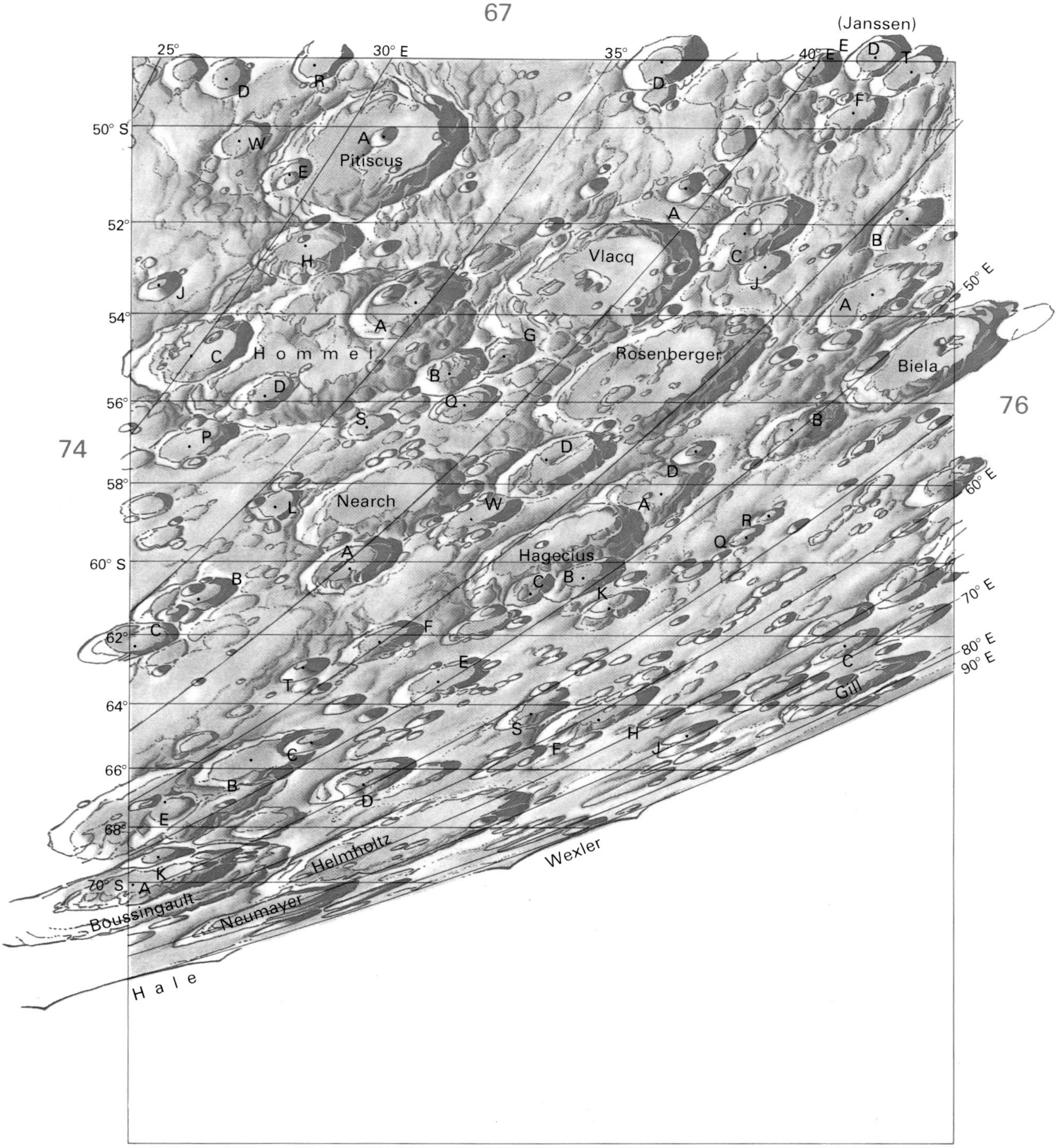

67

(Janssen)

25° 30° E 35° 40° E 50° E 60° E 70° E 80° E 90° E

50° S 52° 54° 56° 58° 60° S 62° 64° 66° 68° 70° S

74

76

Pitiscus

Vlacq

Hommel

Rosenberger

Biela

Nearch

Hagecius

Gill

Helmholtz

Wexler

Boussingault

Neumayer

Hale

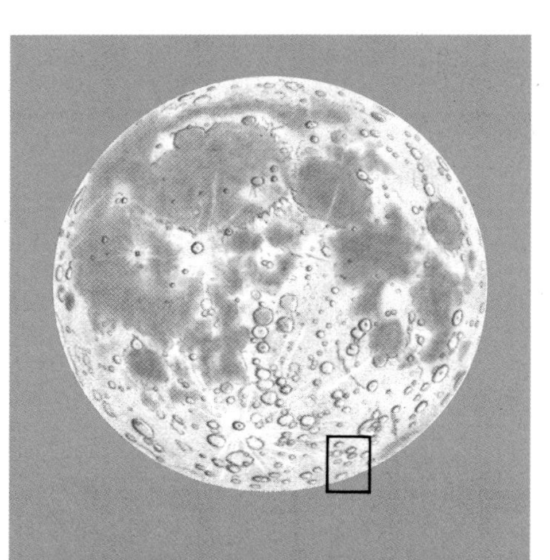

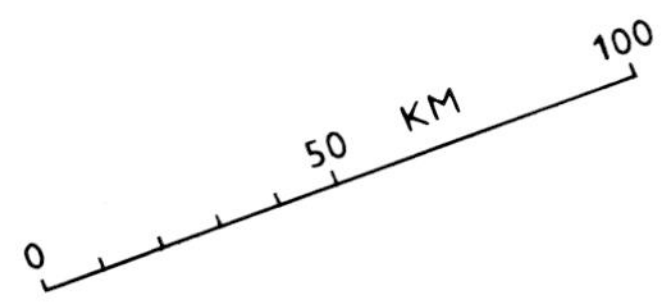

76. WATT

The south-eastern limb of the Moon. The numerous dark patches on the limb itself edge form Mare Australe. M. Australe, as determined from photographs by the *Lunar Orbiters,* is circular with a diameter of about 900 km; from an evolutionary point of view it is evidently a very old lunar basin.

Australe, Mare (Southern Sea) [46°S, 91°E]
Irregular lunar mare, which stretches to the far-side of the Moon; its surface area is 151 000 sq km *and in many places it is covered by craters and light areas.*

Brisbane [49.1°S, 68.5°E] Sir Thomas Brisbane, 1770–1860. Scottish soldier, politician and astronomer.
Crater (45 km).

Hanno [56.3°S, 71.2°E] Hanno, *c.* 500 BC. Carthaginian navigator, who sailed through the Straits of Gibraltar southward to the West-African coast.
Crater (56 km).

Lyot [50.2°S, 84.1°E] Bernard F. Lyot, 1897–1952. French astronomer, invented the solar coronograph and monochromatic polarising filter ('Lyot filter').
Flooded walled plain with a decayed border and a dark circular floor (141 km).

Petrov [61.4°S, 88.0°E] Jevgenii S. Petrov, 1900–1942. Soviet scientist in the field of rocket technology.
Flooded crater (49 km).

Pontécoulant [58.7°S, 66.0°E] Philippe G. Le Doulcet, comte de Pontécoulant, 1795–1874. French mathematician and astronomer. Forecast the return of Halley's comet in 1835 to within 3 days.
Prominent crater (91 km).

Steinheil [48.6°S, 46.5°E] Karl A. von Steinheil, 1801–1870. German mathematician, physicist, optician and astronomer.
Prominent crater with a flooded floor (67 km).

Watt [49.5°S, 48.6°E] James Watt, 1736–1819. Scottish engineer, invented the improved steam engine. Watt's governor was used for controlling the mechanical drives of telescopes.
Crater (66 km); *the western wall adjoins that of the neighbouring crater Steinheil. Together they form a prominent pair of large craters.*

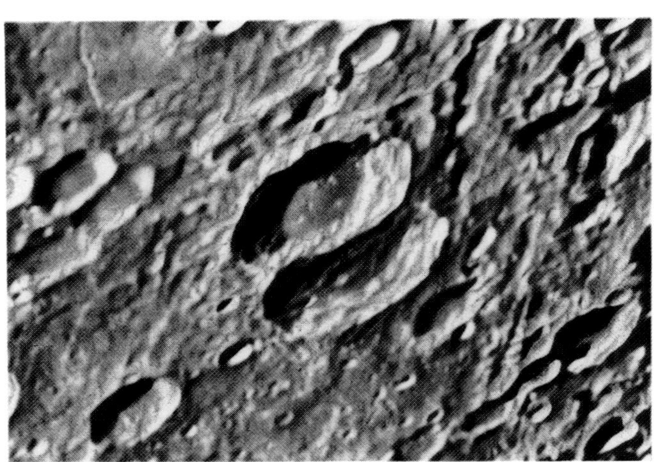

The prominent pair of large craters ***Steinheil*** *and* ***Watt*** *is a good example of the relative age of lunar formations. Watt is older because part of its wall is overlapped by Steinheil.*

68

75

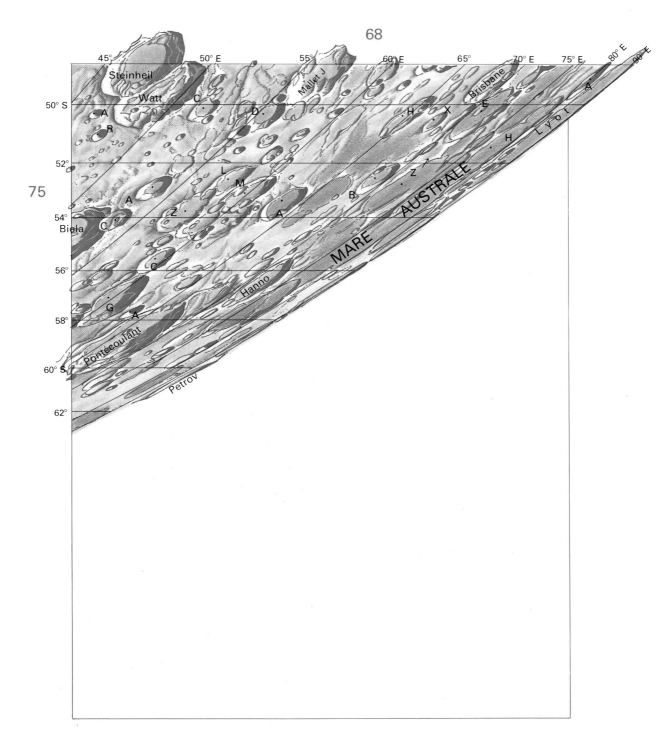

45°
50° E
55°
60° E
65°
70° E
75° E
80° E
90° E
50° S
52°
54°
56°
58°
60° S
62°
Steinheil
Watt
Mallet J
Brisbane
Lyot
Biela
MARE AUSTRALE
Hanno
Pontécoulant
Petrov
A
R
C
D
H
X
E
L
M
Z
B
G

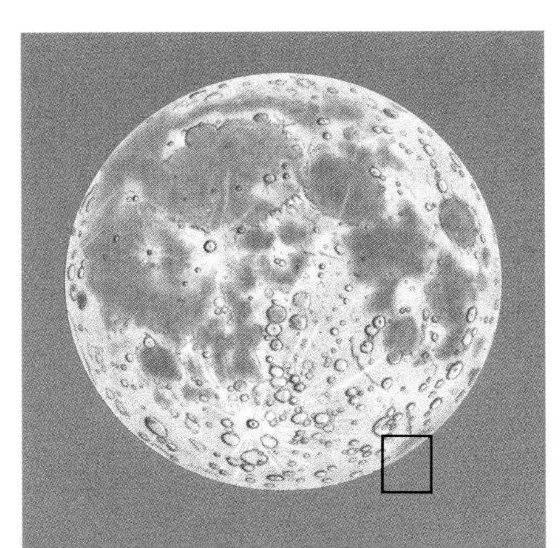

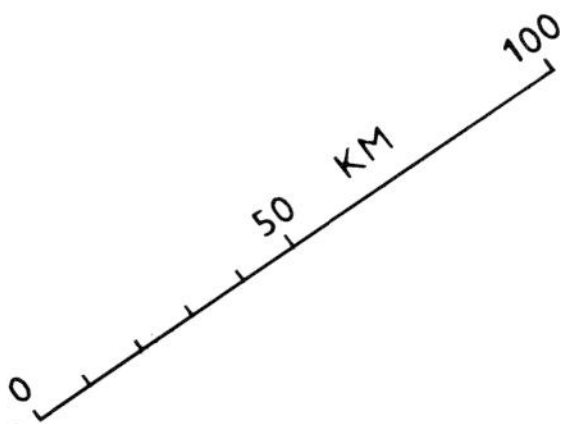

Maps of libration zones

Where should the boundary that separates the near- and far-sides of the Moon be drawn? We can do no better than to commence with the meridians 90°E and 90°W, which constitute the edge of the lunar disk as it is drawn in the orthographic projection of the near-side. This is the aspect that is depicted in the 76 sections that form the major part of this atlas. Such an approach, however, is far too simplistic, for it means, in effect, that the area visible from the Earth is equal to that of the averted far-side, and, as we noted earlier, this is certainly not the case! In fact, to see an 'orthogonal' Moon as it is drawn in the atlas would be an extremely rare event because the visible lunar hemisphere is continually changing its orientation with respect to the Earth.

The librational oscillations in latitude and longitude (p. 8) create a lunar face that 'nods its assent' while, paradoxically, dissentingly 'shaking it from side-to-side'. This effect causes slender crescents of the far-side to come into view from time to time at the expense of similar areas of the near-side, which temporarily disappear. If, for example, the eastern limb of the Moon is turned towards the Earth, the boundary of the visible part can slide over to the very eastern shores of Mare Marginis and Mare Smythii, and almost the whole areas of these maria will come into view (albeit considerably foreshortened). As one would expect, a similar situation occurs when the western limb of the Moon swings towards the Earth and brings into view, onto what is then the near-side, Mare Orientale. This advantage is at the expense of the eastern limb where the limiting boundary moves across Mare Marginis and Mare Smythii, which are then mainly confined to the far side. While these extreme movements are going on it is instructive to note the changes in the oval shape of Mare Crisium and its position in relation to the eastern limb of the visible lunar hemisphere.

The maximum angular values of librations in longitude and latitude are 7°54′ and 6°50′ respectively, and to these must be added 1° to take in the effects of parallactic libration, which is caused by the motion of the observer as the Earth rotates. It turns out that the libration zones, which go through the endless cycle of visibility and invisibility as observed from the Earth, amount to 18% of the total lunar surface area. From this we may conclude that 41% of the near-side is always visible (phase permitting) and 41% is permanently out of sight on the averted hemisphere.

Adding the 18% of the librational areas to the 41% that is always visible gives us, theoretically, a total of 59% of the lunar surface available for observation from the Earth. In practice, however, the peripheral areas are extremely difficult to observe with Earth-based telescopes, so the theoretical 59% has to be reduced to approximately 50% in practice. This is confirmed by the efforts of generations of selenographers, who attempted to map the libration zones. All maps before 1967 are unreliable in the vicinities of meridians 90°E and 90°W, and beyond them, in the librational 'crescents' of the averted hemisphere, the mapping is incomplete and subject to numerous errors. Fortunately, our knowledge of the libration zones is now much better, thanks to the photographic reconnaissance of the lunar surface that was performed by *Lunar Orbiters* during the 1960s.

Why are observations of the libration zones so difficult? First of all, the geometrical determinations of the boundaries of these zones are based on a simplified assumption that the Moon is a smooth sphere; and anyone can see that it is definitely not so! Added to this is the problem of perspective or foreshortening, which complicates the interpretation of observations. For example, in a libration zone the walls of a circular crater appear as a narrow ellipse – or are they two disconnected parallel mountain ridges? Countless numbers of high mountain walls and ridges add their confusion in the peripheral areas by obscuring more remote landscapes, and further details disappear into the shadows, especially in the polar regions. No, the life of an observer of the lunar libration areas is sometimes not a happy one! Up to the present time, even the images sent back to Earth from the orbiting lunar probes have not provided sufficient information to permit cartographers to fill in the blank areas that still exist in their charts of the south polar region. Finally, we have to take into account the fact that a successful observation of a libration zone depends not only on a favourable orientation of the Moon, but also on suitable solar illumination in the zone of interest. And even the efforts of the observer will be to no avail if he or she is troubled by cloudy skies or turbulent atmospheric conditions. It is therefore hardly surprising that successful observations of the libration zones tend to occur at rare intervals; this is why they are of such special interest to diligent amateur astronomers.

The undistorted shapes of lunar formations lying within the libration zones are illustrated in maps I–VIII on pp. 182–189, which show 'sideways' views of the Moon, such as would be obtained by an observer situated over the 90°E and 90°W meridians. Thus each of the eight sections is intersected centrally by a 90° meridian. On both sides of this line the lunar surface is mapped to the theoretical boundaries of the libration zones. The sectors close to the poles are mapped in stereographic projection, while the sectors close to the equator are drawn in the Mercator projection. Both of these cartographical projections are conformal, so that a circular crater is seen on the map as circular, but the scales of such maps vary with selenographic latitude. In the case of the stereographic projection, the scale increases away from the poles to (in this case) a limiting latitude of 45°. The scale of the Mercator projection increases away from the equator to meet the polar projection maps at +45° and −45° latitude. These changes in scale result in some distortions, particularly in those areas close to selenographic latitudes +45° and −45°, where individual areas have to be depicted larger than they really are. However, the overall maximum libration does not exceed a deviation of 11.5° from the mean position.

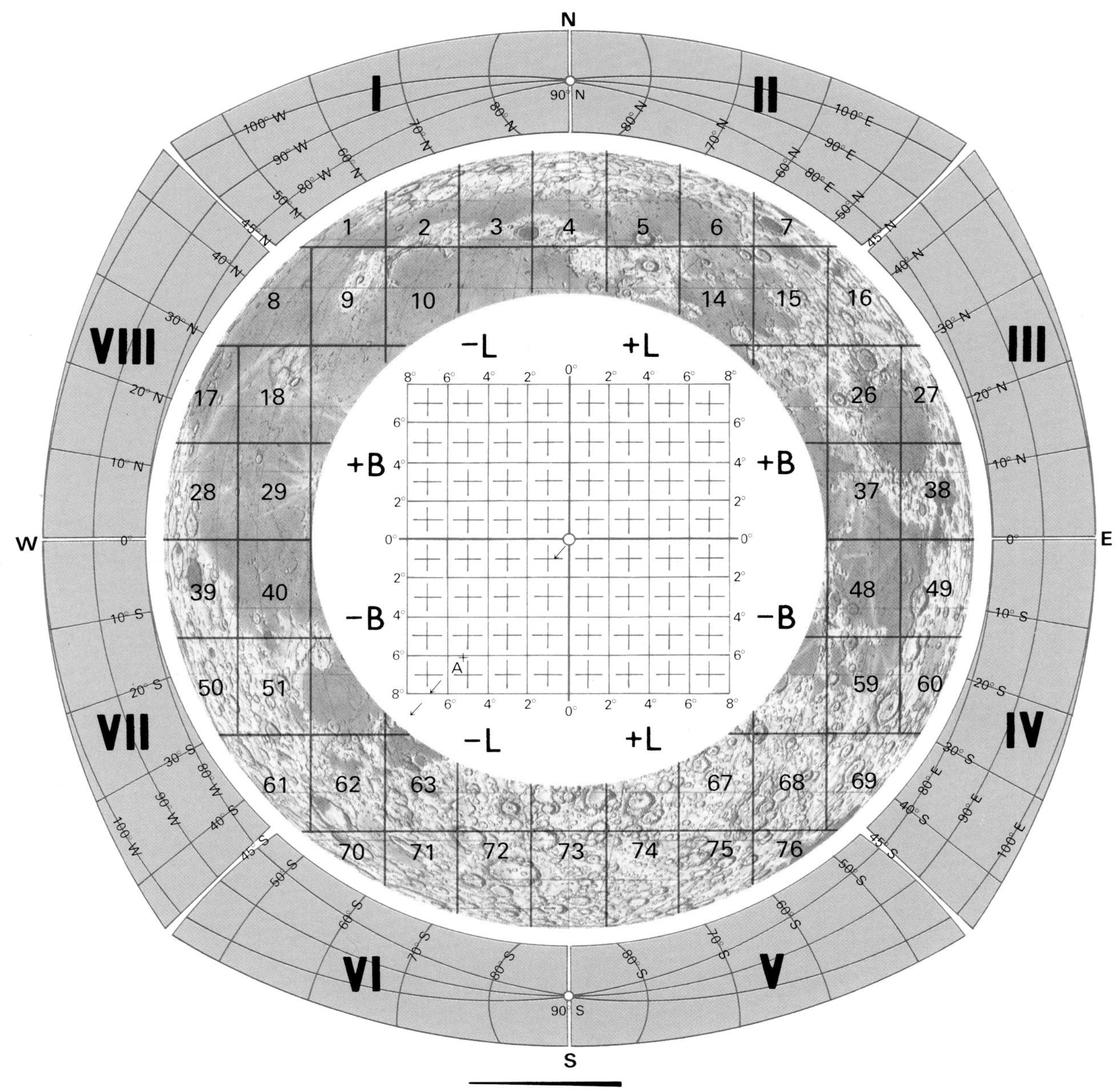

Explanatory note

The diagram shows the relationship between the maps of the libration zones and the map of the near-side of the Moon in 76 sections (p. 29–179). The identities of adjacent maps to a particular map are indicated by numbers along the edges. The nomenclature uses the names that have been adopted in recent years by the IAU for formations in those areas that lie beyond the 90° meridians, in the direction of the averted hemisphere.

The central diagram facilitates the selection of maps according to the librations in longitude, *L*, and latitude, *B*. Its purpose is to determine the direction of the maximum inclination of the Moon's edge towards the observer for a given libration. For example, on 24 February 1986 the values of libration in longitude and latitude were: $L = -5.2°$ and $B = -6.0°$. These co-ordinates, transferred to the diagram, define the position of point A. The line connecting the centre of the diagram with point A indicates the south-western edge of the Moon (sections VI and VII). In this specific case, Full Moon occurred on 24 February, thus providing ideal illumination at the time of favourable libration for observations sections VI and VII.

The libration constants (*L*, *B*) are usually published in astronomical almanacs, along with details of the lunar phases for each day of the year. These enable the observer to anticipate those dates that would be particularly favourable for librational studies.

Libration Zone I. **BRIANCHON**

North-north-western sector

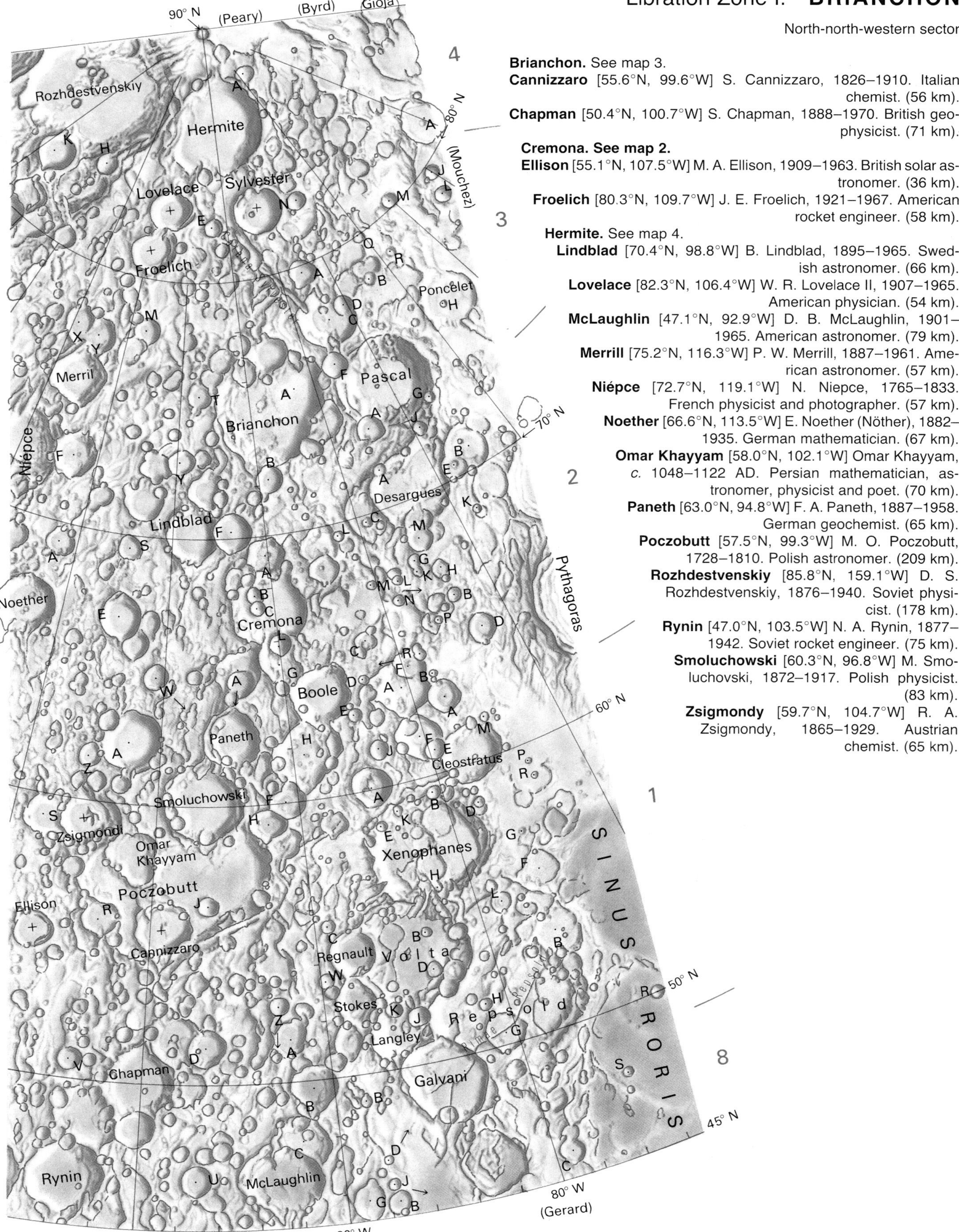

Brianchon. See map 3.

Cannizzaro [55.6°N, 99.6°W] S. Cannizzaro, 1826–1910. Italian chemist. (56 km).

Chapman [50.4°N, 100.7°W] S. Chapman, 1888–1970. British geophysicist. (71 km).

Cremona. See map 2.

Ellison [55.1°N, 107.5°W] M. A. Ellison, 1909–1963. British solar astronomer. (36 km).

Froelich [80.3°N, 109.7°W] J. E. Froelich, 1921–1967. American rocket engineer. (58 km).

Hermite. See map 4.

Lindblad [70.4°N, 98.8°W] B. Lindblad, 1895–1965. Swedish astronomer. (66 km).

Lovelace [82.3°N, 106.4°W] W. R. Lovelace II, 1907–1965. American physician. (54 km).

McLaughlin [47.1°N, 92.9°W] D. B. McLaughlin, 1901–1965. American astronomer. (79 km).

Merrill [75.2°N, 116.3°W] P. W. Merrill, 1887–1961. American astronomer. (57 km).

Niépce [72.7°N, 119.1°W] N. Niepce, 1765–1833. French physicist and photographer. (57 km).

Noether [66.6°N, 113.5°W] E. Noether (Nöther), 1882–1935. German mathematician. (67 km).

Omar Khayyam [58.0°N, 102.1°W] Omar Khayyam, *c.* 1048–1122 AD. Persian mathematician, astronomer, physicist and poet. (70 km).

Paneth [63.0°N, 94.8°W] F. A. Paneth, 1887–1958. German geochemist. (65 km).

Poczobutt [57.5°N, 99.3°W] M. O. Poczobutt, 1728–1810. Polish astronomer. (209 km).

Rozhdestvenskiy [85.8°N, 159.1°W] D. S. Rozhdestvenskiy, 1876–1940. Soviet physicist. (178 km).

Rynin [47.0°N, 103.5°W] N. A. Rynin, 1877–1942. Soviet rocket engineer. (75 km).

Smoluchowski [60.3°N, 96.8°W] M. Smoluchovski, 1872–1917. Polish physicist. (83 km).

Zsigmondy [59.7°N, 104.7°W] R. A. Zsigmondy, 1865–1929. Austrian chemist. (65 km).

Libration Zone II. **NANSEN**

North-north-eastern sector

Belkovich. See map 7.
Compton [56.0°N, 105.0°E] A. H. Compton, 1892–1962. American physicist. (162 km).
Dugan [64.2°N, 103.3°E] R. S. Dugan, 1878–1940. American astronomer. (50 km).
Fabry. See map III.
Nansen [81.3°N, 95.3°E] F. Nansen, 1861–1930. Norwegian explorer. (122 km).
Plaskett [82.3°N, 176.2°E] J. S. Plaskett, 1865–1941. Canadian astronomer. (110 km).
Schwarzschild [70.6°N, 119.6°E] K. Schwarzschild, 1873–1916. German astronomer. (235 km).
Shi Shen [76.0°N, 104.1°E] Shi(h)Shen, *c.* 300 BC. Chinese astronomer. (43 km).

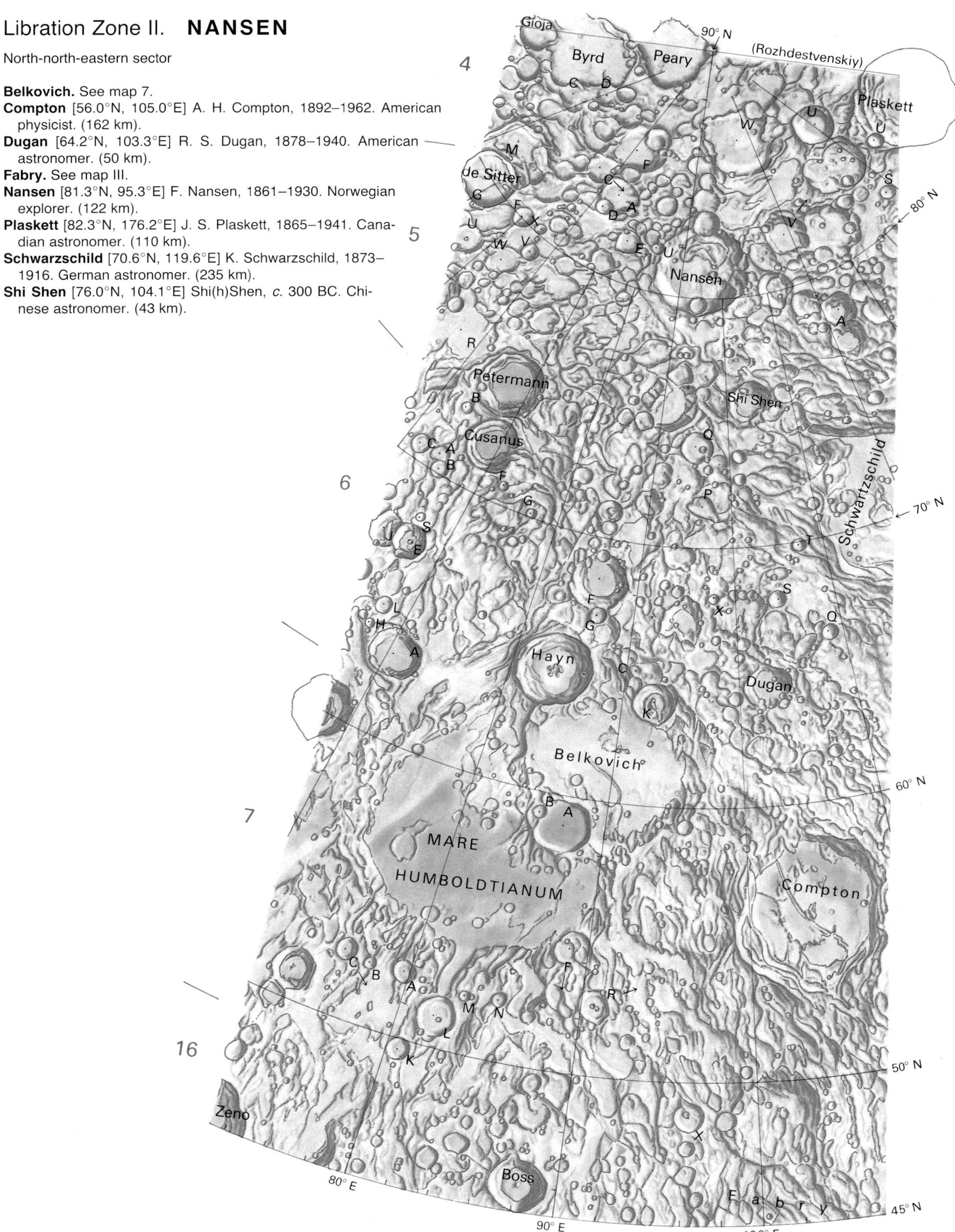

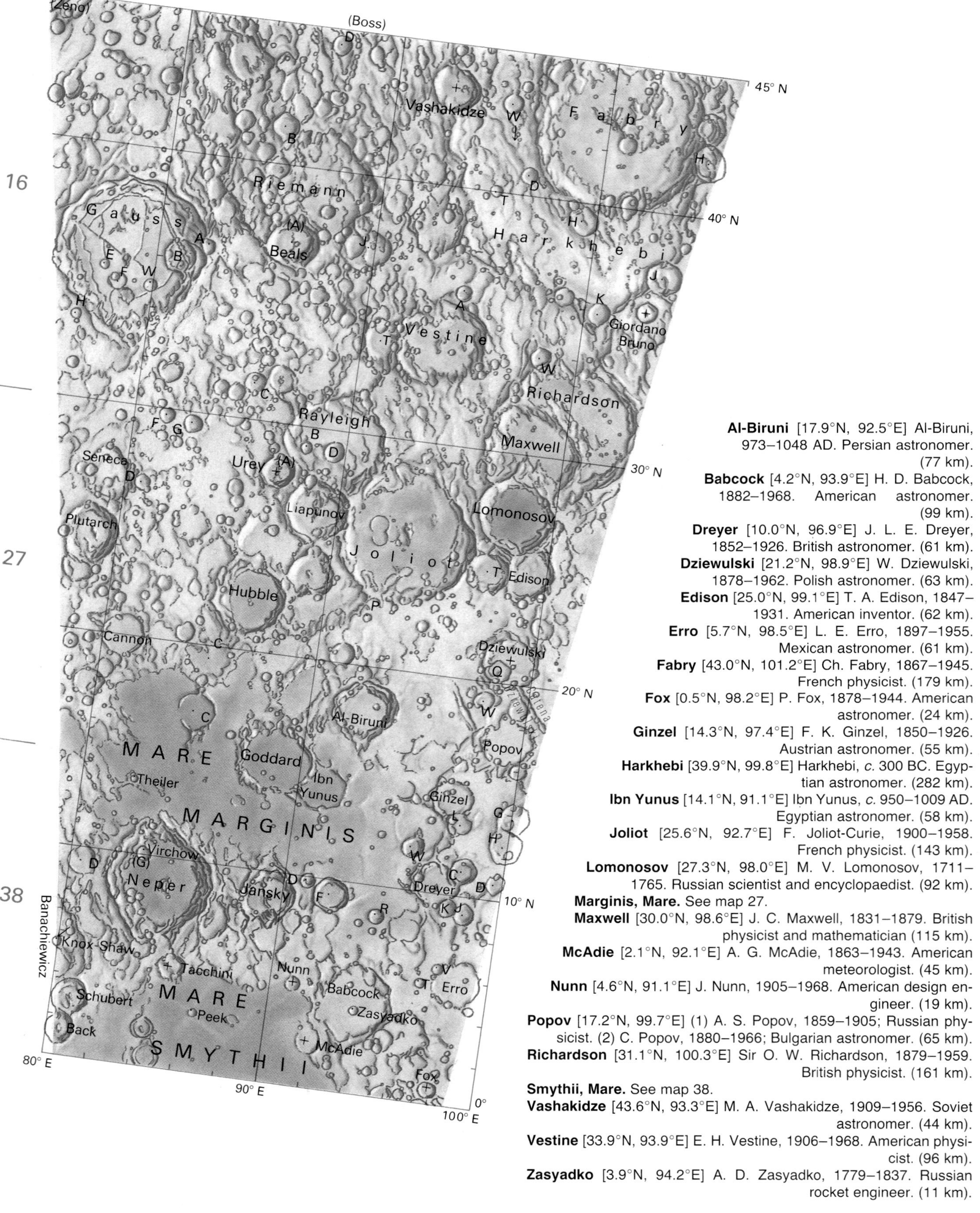

Al-Biruni [17.9°N, 92.5°E] Al-Biruni, 973–1048 AD. Persian astronomer. (77 km).

Babcock [4.2°N, 93.9°E] H. D. Babcock, 1882–1968. American astronomer. (99 km).

Dreyer [10.0°N, 96.9°E] J. L. E. Dreyer, 1852–1926. British astronomer. (61 km).

Dziewulski [21.2°N, 98.9°E] W. Dziewulski, 1878–1962. Polish astronomer. (63 km).

Edison [25.0°N, 99.1°E] T. A. Edison, 1847–1931. American inventor. (62 km).

Erro [5.7°N, 98.5°E] L. E. Erro, 1897–1955. Mexican astronomer. (61 km).

Fabry [43.0°N, 101.2°E] Ch. Fabry, 1867–1945. French physicist. (179 km).

Fox [0.5°N, 98.2°E] P. Fox, 1878–1944. American astronomer. (24 km).

Ginzel [14.3°N, 97.4°E] F. K. Ginzel, 1850–1926. Austrian astronomer. (55 km).

Harkhebi [39.9°N, 99.8°E] Harkhebi, *c.* 300 BC. Egyptian astronomer. (282 km).

Ibn Yunus [14.1°N, 91.1°E] Ibn Yunus, *c.* 950–1009 AD. Egyptian astronomer. (58 km).

Joliot [25.6°N, 92.7°E] F. Joliot-Curie, 1900–1958. French physicist. (143 km).

Lomonosov [27.3°N, 98.0°E] M. V. Lomonosov, 1711–1765. Russian scientist and encyclopaedist. (92 km).

Marginis, Mare. See map 27.

Maxwell [30.0°N, 98.6°E] J. C. Maxwell, 1831–1879. British physicist and mathematician (115 km).

McAdie [2.1°N, 92.1°E] A. G. McAdie, 1863–1943. American meteorologist. (45 km).

Nunn [4.6°N, 91.1°E] J. Nunn, 1905–1968. American design engineer. (19 km).

Popov [17.2°N, 99.7°E] (1) A. S. Popov, 1859–1905; Russian physicist. (2) C. Popov, 1880–1966; Bulgarian astronomer. (65 km).

Richardson [31.1°N, 100.3°E] Sir O. W. Richardson, 1879–1959. British physicist. (161 km).

Smythii, Mare. See map 38.

Vashakidze [43.6°N, 93.3°E] M. A. Vashakidze, 1909–1956. Soviet astronomer. (44 km).

Vestine [33.9°N, 93.9°E] E. H. Vestine, 1906–1968. American physicist. (96 km).

Zasyadko [3.9°N, 94.2°E] A. D. Zasyadko, 1779–1837. Russian rocket engineer. (11 km).

Libration Zone IV. **CURIE**

East-south-eastern sector

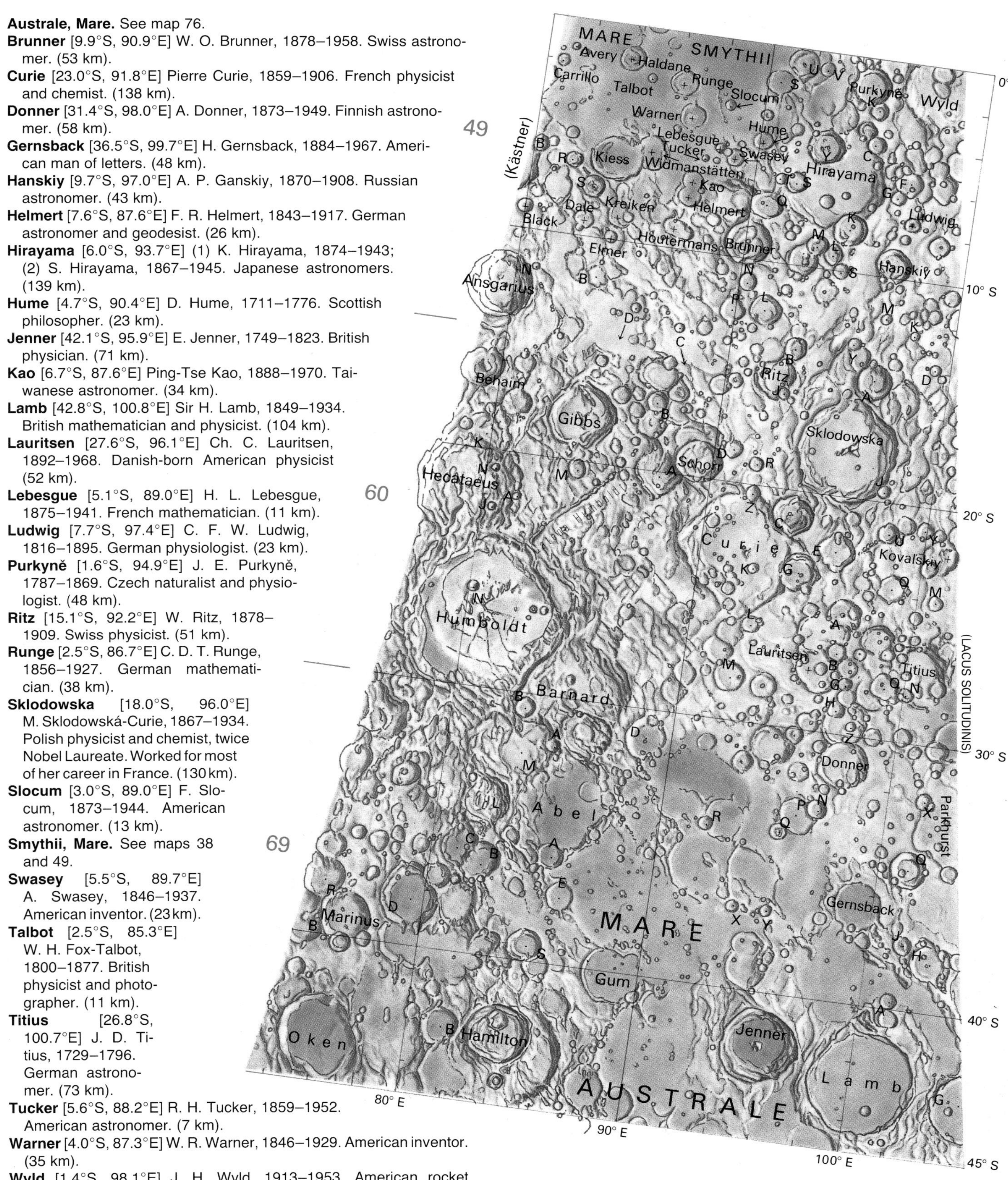

Australe, Mare. See map 76.
Brunner [9.9°S, 90.9°E] W. O. Brunner, 1878–1958. Swiss astronomer. (53 km).
Curie [23.0°S, 91.8°E] Pierre Curie, 1859–1906. French physicist and chemist. (138 km).
Donner [31.4°S, 98.0°E] A. Donner, 1873–1949. Finnish astronomer. (58 km).
Gernsback [36.5°S, 99.7°E] H. Gernsback, 1884–1967. American man of letters. (48 km).
Hanskiy [9.7°S, 97.0°E] A. P. Ganskiy, 1870–1908. Russian astronomer. (43 km).
Helmert [7.6°S, 87.6°E] F. R. Helmert, 1843–1917. German astronomer and geodesist. (26 km).
Hirayama [6.0°S, 93.7°E] (1) K. Hirayama, 1874–1943; (2) S. Hirayama, 1867–1945. Japanese astronomers. (139 km).
Hume [4.7°S, 90.4°E] D. Hume, 1711–1776. Scottish philosopher. (23 km).
Jenner [42.1°S, 95.9°E] E. Jenner, 1749–1823. British physician. (71 km).
Kao [6.7°S, 87.6°E] Ping-Tse Kao, 1888–1970. Taiwanese astronomer. (34 km).
Lamb [42.8°S, 100.8°E] Sir H. Lamb, 1849–1934. British mathematician and physicist. (104 km).
Lauritsen [27.6°S, 96.1°E] Ch. C. Lauritsen, 1892–1968. Danish-born American physicist (52 km).
Lebesgue [5.1°S, 89.0°E] H. L. Lebesgue, 1875–1941. French mathematician. (11 km).
Ludwig [7.7°S, 97.4°E] C. F. W. Ludwig, 1816–1895. German physiologist. (23 km).
Purkyně [1.6°S, 94.9°E] J. E. Purkyně, 1787–1869. Czech naturalist and physiologist. (48 km).
Ritz [15.1°S, 92.2°E] W. Ritz, 1878–1909. Swiss physicist. (51 km).
Runge [2.5°S, 86.7°E] C. D. T. Runge, 1856–1927. German mathematician. (38 km).
Sklodowska [18.0°S, 96.0°E] M. Sklodowská-Curie, 1867–1934. Polish physicist and chemist, twice Nobel Laureate. Worked for most of her career in France. (130 km).
Slocum [3.0°S, 89.0°E] F. Slocum, 1873–1944. American astronomer. (13 km).
Smythii, Mare. See maps 38 and 49.
Swasey [5.5°S, 89.7°E] A. Swasey, 1846–1937. American inventor. (23 km).
Talbot [2.5°S, 85.3°E] W. H. Fox-Talbot, 1800–1877. British physicist and photographer. (11 km).
Titius [26.8°S, 100.7°E] J. D. Titius, 1729–1796. German astronomer. (73 km).
Tucker [5.6°S, 88.2°E] R. H. Tucker, 1859–1952. American astronomer. (7 km).
Warner [4.0°S, 87.3°E] W. R. Warner, 1846–1929. American inventor. (35 km).
Wyld [1.4°S, 98.1°E] J. H. Wyld, 1913–1953. American rocket engineer. (93 km).

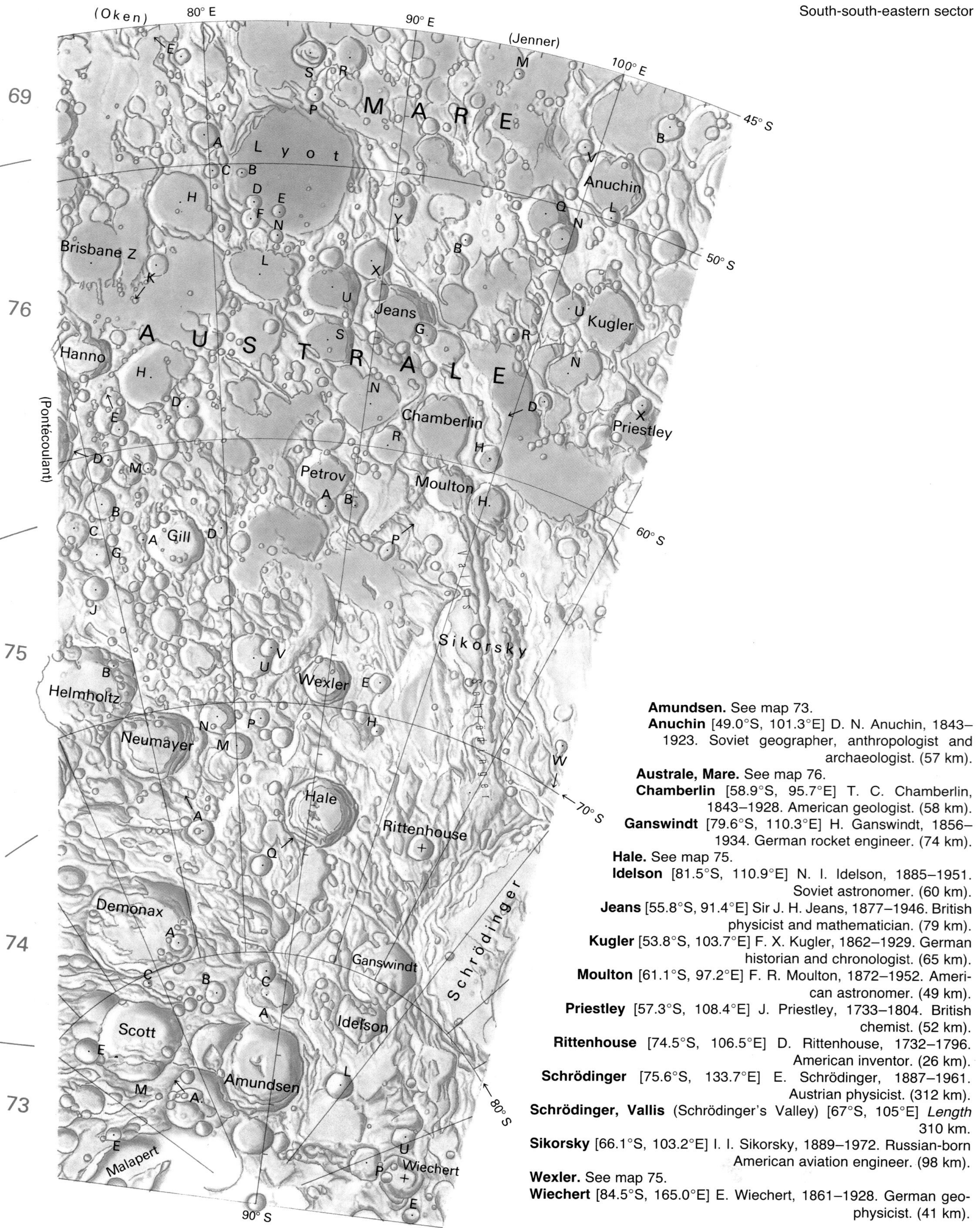

Amundsen. See map 73.
Anuchin [49.0°S, 101.3°E] D. N. Anuchin, 1843–1923. Soviet geographer, anthropologist and archaeologist. (57 km).
Australe, Mare. See map 76.
Chamberlin [58.9°S, 95.7°E] T. C. Chamberlin, 1843–1928. American geologist. (58 km).
Ganswindt [79.6°S, 110.3°E] H. Ganswindt, 1856–1934. German rocket engineer. (74 km).
Hale. See map 75.
Idelson [81.5°S, 110.9°E] N. I. Idelson, 1885–1951. Soviet astronomer. (60 km).
Jeans [55.8°S, 91.4°E] Sir J. H. Jeans, 1877–1946. British physicist and mathematician. (79 km).
Kugler [53.8°S, 103.7°E] F. X. Kugler, 1862–1929. German historian and chronologist. (65 km).
Moulton [61.1°S, 97.2°E] F. R. Moulton, 1872–1952. American astronomer. (49 km).
Priestley [57.3°S, 108.4°E] J. Priestley, 1733–1804. British chemist. (52 km).
Rittenhouse [74.5°S, 106.5°E] D. Rittenhouse, 1732–1796. American inventor. (26 km).
Schrödinger [75.6°S, 133.7°E] E. Schrödinger, 1887–1961. Austrian physicist. (312 km).
Schrödinger, Vallis (Schrödinger's Valley) [67°S, 105°E] *Length* 310 km.
Sikorsky [66.1°S, 103.2°E] I. I. Sikorsky, 1889–1972. Russian-born American aviation engineer. (98 km).
Wexler. See map 75.
Wiechert [84.5°S, 165.0°E] E. Wiechert, 1861–1928. German geophysicist. (41 km).

Libration Zone VI. **HAUSEN**

South-south-western sector

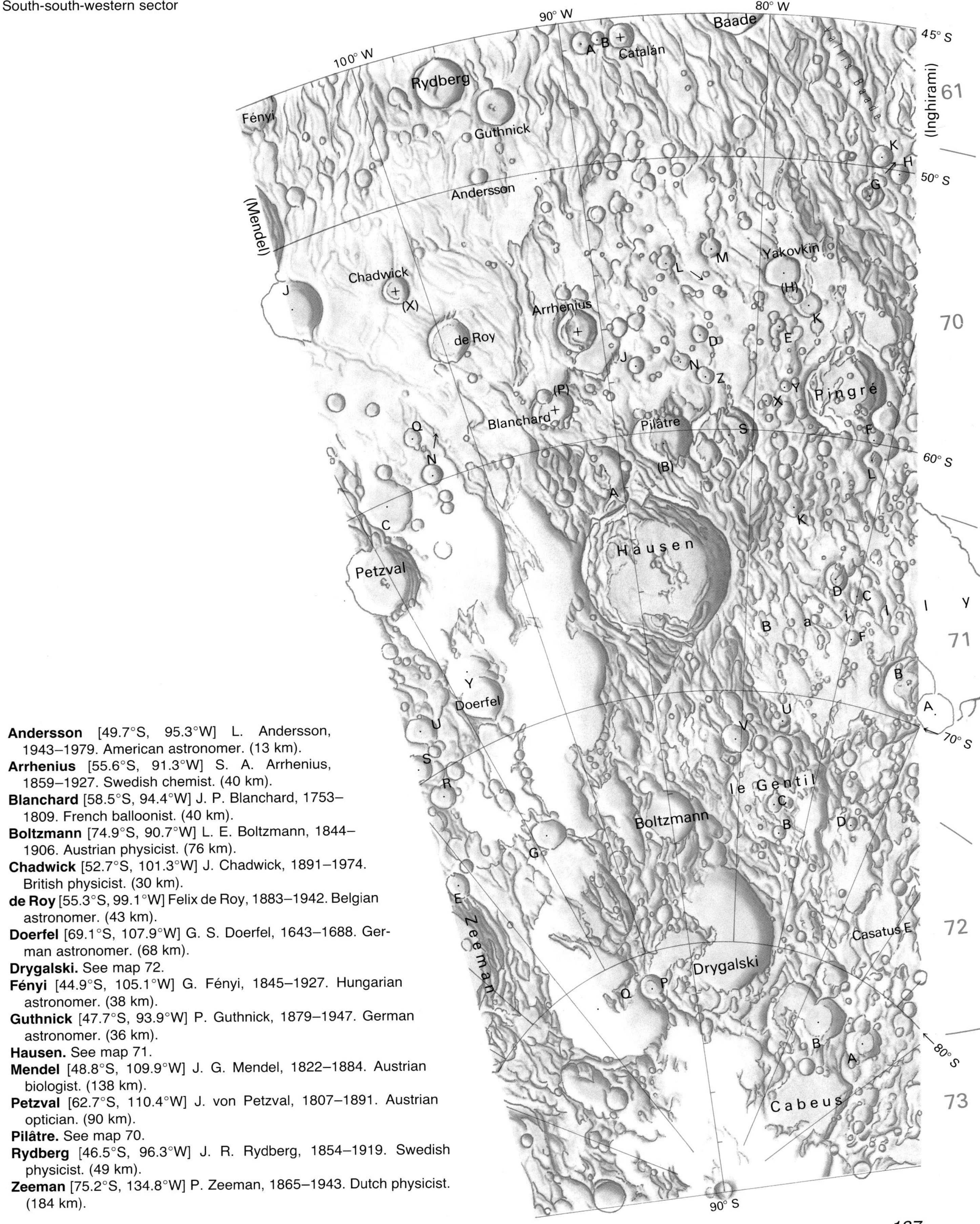

Andersson [49.7°S, 95.3°W] L. Andersson, 1943–1979. American astronomer. (13 km).

Arrhenius [55.6°S, 91.3°W] S. A. Arrhenius, 1859–1927. Swedish chemist. (40 km).

Blanchard [58.5°S, 94.4°W] J. P. Blanchard, 1753–1809. French balloonist. (40 km).

Boltzmann [74.9°S, 90.7°W] L. E. Boltzmann, 1844–1906. Austrian physicist. (76 km).

Chadwick [52.7°S, 101.3°W] J. Chadwick, 1891–1974. British physicist. (30 km).

de Roy [55.3°S, 99.1°W] Felix de Roy, 1883–1942. Belgian astronomer. (43 km).

Doerfel [69.1°S, 107.9°W] G. S. Doerfel, 1643–1688. German astronomer. (68 km).

Drygalski. See map 72.

Fényi [44.9°S, 105.1°W] G. Fényi, 1845–1927. Hungarian astronomer. (38 km).

Guthnick [47.7°S, 93.9°W] P. Guthnick, 1879–1947. German astronomer. (36 km).

Hausen. See map 71.

Mendel [48.8°S, 109.9°W] J. G. Mendel, 1822–1884. Austrian biologist. (138 km).

Petzval [62.7°S, 110.4°W] J. von Petzval, 1807–1891. Austrian optician. (90 km).

Pilâtre. See map 70.

Rydberg [46.5°S, 96.3°W] J. R. Rydberg, 1854–1919. Swedish physicist. (49 km).

Zeeman [75.2°S, 134.8°W] P. Zeeman, 1865–1943. Dutch physicist. (184 km).

Libration Zone VII. **MARE ORIENTALE**

West-south-western sector

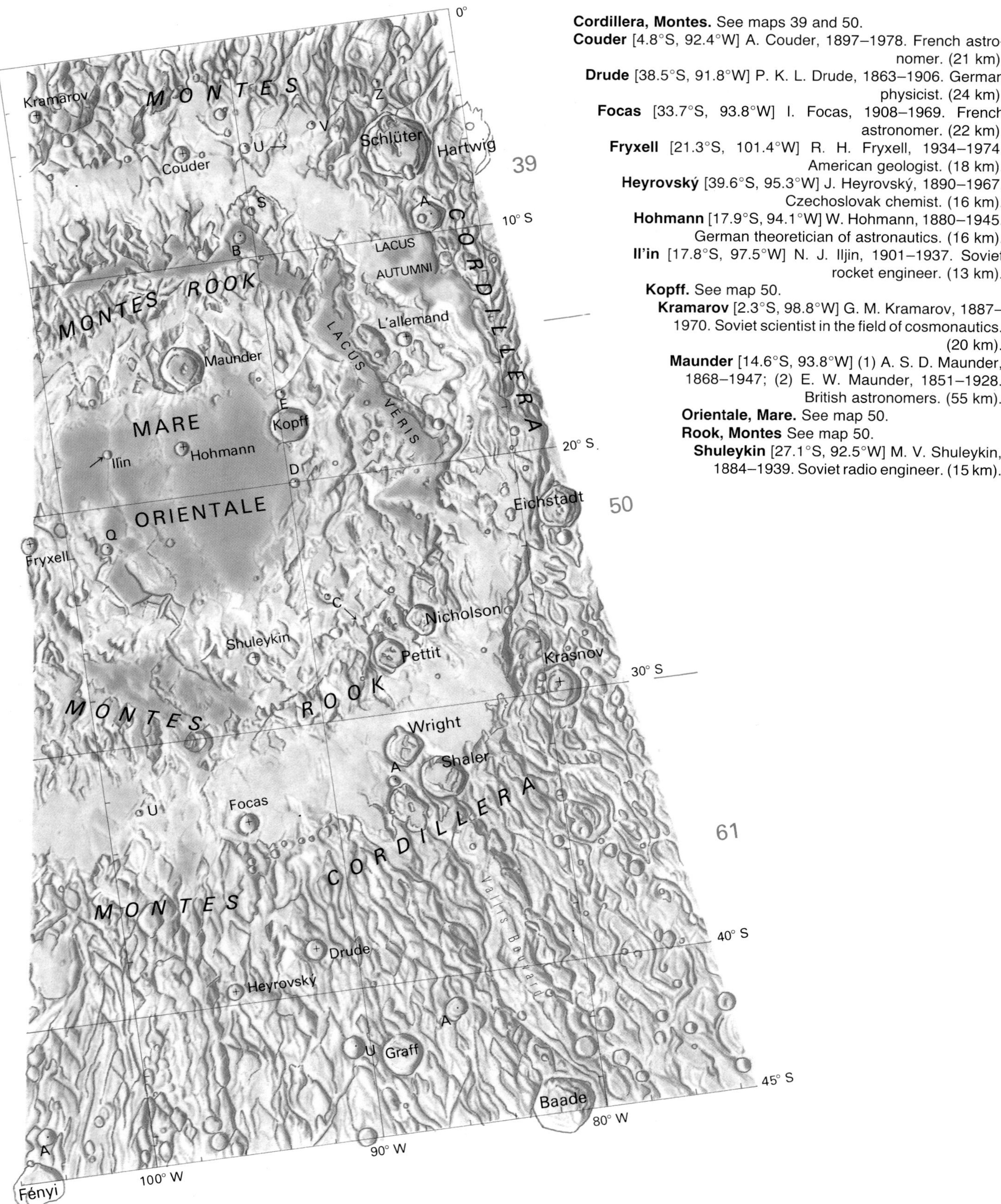

Cordillera, Montes. See maps 39 and 50.

Couder [4.8°S, 92.4°W] A. Couder, 1897–1978. French astronomer. (21 km).

Drude [38.5°S, 91.8°W] P. K. L. Drude, 1863–1906. German physicist. (24 km).

Focas [33.7°S, 93.8°W] I. Focas, 1908–1969. French astronomer. (22 km).

Fryxell [21.3°S, 101.4°W] R. H. Fryxell, 1934–1974. American geologist. (18 km).

Heyrovský [39.6°S, 95.3°W] J. Heyrovský, 1890–1967. Czechoslovak chemist. (16 km).

Hohmann [17.9°S, 94.1°W] W. Hohmann, 1880–1945. German theoretician of astronautics. (16 km).

Il'in [17.8°S, 97.5°W] N. J. Iljin, 1901–1937. Soviet rocket engineer. (13 km).

Kopff. See map 50.

Kramarov [2.3°S, 98.8°W] G. M. Kramarov, 1887–1970. Soviet scientist in the field of cosmonautics. (20 km).

Maunder [14.6°S, 93.8°W] (1) A. S. D. Maunder, 1868–1947; (2) E. W. Maunder, 1851–1928. British astronomers. (55 km).

Orientale, Mare. See map 50.

Rook, Montes See map 50.

Shuleykin [27.1°S, 92.5°W] M. V. Shuleykin, 1884–1939. Soviet radio engineer. (15 km).

Libration Zone VIII. **RÖNTGEN**

West-north-western sector

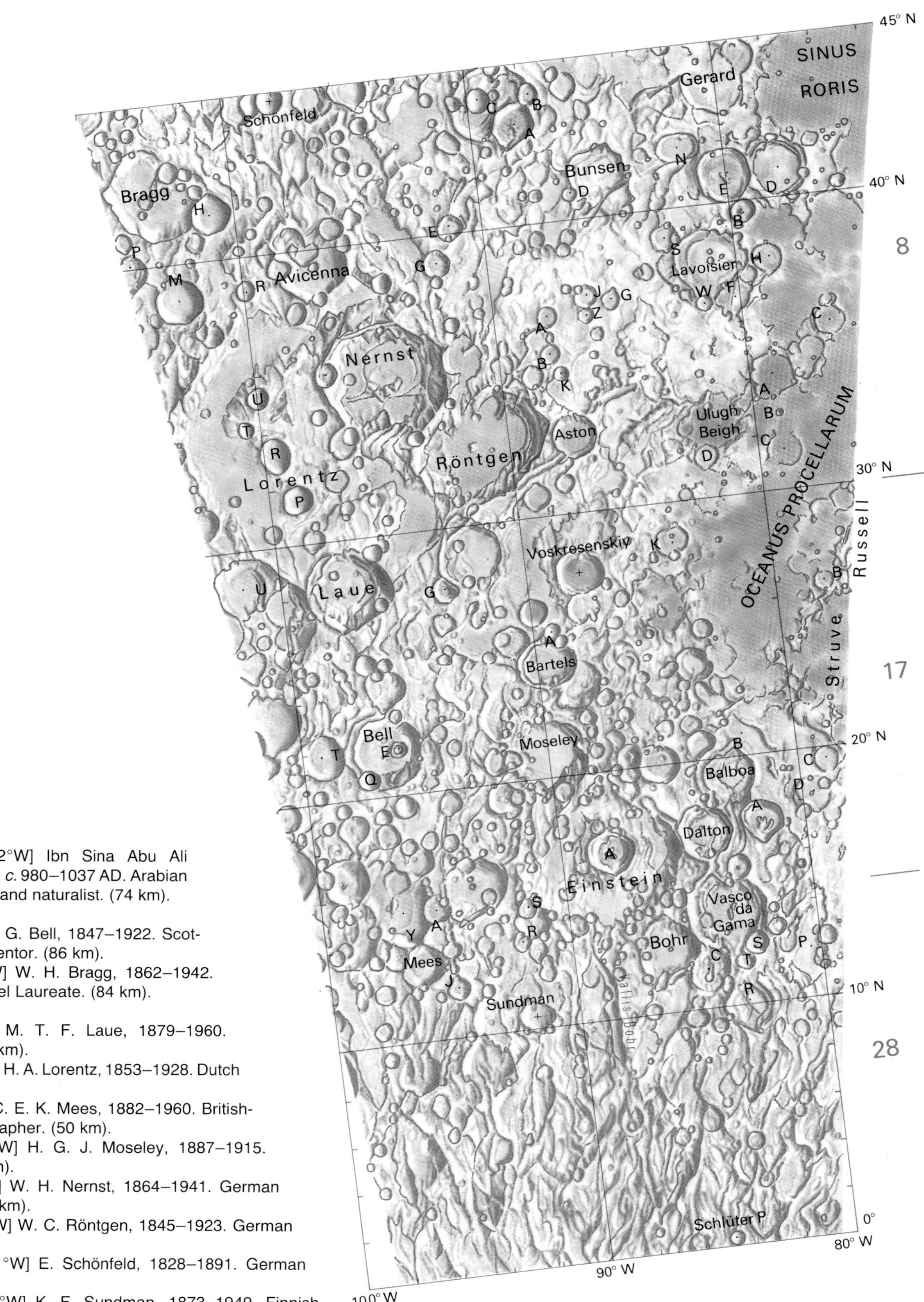

Avicenna [39.7°N, 97.2°W] Ibn Sina Abu Ali Al-Husain ibn Abdullah, *c.* 980–1037 AD. Arabian physician, philosopher and naturalist. (74 km).

Bartels. See map 17.

Bell [21.8°N, 96.4°W] A. G. Bell, 1847–1922. Scottish-born American inventor. (86 km).

Bragg [42.5°N, 102.9°W] W. H. Bragg, 1862–1942. Scottish physicist, Nobel Laureate. (84 km).

Einstein. See map 17.

Laue [28.0°N, 96.7°W] M. T. F. Laue, 1879–1960. German physicist. (87 km).

Lorentz [34.3°N, 96.9°W] H. A. Lorentz, 1853–1928. Dutch physicist. (371 km).

Mees [13.6°N, 96.1°W] C. E. K. Mees, 1882–1960. British-born American photographer. (50 km).

Moseley [20.9°N, 90.1°W] H. G. J. Moseley, 1887–1915. British physicist. (90 km).

Nernst [35.4°N, 94.5°W] W. H. Nernst, 1864–1941. German physical chemist. (118 km).

Röntgen [33.0°N, 91.4°W] W. C. Röntgen, 1845–1923. German physicist. (126 km).

Schönfeld [44.8°N, 98.1°W] E. Schönfeld, 1828–1891. German astronomer. (25 km).

Sundman [10.8°N, 91.6°W] K. F. Sundman, 1873–1949. Finnish astronomer. (40 km).

General map of the Moon

A comparison of the near- and far-side of the Moon reveals considerable differences in the distribution of mare and 'continents'.

On the near-side of the Moon 31.2 per cent of the surface area is covered by maria, while on the far-side, surprisingly, maria can be found on only 2.5 per cent of the surface area: they are Mare Orientale (Eastern Sea), Mare Ingenii (Sea of Longing), Mare Moscoviense (Moscow Sea) and parts of Mare Australe, Mare Marginis and Mare Smythii. Also on the far-side of the Moon there are 621 named craters (1988).

The Mare Orientale basin is the most prominent formation on the lunar far-side; occasionally its outer wall – Montes Cordillera – can be observed from the Earth, on the western limb of the Moon. Mare Moscoviense and the crater Tsiolkovsky, whose floor is filled with very dark mare material, are also quite prominent formations of the far-side.

Near-side of the Moon

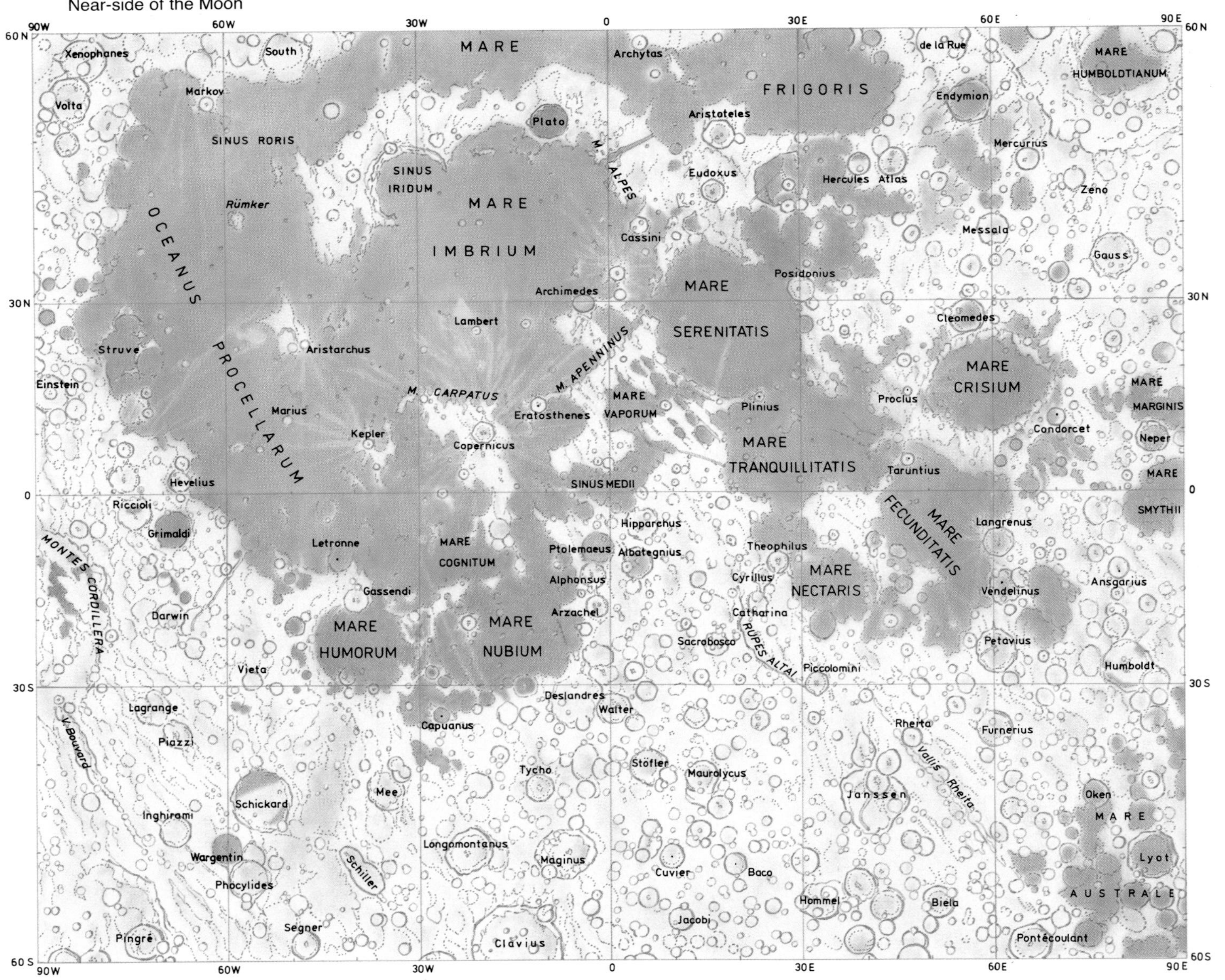

North polar region

180
150 W
150 E
120 W
120 E
90 W
90 E
60 W
60 E
30 W
30 E
0
80 N
70 N
60 N
Rowland
Birkhoff
Sommerfeld
Avogadro
Gamov
Compton
Nöther
Plaskett
Schwarzschild
Brianchon
Nansen
Belkovich
Peary
Byrd
Xenophanes
MARE
HUMBOLDTIANUM
Pascal
Pythagoras
Meton
Anaxagoras
Herschel, J.
de la Rue
MARE
FRIGORIS

South polar region

180
150 E
150 W
120 E
120 W
90 E
90 W
60 E
60 W
30 E
30 W
0
80 S
70 S
60 S
Poincaré
Lemaître
Planck
Lyman
Antoniadi
Fizeau
Schrödinger
Zeeman
Petzval
MARE
AUSTRALE
Amundsen
Drygalski
Hausen
Scott
Bailly
Boussingault
Newton
Moretus
Hagecius
Curtius
Clavius

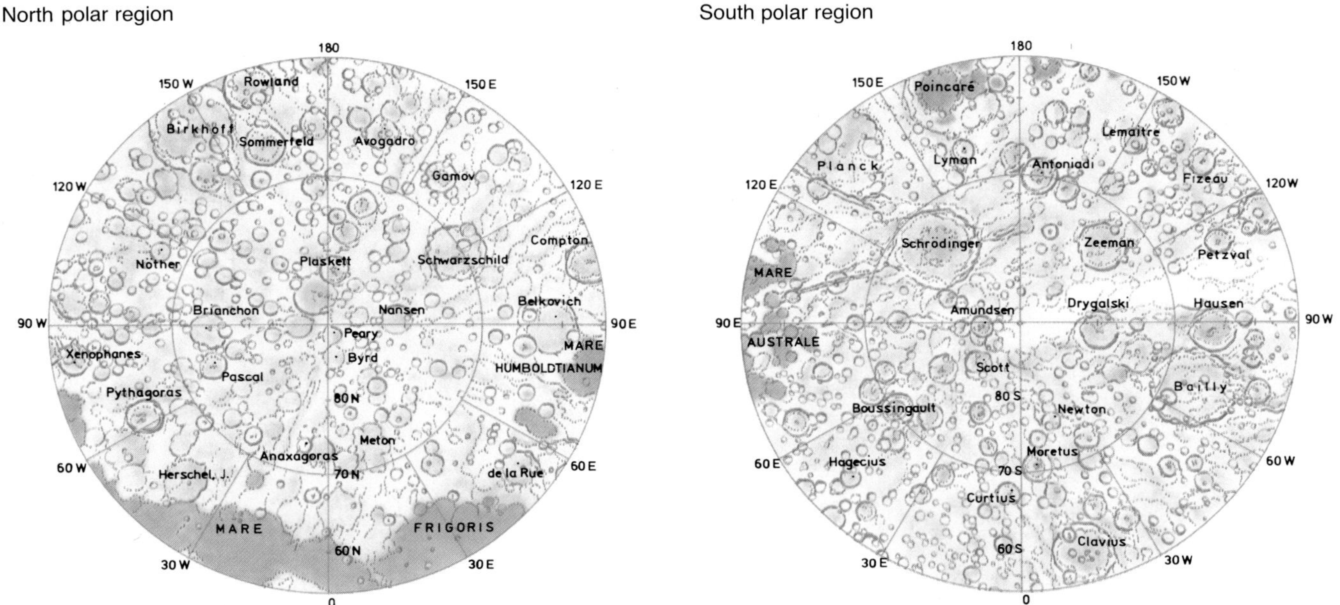

Far-side of the Moon

90 E
120 E
150 E
180
150 W
120 W
90 W
60 N
30 N
0
30 S
60 S
Olivier
Birkhoff
Rowland
Poczobutt
Compton
Coulomb
D'Alembert
Carnot
Debye
Milikan
Campbell
Stefan
Fabry
Fowler
H.G.Wells
Kulik
Landau
Giordano
Bruno
Lorentz
Kovalevskaya
MARE
MOSCOVIENSE
Trumpler
Joliot
Lomonosov
Fitzgerald
Vernadskij
Poynting
Mach
Fersman
Robertson
Anderson
MARE
MARGINIS
Fleming
Mendeleev
Mandel'shtam
Tsander
Michelson
Hertzsprung
Bečvář
Daedalus
Icarus
Korolev
Keeler
Lowell
Pasteur
Doppler
Galois
Aitken
MARE
ORIENTALE
Sklodowska
Gagarin
Houzeau
Tsiolkovskij
Van de
Graaff
Pavlov
Milne
MARE
INGENII
Chebyshev
Jules Verne
Oppenheimer
Apollo
Brouwer
MARE
Leibnitz
Roche
AUSTRALE
Koch
Von Kármán
Pauli
Mendel
Lebedev
Bose
Lippmann
Minkowski
Jeans
Planck
Poincaré
Fizeau

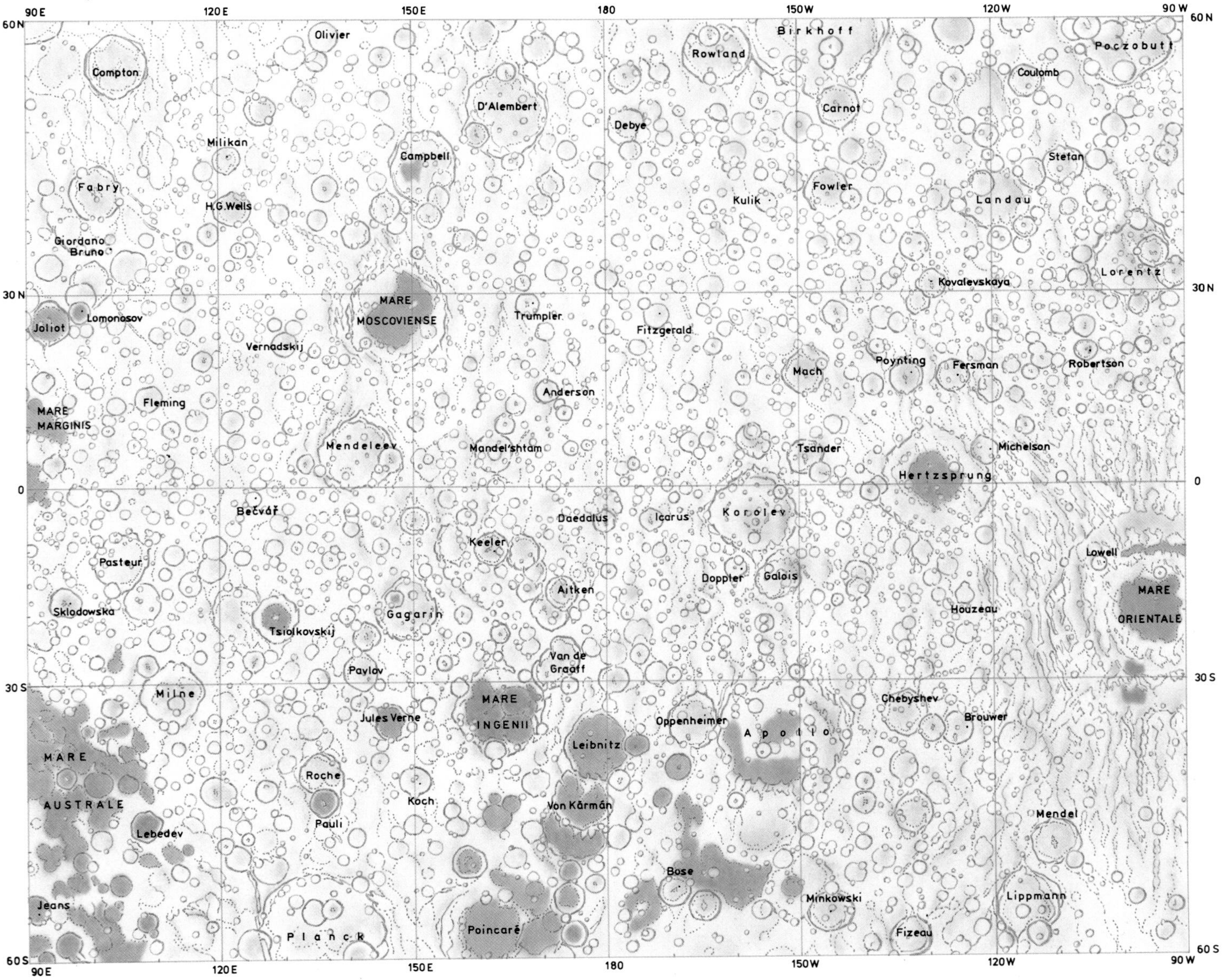

Flights to the Moon

The map opposite calls attention to the initial period of very intensive exploration of the Moon from 1959 to 1976. On the map the places of impacts and softlandings of successful automatic probes and manned space vehicles are indicated. During that period two long-term Soviet and American programmes were in progress. The Soviet programme was based on the use of sophisticated, automatic space probes. The American programme, from the very beginning, was aimed at putting a man on the Moon. Both programmes were complementary to each other and produced a rich harvest of new data about the Moon.

The Soviet programme comprised three generations of *Luna* probes. The simple probes of the first generation by-passed the Moon (*Luna 1,* 1959), and then, for the first time, crash-landed on the Moon (*Luna 2,* 1959) and photographed part of the far-side (*Luna 3,* 1959). The probes of the second series (*Lunas 5–14,* 1965–1968), after several abortive attempts, achieved the first soft-landing on the Moon (*Luna 9,* 1966) and established the first artificial lunar satellite (*Luna 10,* 1966). Of specific importance in the Soviet programme were the probes of the *Zond* series, which orbited the Moon, took photographs and measurements from close range and in some cases returned to the Earth (*Zond 3,* 1965; *Zonds 5–7,* 1968–1969). The probes belonging to the third series were inaugurated by an experiment with *Luna 15* in 1969. Its basic form consisted of a multipurpose transport platform, which made a soft landing on the Moon by means of a four-legged undercarriage. This landing section transported to the Moon either a Moon–Earth rocket equipped for the automatic collection and return of rock samples (*Luna 16,* 1970; *Luna 20,* 1972; and *Luna 24,* 1976) or a remote-controlled mobile laboratory (*Luna 17* with vehicle *Lunokhod 1,* 1970, and *Luna 21* with *Lunokhod 2,* 1973).

The American programme began with a series of probes called *Ranger,* which were crash-landed on pre-selected areas of the lunar surface. Shortly before the final descent, their T.V. cameras photographed the Moon and these photographs were immediately transmitted to the Earth (*Rangers 7, 8* and *9* in 1964–1965). Between 1966 and 1968 two further programmes were carried out, in preparation for the landing of astronauts on the Moon: these involved direct research into the nature of the lunar surface by the soft-landing *Surveyor* probes and the mapping of the Moon in fine detail by the *Lunar Orbiter* artificial lunar satellites. In all, seven *Surveyor* probes were launched, but two attempts were unsuccessful. They were equipped with T.V. cameras remotely controlled from Earth, mechanical surface-samplers (*Surveyors 3* and *7*), and facilities for chemical analysis using alpha-particle scattering (*Surveyors 5, 6* and *7*). Five successful *Surveyor* probes transmitted a total of 87,674 photographs of the lunar surface to Earth.

Some very important work was completed by the five *Lunar Orbiter* satellites. Their original task was to photograph, especially, ten areas chosen for possible landing sites for the *Apollo* expeditions. This task was completed by the first three *Orbiters,* however, and their programme was therefore extended to the global mapping of the Moon's surface. The American programme culminated in a series of manned flights to the Moon, known as the *Apollo* missions. These were preceeded by orbital flights around the Moon (*Apollo 8,* 21–27 December 1968; and *Apollo 10,* 20–26 May 1969). These were followed by the first manned landing on the Moon (*Apollo 11,* 20 July 1969) and another five successful expeditions (*Apollos 12, 14, 15, 16* and *17,* between 1969 and 1972). The *Apollo* spacecraft consisted of three main sections: a command module, a service module and a lunar module. The crew of the spacecraft consisted of three astronauts, who spent most of their time aboard the command module. Following insertion into lunar orbit two astronauts transferred to the lunar module, which was then separated from the command module to soft-land on the Moon. The third member of the crew remained in the command module, which, connected to the service module, continued to orbit the Moon. The activity of the astronauts on the lunar surface was concentrated on geological survey work, the collection of rock samples, photographic documentation and the installation of scientific apparatus (including the geophysical station 'ALSEP').

The results of the combined attack on unravelling the mysteries of the Moon were admirable. In a mere decade incomparably more information about the Moon was collected than ever before and the complete analysis will take several decades. This was the main reason why in the mid-1970s all flights to the Moon ceased. The experience gathered will be used in the planning of new flights, in support of the construction of permanently manned bases on the Moon.

Note: In addition to the list of successful lunar landings (p. 193), some 26 rockets, probes, etc., have crash-landed on the Moon.

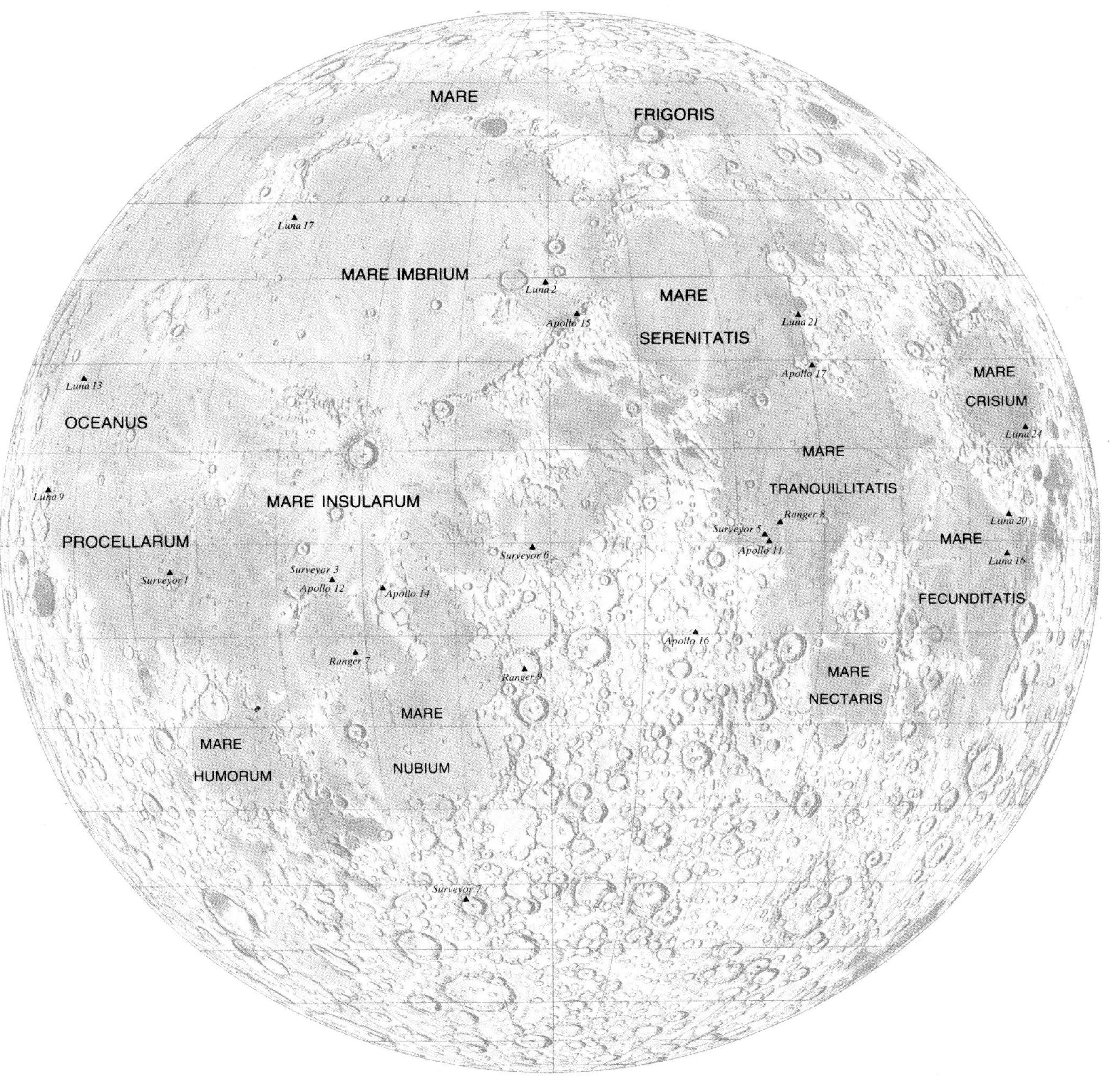

Probe	Date of reaching the Moon	Results
Luna 2	13. 9. 1959	The first probe to reach the Moon.
Ranger 7	31. 7. 1964	4308 photographs of the Moon, details down to 1 m and below (crash-landing).
Ranger 8	20. 2. 1965	7137 photographs (crash-landing).
Ranger 9	24. 3. 1965	5814 photographs, details down to 25 cm (crash-landing).
Luna 9	3. 2. 1966	The first soft-landing, 4 panoramic photographs.
Surveyor 1	2. 6. 1966	Soft-landing, 11 240 photographs.
Luna 13	24. 12. 1966	Three panoramic photographs, mechanical soil-probe.
Surveyor 3	20. 4. 1967	6326 photographs, later the *Apollo 12* expedition landed here.
Surveyor 5	11. 9. 1967	19 118 photographs; properties of lunar surface analysed.
Surveyor 6	10. 11. 1967	29 952 photographs.
Surveyor 7	19. 1. 1968	21 038 photographs, mechanical scoop, chemical analysis.
Apollo 11	20. 7. 1969	The first men on the Moon (Armstrong and Aldrin; Collins in orbit).
Apollo 12	19. 11. 1969	Conrad and Bean on the Moon, Gordon in orbit.
Luna 16	21. 9. 1970	Automatic collection of lunar rock sample, brought back to Earth.
Luna 17	17. 11. 1970	Mobile laboratory *Lunokhod 1* (drove 10 540 m across the lunar surface).
Apollo 14	5. 2. 1971	Shepard and Mitchell on the Moon, Roosa in orbit.
Apollo 15	30. 7. 1971	Scott and Irwin on the Moon, Worden in orbit.
Luna 20	21. 2. 1972	Automatic collection of rock samples, brought back to Earth.
Apollo 16	21. 4. 1972	Young and Duke on the Moon, Mattingly in orbit.
Apollo 17	11. 12. 1972	Cernan and Schmitt on the Moon, Evans in orbit.
Luna 21	15. 1. 1973	Mobile laboratory *Lunokhod 2* (covered a distance of 37 km).
Luna 24	18. 8. 1976	Core sample taken automatically, down to a depth of 2 m.

Fifty views of the Moon

The wealth of observable details on the Moon's surface is demonstrated here by a selection of photographs of diverse formations, such as typical walled plains, craters, valleys, rilles, wrinkle ridges, domes, mountain ranges and also various peculiar and rare features. These photographs may help amateur observers to choose suitable objects for study, depending on the resolving power of their telescopes.

There follow 50 examples of interesting lunar formations. Each is briefly described and its dimensions are given. Reference numbers, adjacent to the names of formations, indicate the map or maps on which they are shown, 1–76 being the sections depicting the near-side, I–VIII those for the libration zones. Approximate times of sunrise and sunset are given on the assumption that all librations are zero: in practice there can be deviations of up to plus or minus three-quarters of a day to take into account. In each descriptive paragraph the abbreviation *col.* and its value is a reference to co-longitude; the illustrations show each lunar feature as it would appear under the lighting condition imposed by the given value of co-longitude.

For the most part, the north is uppermost in the following illustrations. The positions of the 50 objects are shown on the schematic map on the back endpaper.

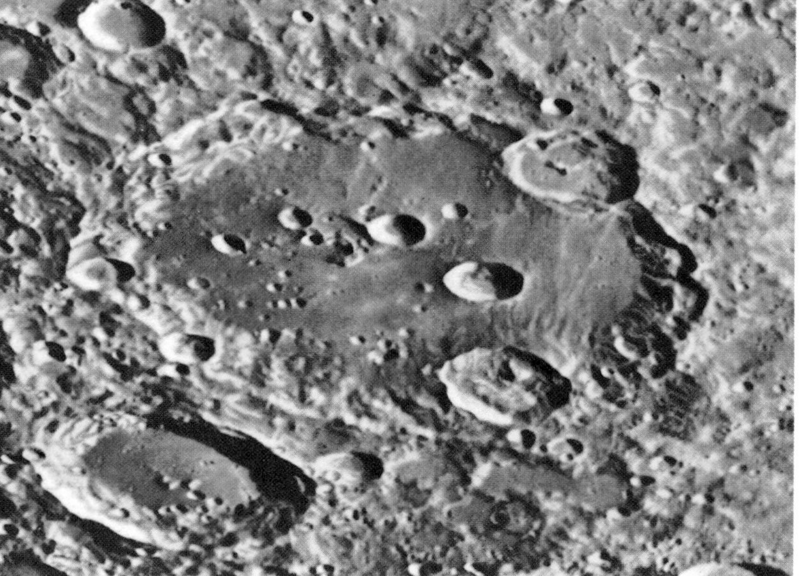

1. **Clavius** (maps 72, 73) *col. 44°*

One of the largest walled plains, polygonal outline, 225 km in diameter. Central mountain massif with numerous craters on its floor. The wall is eroded, and does not overlook the surrounding terrain. There is a fine arc of lesser craters crossing the floor. Sunrise is 1–2 days after First Quarter, and sunset 1–2 days after Last Quarter.

2. **Janssen** (maps 67, 68) *col. 128°*

A large walled plain with considerably eroded walls, 190 km in diameter. On the floor, there is a mountain massif and a system of rilles. The largest of the rilles is visible through even a small telescope; its length is 110 km and it is up to 6 km wide. Sunrise is 4 days after New Moon; sunset occurs 4 days after Full Moon.

3. **Plato** (maps 3, 4) *col. 24°*

Prominent walled plain with a dark floor, visible through even binoculars. It is 101 km in diameter and has an average depth of about 1000 m. Individual mountain peaks, on the eastern and western walls, rise to more than 2000 m above the floor, on which there are four craterlets about 2 km across (a test for larger telescopes). Sunrise 0.5 day after First Quarter, sunset 0.5 day after Last Quarter.

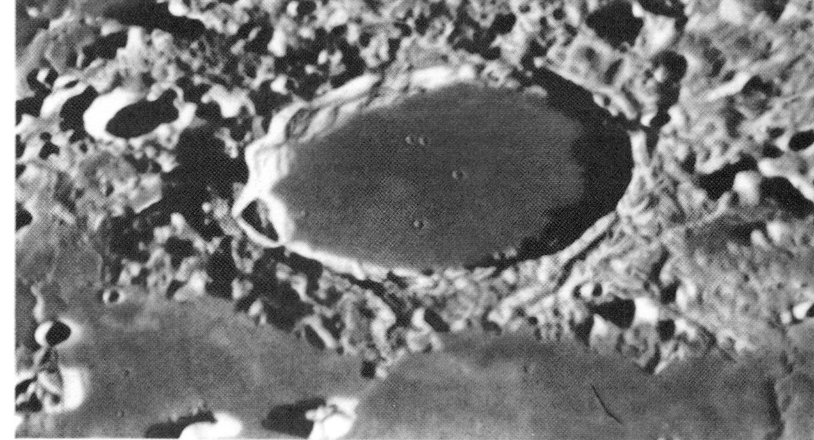

4. **Ptolemaeus** (map 44) *col. 7°*

A large walled plain with a considerably eroded polygonal wall. It is 153 km in diameter and has an average depth of 2400 m. What appears to be a flat floor is covered with numerous craterlets and crater pits and is rippled by shallow circular depressions, which are visible only under low illumination. Sunrise occurs at First Quarter, and sunset at Last Quarter.

5. **Sinus Iridum** (map 10) *col. 44°*

The remains of a 260 km diameter walled plain. Its northern and western ramparts are well preserved and are called Montes Jura – the large crater Bianchini is situated about halfway along this mountain range. The eastern and western ends of the Jura Mountains terminate with the 'capes' Promontorium Laplace and Promontorium Heraclides respectively. The interior of Sinus Iridum is flooded with mare material on which there are several wrinkle ridges and small craters. Sunrise and sunset take place 2–3 days after First Quarter and Last Quarter, respectively.

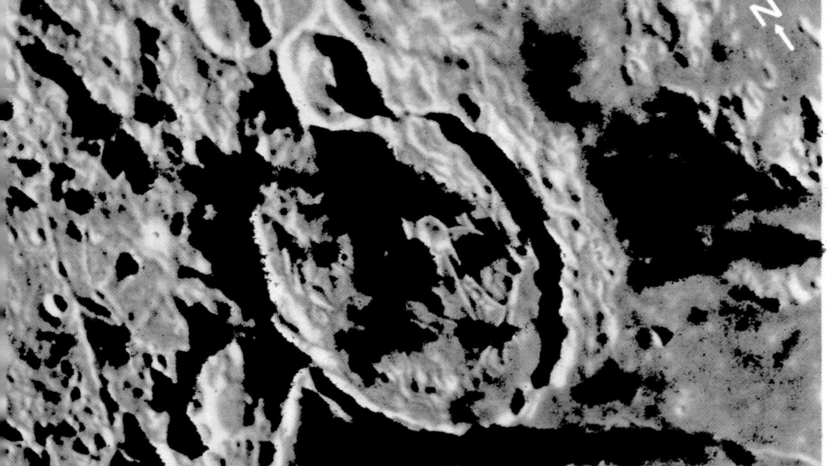

6. **Gassendi** (map 52) *col. 53°*

A characteristic walled plain, 110 km in diameter. The interior contains a group of central peaks, a complex system of rilles and numerous protruding features on the floor. In the south the height of the wall is less than 200 m, to the east and west it reaches more than 2500 m above the floor; in the north it is interrupted by the crater **Gassendi A** (33 km/3600 m). Sunrise is 3 days after First Quarter, and sunset 3 days after Last Quarter.

7. **Copernicus** (map 31) *col. 31°*

Spectacular ring mountain, one of the most prominent centres of bright radiating rays. The terraced walls are elevated 900 m above the surrounding terrain, the depth of the crater is about 3760 m, and its diameter is 93 km. On the inner side of the wall numerous landslides are found. The crater's shape is approximately hexagonal. A group of central mountains rise to 1200 m above the floor. Bright rays can be traced as far as 800 km from the crater. Sunrise is 1.5 days after First Quarter, sunset is 1.5 days after Last Quarter.

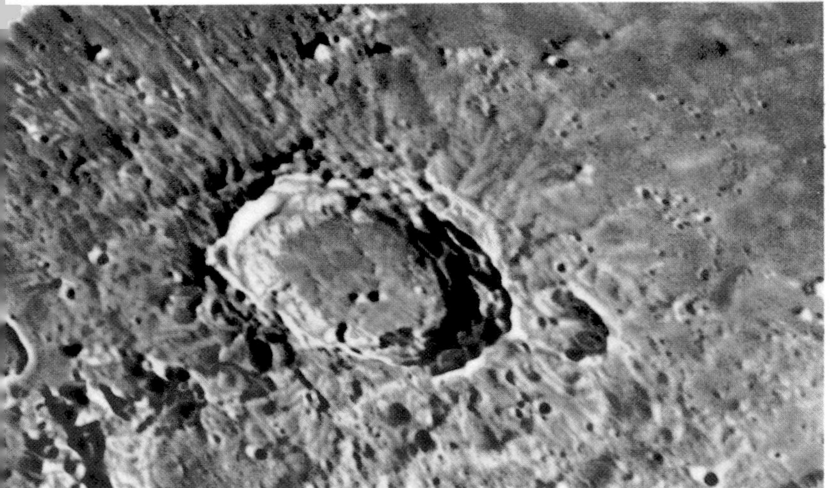

8. **Aristoteles** (map 5) *col. 7°*

A beautiful ring mountain with internal terracing, 87 km in diameter. There is a group of small central peaks on its floor. Outside, radial structure is clearly visible in the ejecta blanket. Aristoteles is one of a prominent pair with the crater Eudoxus (map 13). Sunrise is 1.5 days before First Quarter, and sunset 1.5 days before Last Quarter.

9. **Theophilus** (maps 46, 47) *col. 351°*

One of the most prominent ring mountains with a considerable terraced interior wall. It is 100 km in diameter, and has a depth of 4400 m; the ridge of the wall rises 1200 m above the surrounding terrain. A group of central mountains rises to about 1400 m above the floor. The wall of Theophilus interrupts that of neighbouring Cyrillus, proving that the former is the younger formation. Sunrise is 5 days after New Moon, and sunset 5 days after Full Moon.

10. **Tycho** (map 64)

col. 168°

A very prominent crater, one of the youngest on the Moon: the approximate age is 100 million years. The crater's diameter is 85 km, its depth is 4850 m. There is a central mountain massif, 1600 m high, terraced walls and a very uneven floor. Sunrise 1 day after First Quarter, sunset 1 day after Last Quarter.

11. **Tycho** (map 64)

col. 93°

Under high illumination, and especially around Full Moon, Tycho is the most prominent crater on the Moon, and also the most distinct centre of a bright ray system which can be traced over a distance of more than 1500 km. The crater is surrounded by a ring of dark material over 150 km in diameter. Individual parts of the ray system are observable when Tycho is immersed in shadow on the night side of the Moon.

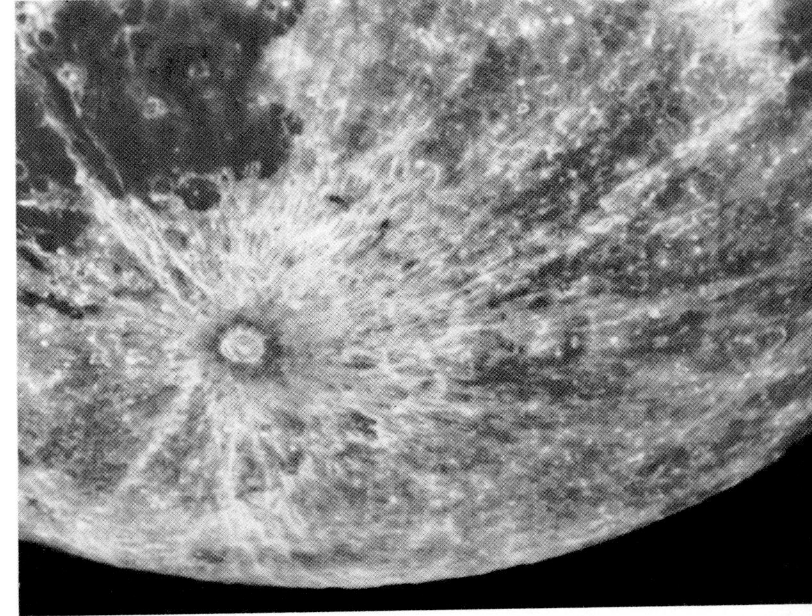

12. **Byrgius A** (maps 50, 51)

col. 134°

A regular, circular crater, 19 km in diameter, with a sharp rim interrupting the eastern wall of the crater Byrgius. Close to the terminator Byrgius A does not stand out at all; however, under high illumination during the period from approximately Full Moon until Last Quarter it is one of the most prominent centres of bright rays.

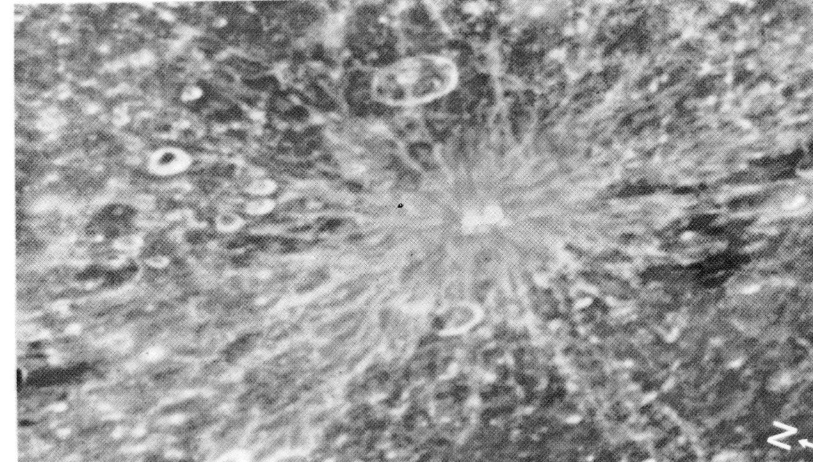

13. **Proclus** (map 26)

col. 93°

A polygonal crater with a sharp-edged outer wall, 28 km in diameter, 2400 m deep. A conspicuous centre of an asymmetric system of bright rays which radiate mostly in three directions. The darker Palus Somnii is bordered by these rays. Sunrise is 3.5 days after New Moon, sunset 3.5 days after Full Moon.

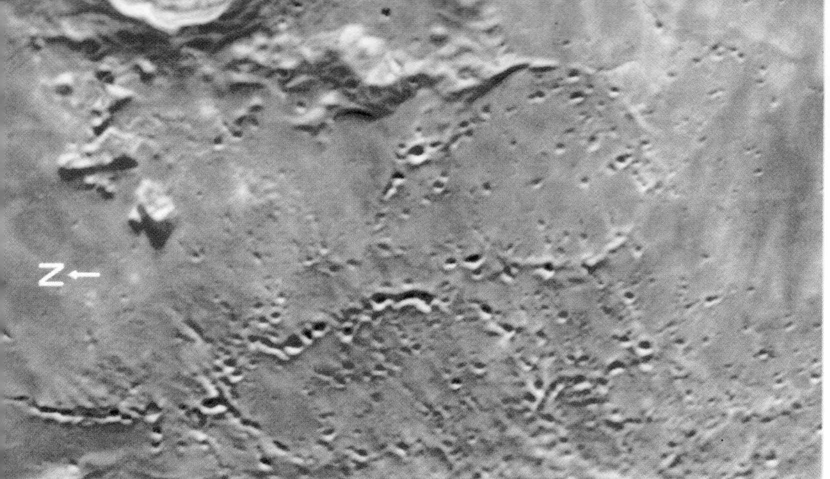

14. Crater chain close to Stadius (maps 20, 32)

col. 24°

An interesting group of craters north-north-west of the flooded crater Stadius and reaching as far as the edge of Mare Imbrium, where it continues in the form of a rille composed of crater pits. They are obviously secondary formations related to the origin of nearby Copernicus. Sunrise, about 1 day after First Quarter, sunset one day after Last Quarter.

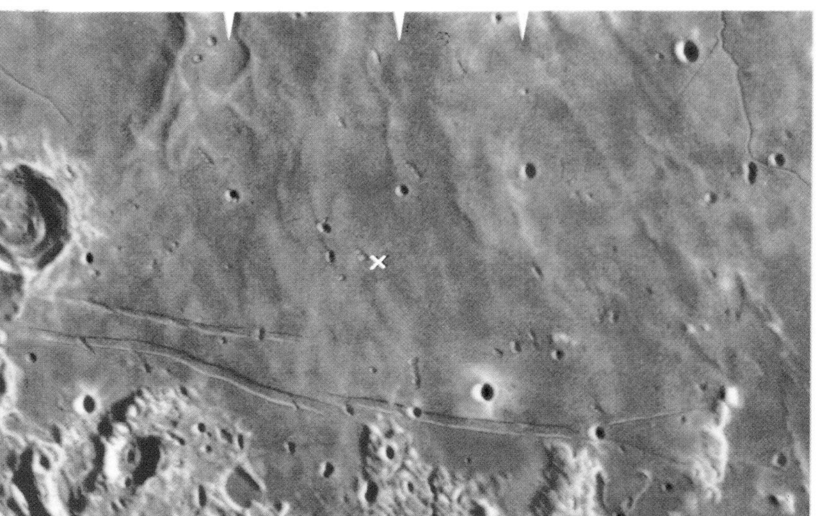

15. Statio Tranquillitatis (map 35)

col. 351°

The soft-landing site of the *Apollo 11* expedition. Close to it are the three small craters Aldrin, Armstrong and Collins (3.4 km, 4.6 km and 2.4 km in diameter), named after the *Apollo 11* crew. These are the only examples on the near-side of the Moon of lunar formations named after living persons. Sunrise 2 days before First Quarter, sunset 2 days before Last Quarter.

16. Mösting A (maps 43, 44)

col. 20°

A small crater, 13 km in diameter, 2700 m deep. It is situated close to the centre of the lunar disk and is easily observed under any angle of illumination. It is well known because of its value as the basic reference point in the system of selenographic co-ordinates. Sunrise and sunset occur shortly after First and Last Quarters, respectively.

17. Censorinus (map 47)

col. 134°

A small circular crater, 3.8 km in diameter, surrounded by very light material. The wall of the crater is not elevated much above its surroundings and the crater itself is observable only when close to the terminator. Under high illumination it is one of the brightest features on the Moon. Sunrise occurs somewhat less than 5 days after New Moon, sunset approximately 3 days before Last Quarter.

18. **Aristarchus** (map 18) *col. 134°*

One of the brightest objects on the Moon and one of the most conspicuous centres of bright rays. The crater's diameter is 40 km, its depth 3000 m. It is so bright that it is clearly visible even on the night-side of the Moon in the so-called earthshine. A number of transient lunar phenomena (T.L.P.s) such as hazes, light-increases, etc., have been observed in the vicinity of Aristarchus. Sunrise occurs almost 4 days after First Quarter, sunset about 4 days after Last Quarter.

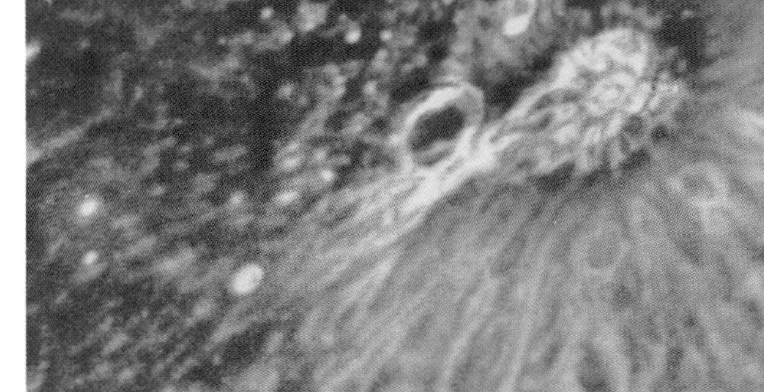

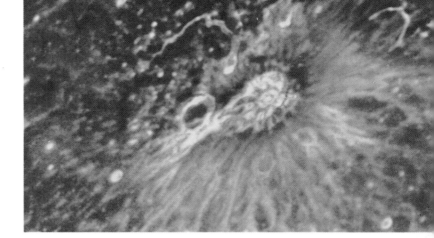

19. **Vallis Schröteri** (map 18) *col. 64°*

A large sinuous valley, resembling a dry river-bed with numerous meanders. It starts in the crater to the west of Aristarchus, then widens to 10 km, several times changes its direction, then narrows down and finally, at a distance of about 160 km from its origin, it disappears. The floor of the valley is flat and a very narrow sinuous rille, which is not visible from the Earth, zigzags along it. Sunrise is 4 days after First Quarter, sunset occurs 4 days after Last Quarter.

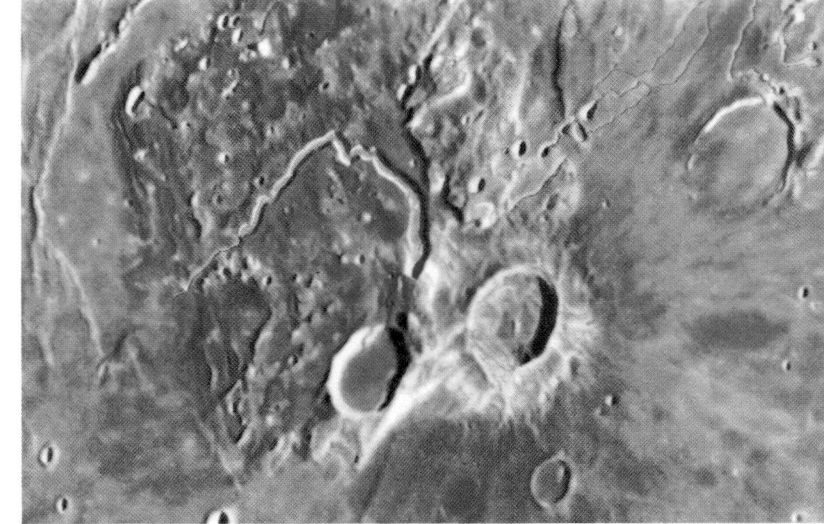

20. **Vallis Alpes** (maps 4, 12) *col. 20°*

One of the best-known lunar valleys, about 180 km long. In the west, it is enclosed by mountains. The floor of the valley is flat, and flooded with mare material. A narrow sinuous rille, which is not observable through smaller telescopes, runs along the middle of the valley. Sunrise and sunset occur just before First and Last Quarters, respectively.

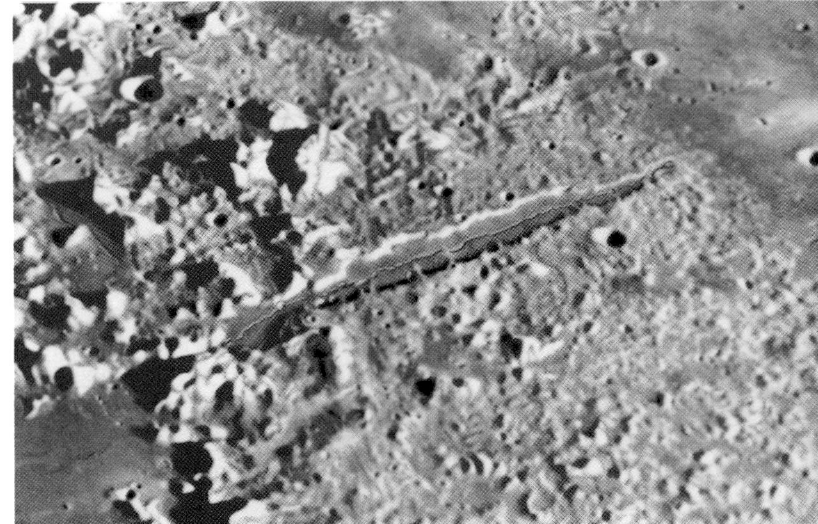

21. **Vallis Rheita** (map 68) *col. 340°*

A large valley, 30 km wide in places, starting near the western wall of the crater Rheita and continuing for a distance of about 500 km, where it narrows down to 10 km. The valley is considerably eroded, and in several places interrupted by larger craters, so that it may appear to end near the crater Young, whereas in fact it continues to the crater Reimarus. Sunrise is 3–4 days after New Moon, sunset occurs 3–4 days after Full Moon.

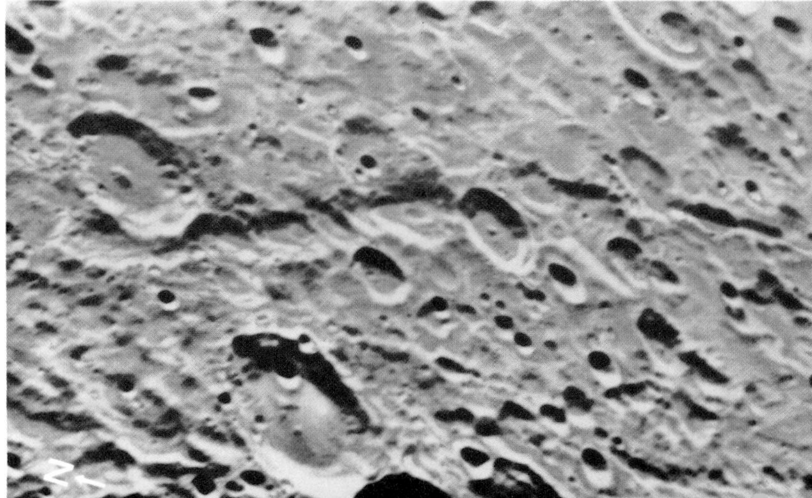

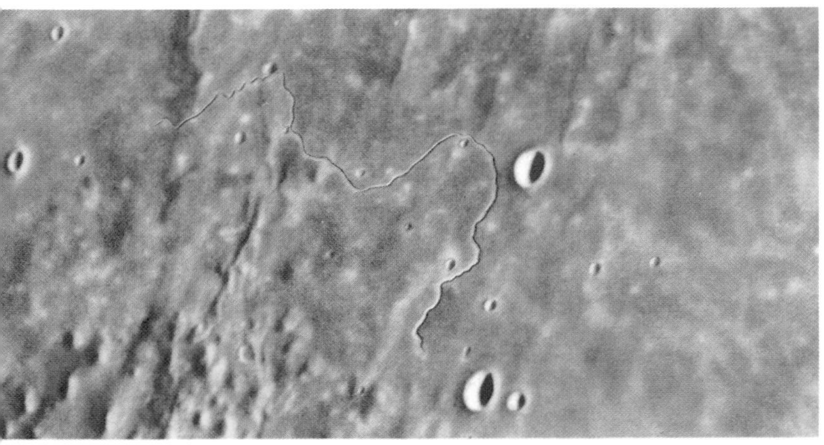

22. **Rima Marius** (map 18) *col. 61°*

A typical sinuous rille, resembling a parched meandering river-bed. Close to the crater Marius C (map 29) the rille is about 2 km wide; at the opposite end, near Marius P, it narrows down to 1 km. The total length is about 250 km. It has been suggested that such rilles may be the remains of lava channels which once contributed to the flooding of the maria. This object repays observation with larger telescopes and will tax the skill of experienced observers. Sunrise is 4 days after First Quarter, sunset occurs 4 days after Last Quarter.

23. **Rimae Triesnecker and Rima Hyginus**

Rimae Triesnecker (*Below, bottom left*) (map 33) *col. 357°*

A complex system of rilles creating the illusion of deep clefts. In fact, these are shallow valleys with flat floors, usually 1–2 km wide. The greatest extent of the system is about 200 km. A very suitable object for larger telescopes. Sunrise is just before First Quarter, sunset occurs just before Last Quarter.

Rima Hyginus (*Below, top left*) (map 34) *col. 357°*

One of the unique lunar formations. The 220 km long rille is divided by the crater Hyginus into two sections of different length. The average width of the rille is 2–3 km, its depth is only a few hundred metres. The rille in some places changes into a line of linked crater pits. A small telescope shows just a simple rille. Sunrise is almost one day before First Quarter, sunset occurs one day before Last Quarter.

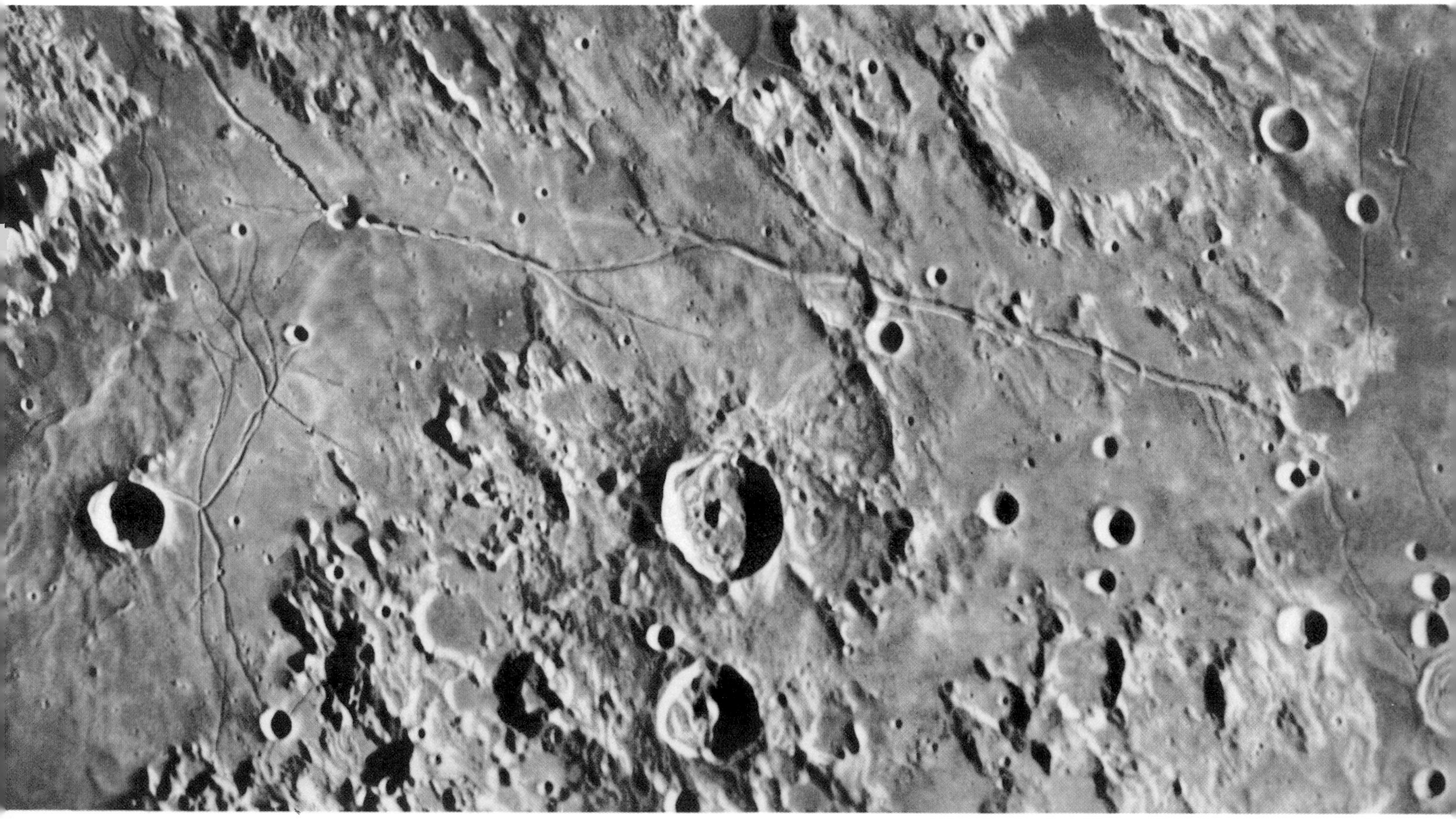

24. **Rima Ariadaeus** (*Above, top right*) (map 34) *col. 351°*

This rille, 220 km long, assumes the form of a shallow valley with a flat floor, 3–5 km wide. The valley is in many places interrupted by mountain ridges intruding from the surrounding terrain. Along the western extension of Rima Ariadaeus a narrow rille branches off to join the rille Rima Hyginus. Close to the terminator, Rima Ariadaeus is easily observable through small telescopes. Sunrise is one day before First Quarter, sunset one day before Last Quarter.

25. **Rimae Hippalus** (maps 52, 53) *col. 40°*

A system of wide, easily observable rilles which, in concentric arcs, follow the eastern edge of the Mare Humorum basin. The greatest span of the system is about 240 km from north to south. The westernmost rille crosses the lava floor of the flooded crater Hippalus. The rille on the extreme east of this system is interrupted by the regular circular craters Campanus A (11 km in diameter) and Agatharchides A (16 km in diameter). Sunrise and sunset occur 2.5 days after First Quarter and Last Quarter, respectively.

26. **'Hyperbolas' near Cauchy** (map 36) *col. 134°*

An extraordinarily interesting landscape. The rilles and faults near the crater Cauchy have an appearance reminiscent of the two branches of a hyperbola. To the north of the crater there are two rilles, Rimae Cauchy, which link up to produce a total length of 210 km. South of Cauchy, on the eastern edge of the Mare Tranquillitatis, a rille proceeds to the north-west, where, in the vicinity of Cauchy C, it continues as the narrow escarpment Rupes Cauchy, 120 km long. If illuminated by the rising Sun (4 days after New Moon) it casts a narrow shadow, while under the setting Sun (4 days after Full Moon) the brightly illuminated slope of the fault can be seen. South of this fault there are two large lunar domes.

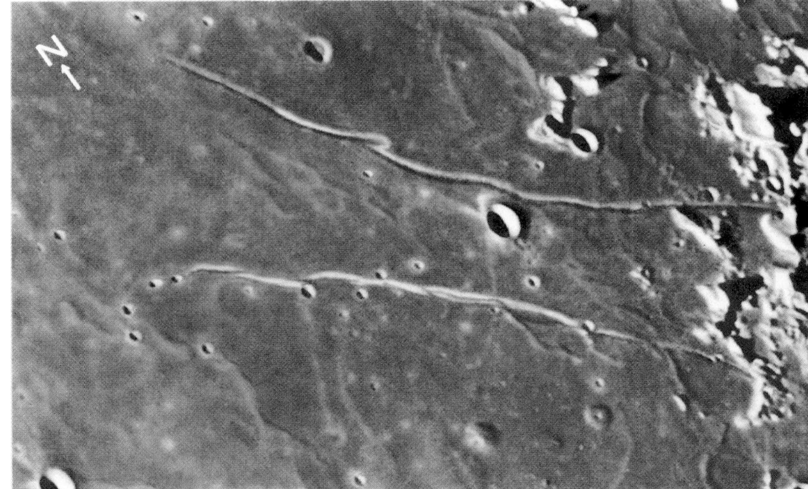

27. **Rupes Recta** (map 54) *col. 20°*

The best known fault on the Moon, easily observable even through a small telescope. The length of the fault is 110 km, its height 240–300 m, and its apparent width 2.5 km. Thus, it is not a steep scarp but a moderate slope. When illuminated by the rising Sun (less than a day after First Quarter) it casts a striking shadow. Before sunset (shortly after Last Quarter) the illuminated slope of the fault shines brightly. Between the small craters Birt E and Birt F, runs Rima Birt, 1.5 km wide and 50 km long.

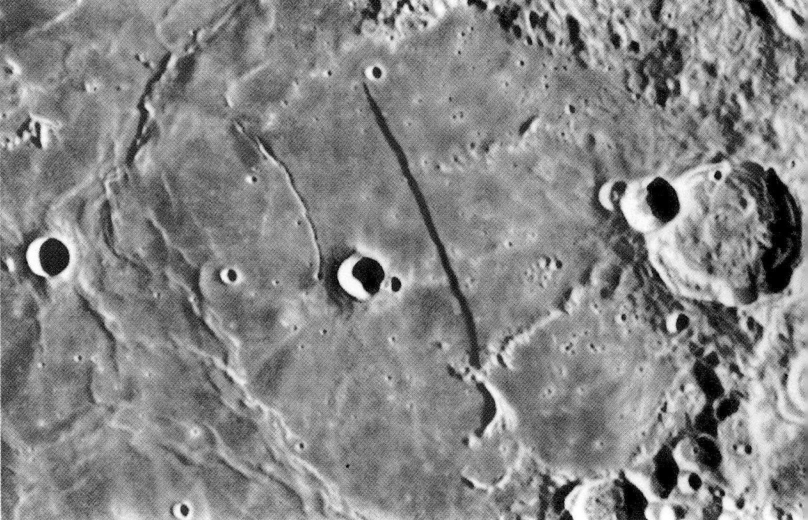

28. **Mons Gruithuisen Gamma** (map 9) *col. 52°*

A mountain massif in the form of a lunar dome with a circular base 20 km in diameter. The foreshortening effect in this area gives the formation the appearance of an upturned bathtub. There is a summit craterlet 900 m in diameter, which is a suitable test object for larger telescopes. Sunrise is 3 days after First Quarter, sunset 3 days after Last Quarter.

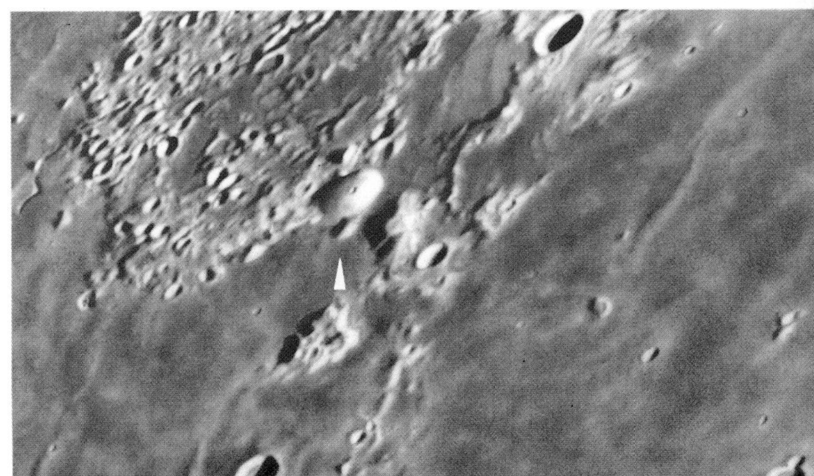

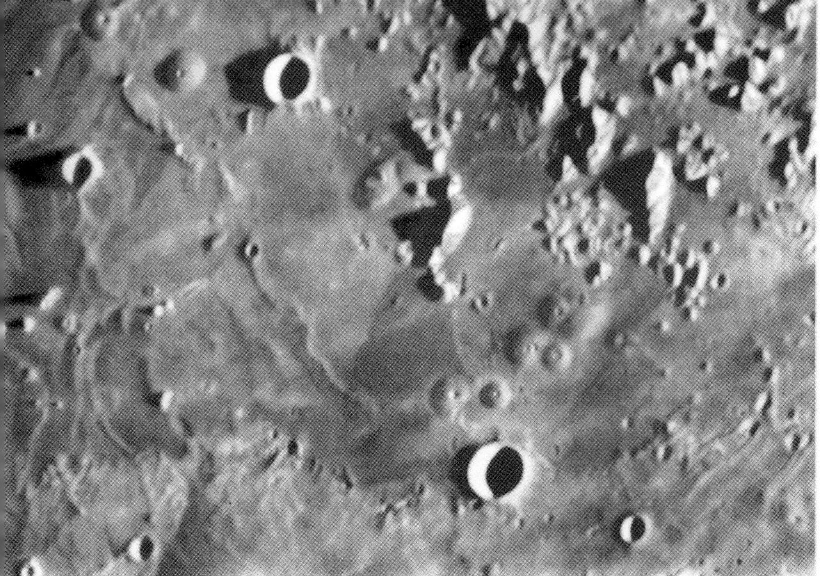

29. **Domes near Hortensius and Milichius** (map 30)

col. 34°

One of the best-known groups of lunar domes, north of Hortensius, and a solitary dome to the west of Milichius. The domes have approximately circular bases, 10–12 km in diameter, and are 300–400 m high. The majority have summit craterlets, about 1 km in diameter. Domes are manifestations of lunar volcanism. They are observable only when close to the terminator. Sunrise is 2.5 days after First Quarter, sunset is 2.5 days after Last Quarter.

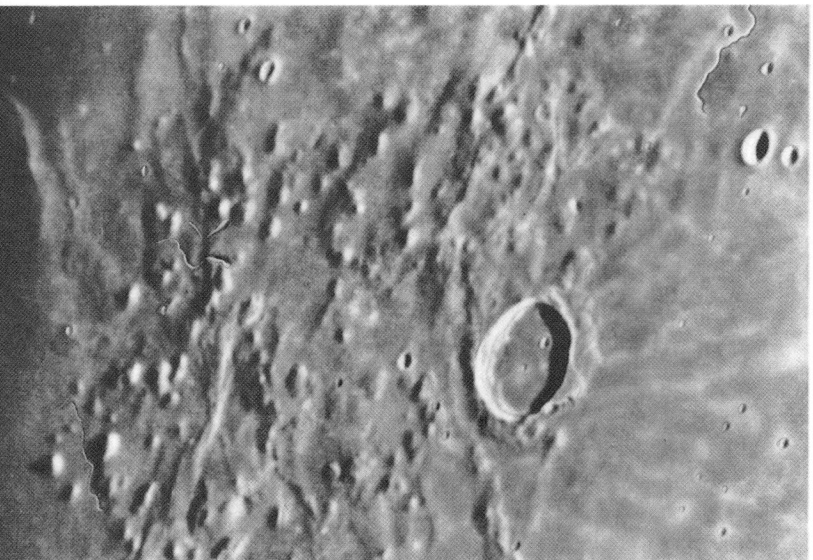

30. **Domes near Marius** (map 29)

col. 61°

An extensive field of low domes in the vicinity of Marius (41 km in diameter) is an apparent manifestation of former volcanic activity in this part of Oceanus Procellarum. This area was one of the selected (but never used) landing sites for the *Apollo* missions. The domes are observable only when close to the terminator, i.e. at sunrise (4 days after First Quarter) or sunset (4 days after Last Quarter).

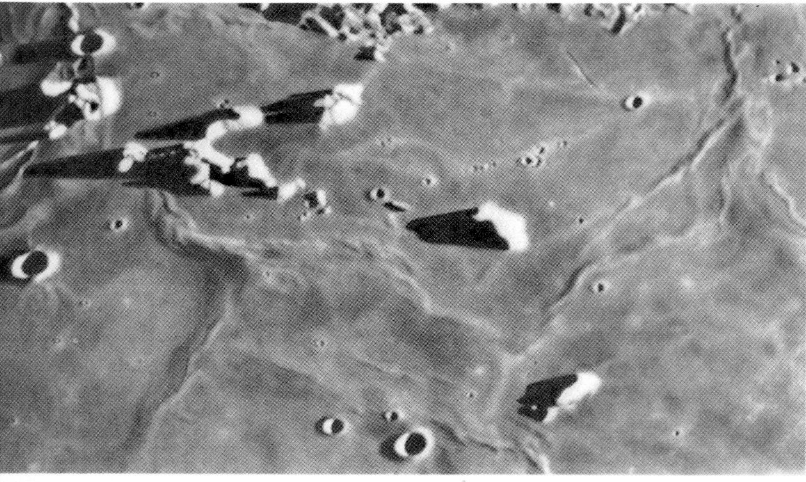

31. **Mons Pico and Montes Teneriffe** (map 11)

col. 20°

Mons Pico is an isolated mountain with a base area of 15×25 km and a height of 2400 m. The peaks of Montes Teneriffe also reach a height of 2400 m above the surrounding surface of Mare Imbrium. In the neighbourhood there are prominent wrinkle ridges. Sunrise is 1 day after First Quarter, sunset 1 day after Last Quarter.

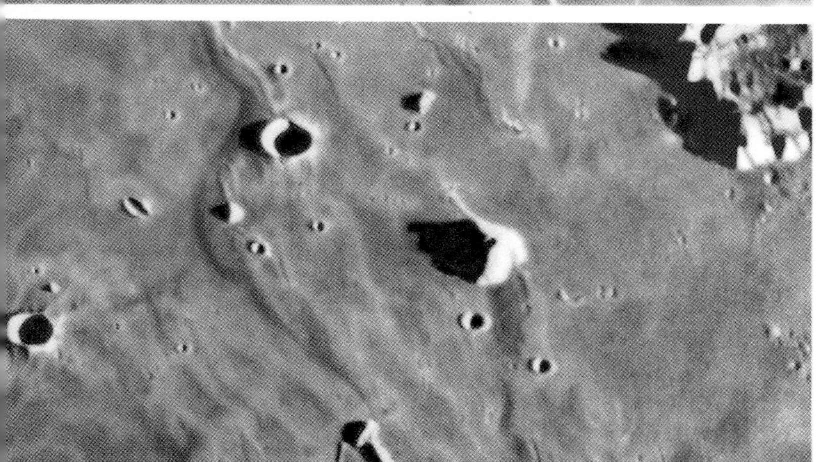

32. **Mons Piton** (map 12)

col. 10°

A characteristic isolated mountain massif reaching a height of 2250 m above the surface of Mare Imbrium. The diameter at the base of the massif (25 km) is eleven times its height. Thus, it is a relatively flat formation with gentle slopes, despite the shadows cast under oblique illumination, which give a false impression of a very steep mountain. Sunrise and sunset take place at the quarter phases.

33. **Montes Alpes** (maps 4, 12) *col. 10°*

A typical lunar mountain range composed of separate mountainous areas with peaks rising about 1800–2400 m above the adjacent Mare Imbrium. The highest mountain, Mons Blanc, is 3600 m high. The lunar Alps offer dramatic views, especially at sunrise (at First Quarter) when the long tapering shadows give a completely false impression of jagged, towering peaks.

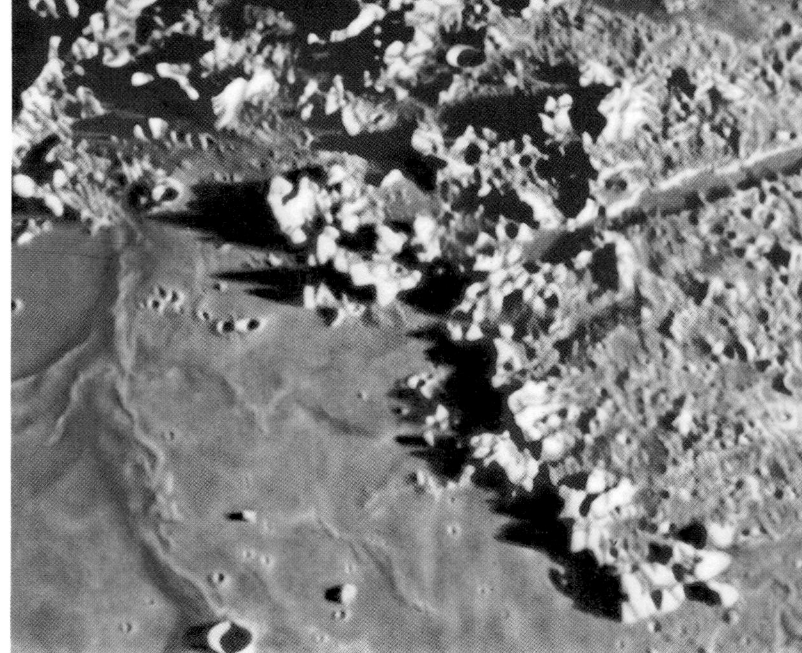

34. **Montes Apenninus, Rima Hadley** (map 22) *col. 156°*

The northern part of the greatest lunar mountain range, which attains a height of 5000 m above the level of adjacent maria. Through a larger telescope, the rille Rima Hadley (width about 1.5 km, depth 300 m), in the vicinity of which the *Apollo 15* mission landed, can be distinguished. The Apennines are a magnificent sight at sunset (at Last Quarter) when their western slopes are illuminated by the Sun.

35. **Rupes Altai** (map 57) *col. 351°*

The Altai fault, formerly called the Altai Scarp, is a substantial remnant of the concentric 'walls' surrounding the Mare Nectaris basin. The length of the arc-shaped fault is 480 km. The slopes descending into the basin shine brightly under the rising Sun (2–3 days before First Quarter) and, vice versa, they disappear into their own shadows under the setting Sun (2–3 days before Last Quarter).

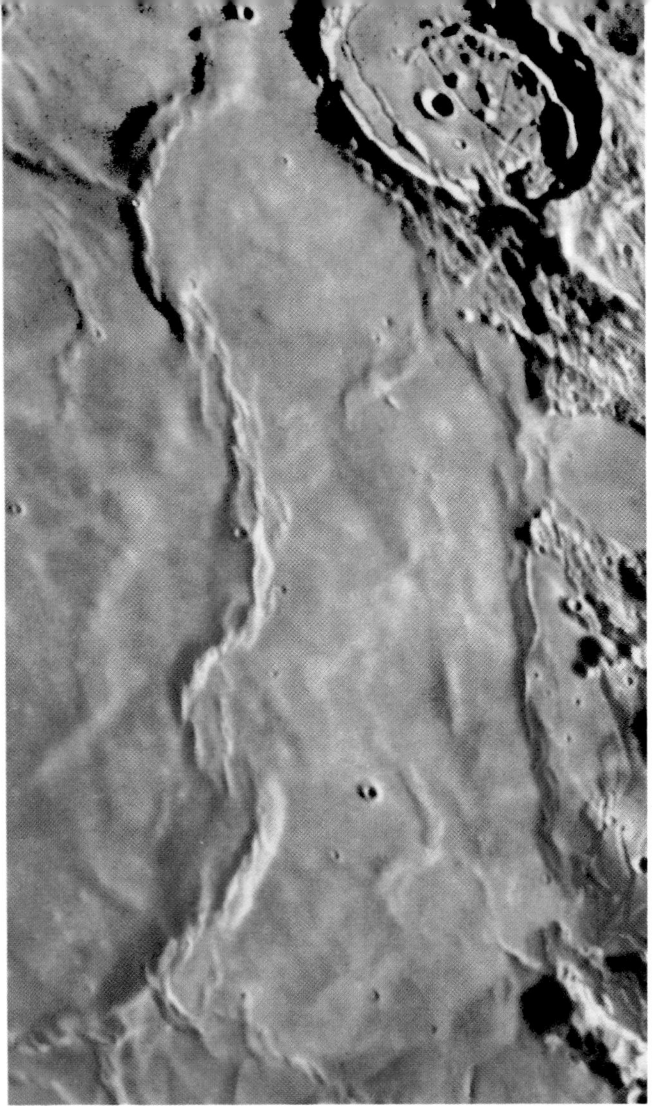

36. **Lacus Mortis** (map 14) *col. 340°*

Lacus Mortis occupies the floor of a considerably eroded walled plain of approximately 150 km diameter. The remains of its walls are preserved, mostly in the west. The observer's attention will be attracted by the crater Bürg (diameter 40 km), which is a centre of wrinkle ridges running to the north and south-west. Other easy objects are two wide rilles from the Rimae Bürg system; the shorter one passes by the southern edge of L. Mortis into a fault which casts a wide shadow at sunrise (2 days before First Quarter) and shines brightly at sunset (2 days before Last Quarter).

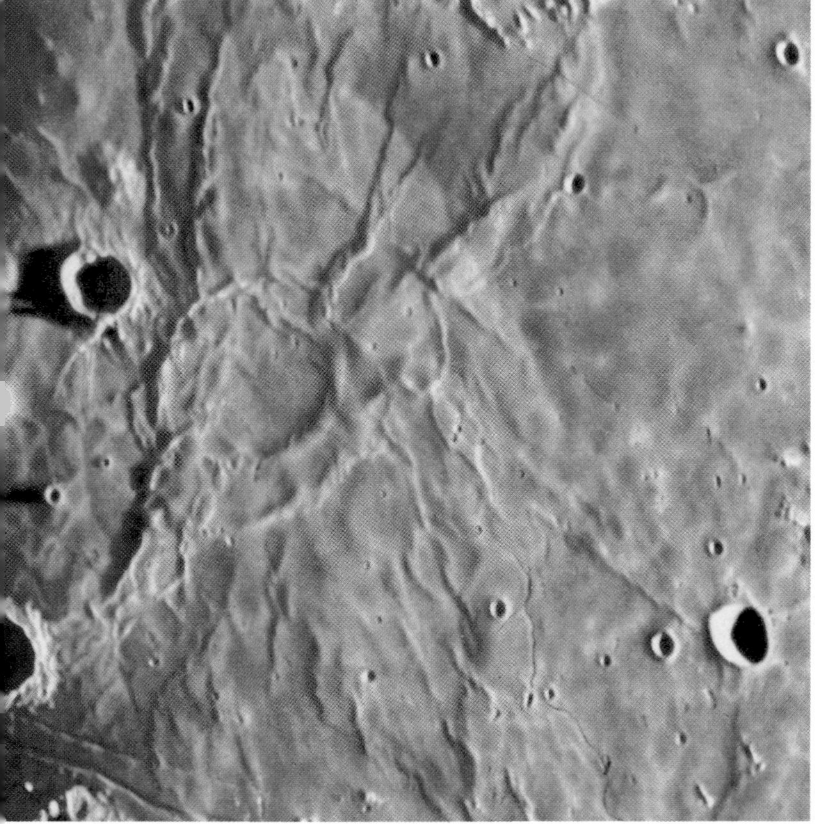

37. **Dorsa Smirnov and Posidonius** (maps 14, 24) *col. 340°*

Parallel to the eastern edge of Mare Serenitatis there is a prominent system of wrinkle ridges. Formerly known as the Serpentine Ridge, its northern part is now called Dorsa Smirnov. The wrinkle ridges are rounded, are of modest height (tens to hundreds of metres) and are observable only when they are close to the terminator. At their highest point, formerly named Posidonius Gamma, lies a small crater about 2 km across. The walled plain Posidonius (95 km in diameter, 2300 m deep) has a complicated system of rilles on its floor. Sunrise is 2 days before First Quarter; sunset takes place 2 days before Last Quarter.

38. **Lamont** (map 35) *col. 340°*

The 'ghost' crater Lamont is a unique formation measuring some 75 km in diameter. The circular walls, which resemble typical wrinkle ridges, range from 5–10 km in width and are up to 200 m in height. An extensive system of wrinkle ridges is associated with Lamont in the western region of Mare Tranquillitatis. Sunrise and sunset occur 2 days before First Quarter and Last Quarter, respectively.

39. **Hesiodus A** (map 54) *col. 40°*

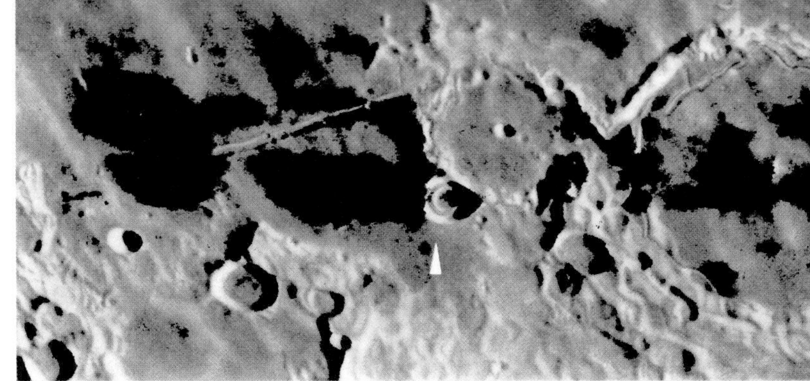

One of the rare examples of a crater with double concentric walls. The diameter of the outer wall of Hesiodus A is 15 km. A similar formation, in an adjacent area, is the crater Marth, 7 km in diameter, which is located in Palus Epidemiarum (map 63). Sunrise is 2–3 days after First Quarter, sunset 2–3 days after Last Quarter.

40. **Wargentin** (map 70) *col. 66°*

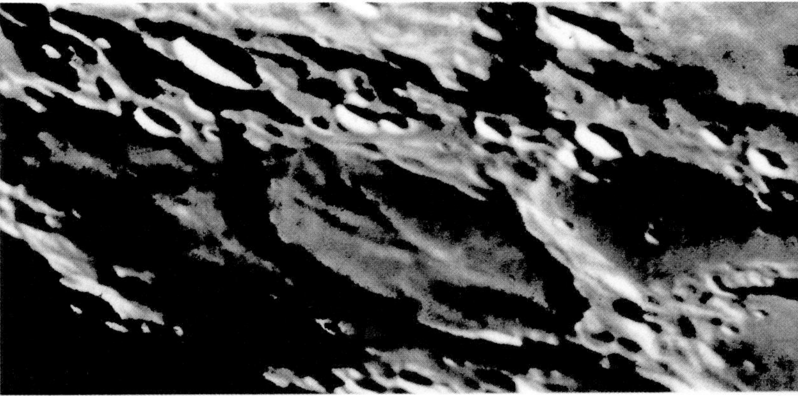

A rarity, breaking the normal rule that a crater's floor always lies below the level of the exterior terrain. The crater Wargentin, 84 km in diameter, is filled up to its brim with solidified lava, forming an elevated plateau with wrinkle ridges on its surface. Sunrise is 2–3 days before Full Moon; sunset 3 days before New Moon.

41. **Linné** (map 23) *col. 156°*

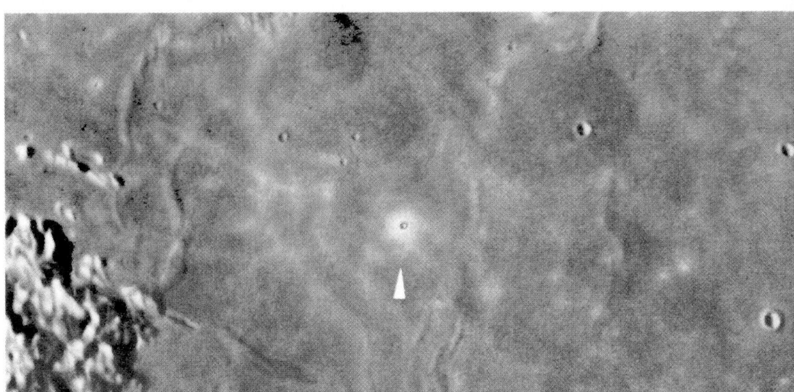

A small, relatively young crater, 2.4 km in diameter, 600 m deep. It is surrounded by bright ejected material; under high illumination it appears through the telescope as a bright spot. In the literature, since the second half of the nineteenth century, numerous mysterious changes and disappearances of Linné have been recorded. These are classic examples of observing errors which occur when lunar details are close to the limit of resolution of a telescope. Sunrise and sunset take place 1 day before First Quarter and Last Quarter, respectively.

42. **Messier and Messier A** (map 48) *col. 118°*

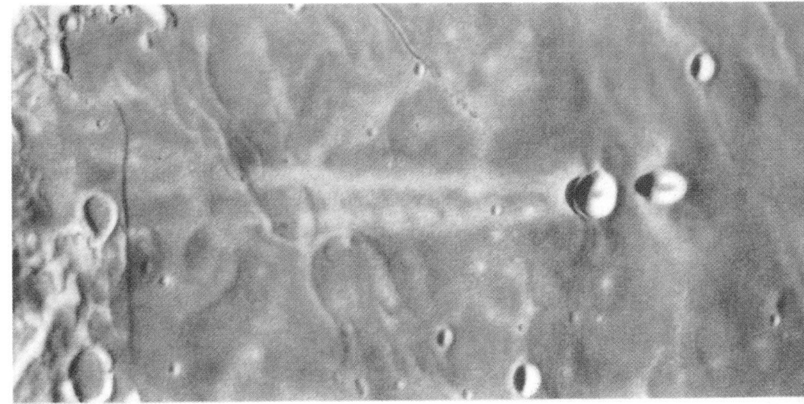

A characteristic pair of craters. Messier is an oval crater elongated in an E–W direction. Its size is 9 × 11 km and it may have originated as the result of an oblique meteorite impact from space. Messier A comprises two circular craters; the younger, eastern crater overlaps part of the smaller, older formation. This twin-crater measures 11 × 13 km. To the west, two straight narrow rays, resembling the tail of a comet, trail across the surface of Mare Fecunditatis to a distance of about 120 km. Sunrise is 3–4 days after New Moon, sunset 3–4 days after Full Moon.

43. **Reiner Gamma** (map 28) *col. 64°*

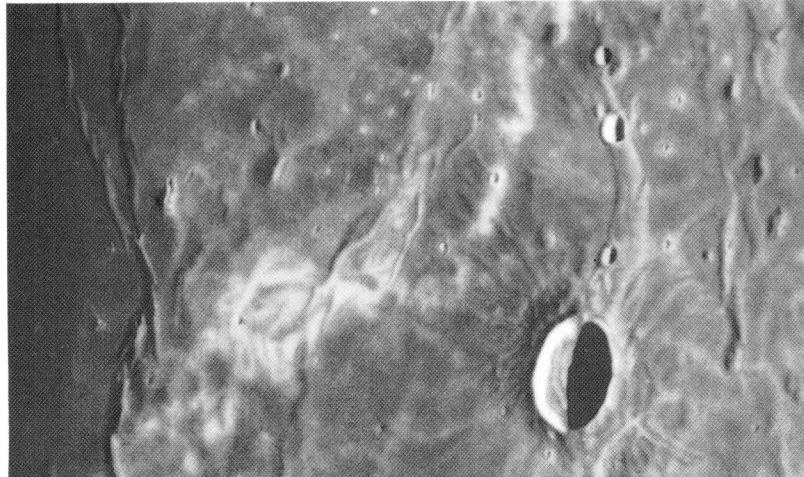

On the near-side of the Moon this is an extraordinary and unique formation in Oceanus Procellarum, some distance west of the crater Reiner. It is formed by extra-bright material on the surface of the mare and even detailed photographs taken by the *Lunar Orbiter* probes did not show any evidence of relief. Sunrise is 5 days after First Quarter, sunset 3 days before New Moon.

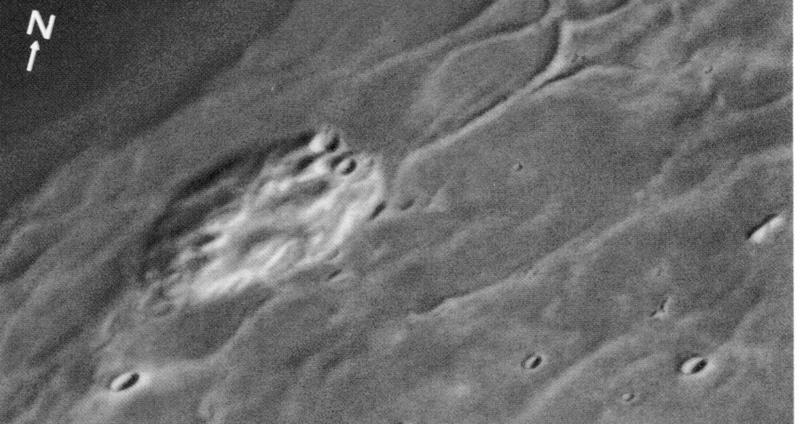

44. **Mons Rümker** (map 8) *col. 64°*

The most extensive complex of domes on the Moon. The diameter of this volcanic formation is about 70 km. Several low, flattish domes add to the irregular shape of this structure and there are dozens of crater pits which are too small to be discerned from Earth. Sunrise is 5 days after First Quarter; sunset occurs 3 days before New Moon.

45. **The 'bridge' over the valley Bullialdus W** (map 53) *col. 198°*

One of the lunar rarities. From the crater Bullialdus (61 km in diameter, 3510 m deep) a shallow valley denoted by the letter W progresses in a north-westerly direction. Approximately 100 km from the crater Bullialdus, the valley is interrupted by a flat, 10 km wide wall which gives the illusion of a 'bridge'. Sunrise: 2 days after First Quarter; sunset: 2 days after Last Quarter.

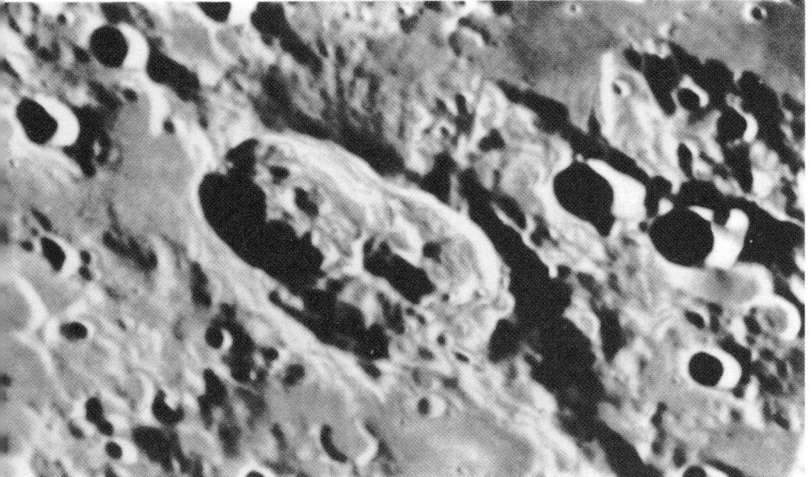

46. **Hainzel – a composite crater** (map 63) *col. 198°*

Three craters overlapping each other to form a single composite formation. The oldest and largest of these is Hainzel (70 km in diameter), Hainzel C is of intermediate age, while the youngest is Hainzel A (53 km in diameter). Sunrise is 3 days after First Quarter, sunset 3 days after Last Quarter.

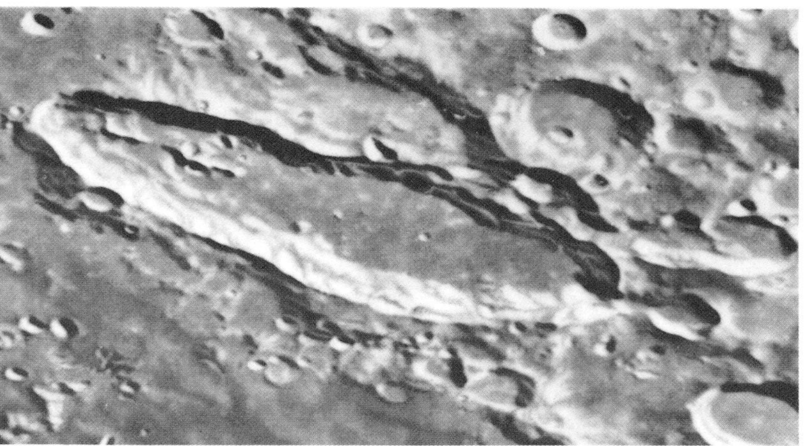

47. **Schiller** (map 71) *col. 61°*

An extraordinarily elongated crater, measuring 179 × 71 km, the shape of its wall resembles a footprint. The floor of the crater is smooth, as if flooded with lava. Non-circular formations have complicated structural histories and Schiller, by virtue of its size, is the most remarkable example on the Moon. Sunrise: 3 days after First Quarter; sunset: 3 days after Last Quarter.

48. **Bailly** (map 71) *col. 81°*

The largest walled plain on the near-side of the Moon, diameter over 303 km. Such a large formation is classified as a lunar basin. It has an uneven floor with numerous craters. Sunrise is 3 days before Full Moon, sunset 3 days before New Moon.

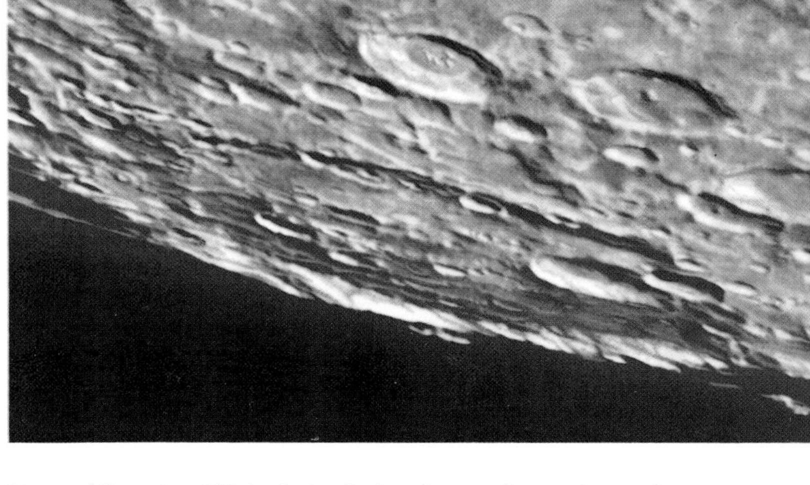

49. **Mare Marginis** (maps 27, 38, III)

(Left), col. 329° *(Right), col. 50°*

Mare Marginis is of elongated shape in a SE–NW direction. It is observable east of Mare Crisium at favourable librations. The undistorted shape of M. Marginis can be seen in the right-hand photograph, which is from *Apollo 16.* The large crater Neper (diameter 137 km) dominates the southern shore of Mare Marginis. Sunrise: shortly after New Moon; sunset follows Full Moon.

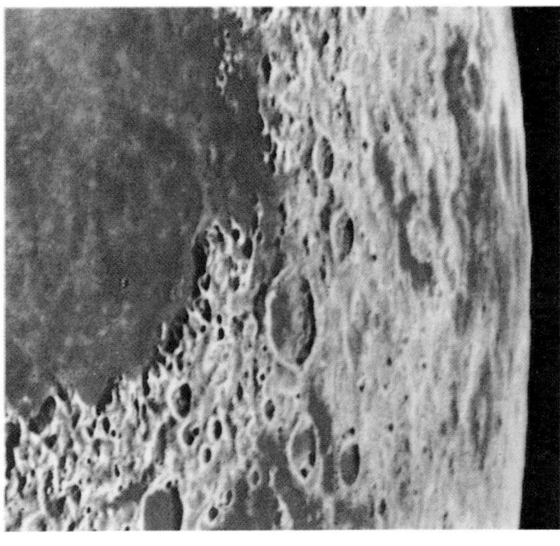

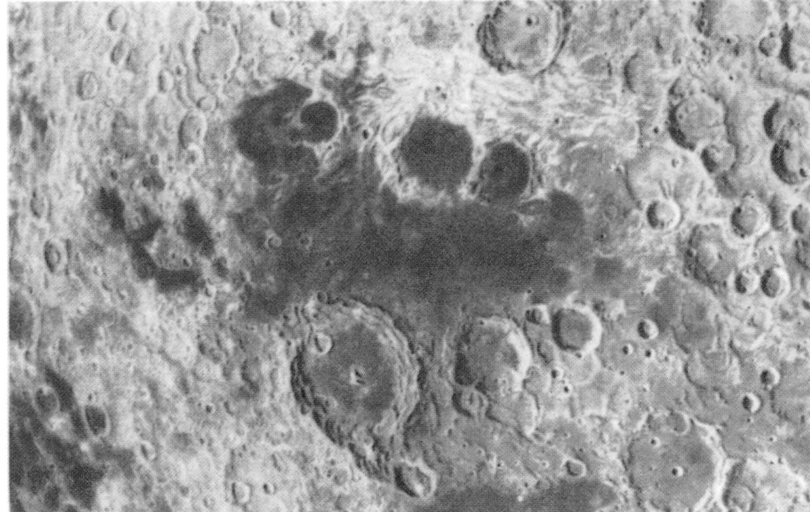

50. **Mare Crisium** (maps 26, 27, 37, 38)

(Left), col 335° *(Right), col. 329°*

A circular structure resembling a giant crater, 570 km in diameter. It is a typical lunar basin of impact origin. In the distant past the floor was submerged under thick layers of lava. These cover deeper and denser material that causes an anomaly in the local gravitational field (as detected by lunar orbiting space probes). These mass-concentrations are called 'mascons'. Numerous wrinkle ridges interrupt the smooth surface of the mare which, being close to the edge of the lunar disk, appears as an ellipse with its major axis running N–S. The shape of this ellipse changes dramatically, especially at the extremes of longitudinal libration. The actual, undistorted shape of the mare is elongated in an E–W direction as shown in the right-hand photograph. Sunrise is 2–4 days after New Moon; sunset occurs 2–4 days after Full Moon.

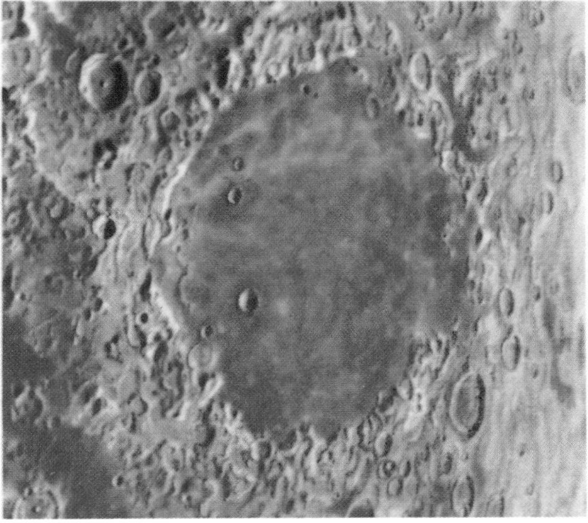

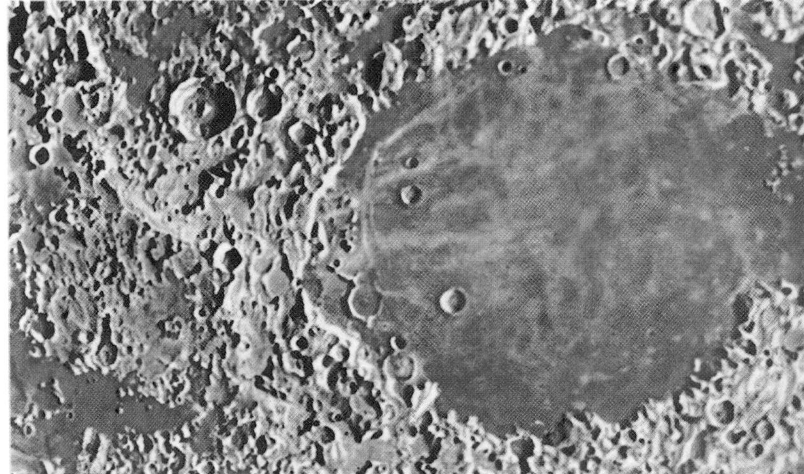

Observation of the Moon

There is no other celestial body that offers such a wealth of detail, changes and opportunities for observation. Initially, let us try to forecast the conditions of visibility: for example, will the Moon be visible this evening, and if so, in which direction will it lie relative to the Sun and what will be its altitude above the horizon? Let us consult the calendar to find the next principal lunar phase. From this we can determine the respective positions of the Sun and Moon, and hazard a guess as to where the Moon will be at a certain hour, and what appearance it will exhibit. We can then verify our assessment by going outside to look at the sky.

For a start, the lunar observer must familiarize himself with the names, shapes and positions of the lunar maria. These duller areas can be distinguished with the naked eye, but a pair of binoculars or small telescope will improve the view and allow positive identifications to be made by referring to the map (Fig. 14). The best views are obtained with a pair of binoculars that is mounted on a fixed tripod; if a tripod is not available, the binoculars can be steadied against a window frame or tree, etc. A typical pair of binoculars will have a magnification of about 7× or 10×, which is probably enough for a hand-held instrument. In the absence of a high-magnification telescope, it is possible to adapt a pair of binoculars by using an additional eyepiece/s to increase the magnification. Suitable eyepieces (or oculars, as they are sometimes called) are those with focal lengths of 15–30 mm and these must be placed some 10 or 15 cm behind the normal binocular eyepice. This of course involves the use of a cardboard or metal tube in which to mount the eyepieces so as to ensure correct alignment of the optics. Experimenting in this way it should be possible to obtain magnifications of 30× to 50× provided that the instrument is properly mounted for stability. Why not try it?

A pair of binoculars is perfectly adequate for getting one's bearings on the Moon and for learning the basic nomenclature. However, for more detailed observations an astronomical telescope is a necessity. This can take the form of a small, portable instrument, or a larger, fixed telescope on a permanent mounting under a dome or perhaps in some sort of housing with a sliding roof. The smaller telescopes are usually refractors, which employ an objective lens and have a choice of eyepieces. Reflecting telescopes, as their name implies, contain metal-coated parabolic mirrors, which range in size upwards from about 15 cm diameter – some of those used by amateur astronomers are very large instruments. A third type of telescope is the catadioptric, which combines both lenses and mirrors in the pursuit of a good quality image; these are very compact and portable. The choice of telescope depends on the requirements of the observer, who must decide which type to make or purchase. If he or she decides to make a telescope, reference should be made to the specialized books and publications devoted to the grinding, polishing and mounting of astronomical telescope mirrors. Here, however, we assume that the observer has a telescope and is preparing to use it to observe the Moon.

Often the new owner of a telescope seems to be obsessed with its magnifying power, as if this alone determines the excellence of the optical system. This is rather like judging the performance of a car by the maximum possible speed shown on the dial of the speedometer! Needless to say, magnifying power is important, but, even more so is the *resolving power* of the telescope, which determines how much fine detail can be discerned on such objects as the Moon and planets. This is the main question to be addressed.

The resolving power of a telescope can be determined by measuring the minimum angular distance between two points that can be separately seen – the limit is reached when the images of the two points coalesce into one. This is expressed in radian seconds (r'') and can be calculated from the simple formula $r'' = 120''/d$, where r'' is the resolving power and d is the diameter of the objective or mirror in millimetres. For example, an objective of 80 mm diameter should have a resolving power of 1.5″.

From the Earth, a lunar crater with a diameter of about 1.9 km may be discerned using a telescope with a resolution of 1″ (sometimes called one second of arc), so, theoretically, we ought to be able to distinguish a 3 km diameter lunar crater with an 80 mm objective. This assumes, of course, that our telescope has a high-quality objective and is being used under ideal conditions, when the atmosphere is clear and still. The actual quality of the atmospheric 'seeing' conditions can be tested empirically by using as test objects small lunar craters, of which dozens are listed in the descriptions accompanying the 76 sections of the map of the Moon.

It is also worth noting that the visibility of detail depends not only on the diameter of the objective or mirror, but also on its type, of which there are several. A very important property of

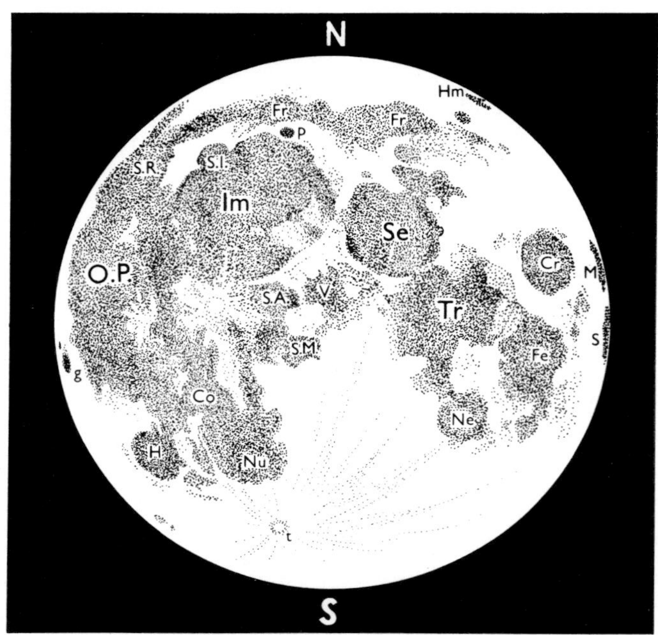

Fig. 14. *A map of the lunar maria and craters visible with the naked eye or using a pair of binoculars. Key:* **N** *= north,* **S** *= south; Co = Mare Cognitum (Known Sea), Cr = Mare Crisium (Sea of Crises), Fe = Mare Fecunditatis (Sea of Fertility), Fr = Mare Frigoris (Sea of Cold), H = Mare Humorum (Sea of Moisture), Hm = Mare Humboldtianum (Humboldt's Sea), Im = Mare Imbrium (Sea of Rains), M = Mare Marginis (Border Sea), Ne = Mare Nectaris (Sea of Nectar), Nu = Mare Nubium (Sea of Clouds), O.P. = Oceanus Procellarum (Ocean of Storms), S = Mare Smythii (Smyth's Sea), Se = Mare Serenitatis (Sea of Serenity), S.A. = Sinus Æstuum (Bay of Billows), S.I. = Sinus Iridum (Bay of Rainbows), S.M. = Sinus Medii (Central Bay), S.R. = Sinus Roris (Bay of Dew), Tr = Mare Tranquillitatis (Sea of Tranquillity), V = Mare Vaporum (Sea of Vapours); g = Grimaldi (crater), p = Plato (crater), t = Tycho (crater).*

a lens (or mirror) is its 'speed', which is defined as the ratio of its diameter to its focal length. (For example, a lens with a diameter of 80 mm and a focal length of 1200 mm has a *focal ratio* or speed of f/15). For lunar and planetary work the most useful telescopes are those with focal ratios of f/12 to f/15. The type of objective and its focal ratio, together with the design of a chosen eyepiece (of which, again, there are many types) all influence the contrast, colour rendition and definiton, which, combined, yield the overall clarity of image.

And what role is played by the magnifying power in all this? Certainly it does not affect the theoretical resolution of a telescope, but by choosing a suitable eyepiece, the performance of the instrument can be optimized to suit prevailing atmospheric conditions. Telescopic magnification, z, is simply the ratio $F{:}f$, where F is the focal length of the objective or mirror, and f the focal length of the eyepiece expressed in the same units. For example, if our telescope has an objective with a focal length of 1 m and an eyepiece with a focal length of 10 mm, then the magnification $z = 1000/10 = 100\times$. From this we can see that magnification depends on focal length only and has no relevance to the diameters of the optical elements.

Common sense suggests that there will be a considerable difference between the views of the Moon obtained through two telescopes using identical magnification but with objectives or mirrors of different diameter. This is true, since more details will be revealed by the instrument with the larger aperture (always assuming, of course, that the quality of the optical components in both telescopes is beyond reproach). For every telescope objective or mirror there is an optimum magnification that can be applied to allow the telescope to attain its maximum performance in the rendering of fine detail. This property is sometimes known as useful magnification and, for smaller instruments, is approximately equal to the diameter of the objective or mirror expressed in millimetres. For example, a telescope with an 80 mm diameter objective has a useful magnification of approximately 80×. It goes without saying that we can employ eyepieces of shorter focal length to increase the magnification to anything we like: but, if we do this, no further detail will be seen and no additional information will be obtained. This situation has its parallel in television when we compare the performance of small- and large-screen receivers tuned to the same station. In the case of a telescope, limitations are imposed by the agitation and turbulence of the Earth's atmosphere; and these effects are amplified by magnification, so that an observer often has little choice but to reduce the magnification to a level below that of which the instrument is capable. Experience shows that about two-thirds of bright starry nights cannot be used for observation of the finest details because of this problem.

Up to now we have considered only the optical parameters, and no matter how good these are, all will be lost if the telescope is inadequately mounted. For a small telescope, the basic requirement is a solid tripod, or some other vibrationless support, which enables the instrument to be rotated around two axes that are perpendicular to each other. It should then be possible to point the telescope to any chosen part of the sky without introducing excessive vibration. Beware of portable camera tripods – they are worse than useless; and even some of the tripods that are supplied with telescopes shake like aspen leaves at the lightest touch or in the slightest breeze. The stability of a telescope mounting is obviously improved by increasing its mass (which decreases its portability). This is a desirable property of the equatorial or 'parallactic' mounting, on which a camera or telescope can be fixed and, driven by a small motor, made to cancel the Earth's rotation and thus follow the stars automatically. The equatorial mounting is necessary – some might say essential – for mounting medium-sized and larger instruments, which are used for detailed observation or photography.

Probably the newcomer to astronomy would prefer to select a telescope that is 'universal', in the sense that it can be used for observing the Moon, planets, stars, etc., and not too expensive. A suitable choice might be a small refractor with an objective 60–80 mm in diameter, working at a focal ratio of about f/10. Some improvement in performance could be expected from a reflecting telescope with a mirror 100–150 mm in diameter. More advanced observers of the Moon and planets might prefer a more expensive, medium-sized refractor with an aperture larger than 100 mm and with a focal ratio in the region of f/12 to f/15. Other options include larger reflectors or catadioptric telescopes with an aperture greater than 150 mm. Whichever instrument is chosen it should be mounted in a position from which it has an unobscured view of the sky, and which is as far away as possible from man-made turbulence from chimneys, etc. It is ironic that on cold nights we can observe the Moon with the naked eye while sitting comfortably beside the window of a warm room. Regrettably, this cannot be done with a telescope!

Theoretically, a high-quality telescope on a stable mounting, under favourable observing conditions, ought to reveal details close to the limit of resolution of the telescope. But this depends on the observer's experience and how he or she interprets the incoming image, which is why it is necessary to train the eye so that it is not deceived. Very often a visitor to an observatory will look at an object through a telescope and imagine that he or she is making an observation; frequently, such a person is disappointed by what they think they see, and how different it is from the miraculous accounts and magnificent photographs that are published in the astronomical literature. Such a visitor does not realize that observing is an art that has to be learned and exercised. Disappointments, with persistence, lead to success!

A trained eye viewing through even a small telescope can see a vast amount of detail, and an experienced observer will use just enough magnification to clarify it. Excessive magnification decreases the brightness of the image and increases the shimmering effect, which degrades the image. Even with a magnification of 30–40 times, an experienced eye will be able to detect fine detail if the observer is familiar with the appearance of the object from his or her former observations, and from the literature, photographs and maps.

For a start, let us concentrate on some of the named formations, which are shown on the map. Observing them is, in itself, an absorbing and long-term programme involving a step-by-step scrutiny of the Moon's surface during different phases, including Full Moon. At Full Moon a telescopic exploration of the Moon's face is not an end in itself, bearing in mind that the absence of shadows removes all evidence of vertical relief; it is, however, the time when lunar eclipses occur and it can be rewarding to observe the contacts of small bright craters with the Earth's shadow as it moves across the lunar disk (see p. 25). If we want to test the performance of our telescope and, at the same time, improve our observational prowess, we can resort to the methods used by the classical selenographers –pencil and paper – and attempt to draw the detail we see as faithfully as possible. How should we proceed? Well, first we must assemble the usual drawing implements: paper, pencils (HB or softer), eraser and a clip-board. In addition we shall require a small lamp to illuminate the paper, without dazzling the observer. We could commence drawing on a blank sheet of paper, but life can be made easier if an outline of the chosen formation is sketched in first by reference to a photograph or map. Such a procedure saves time, for we can then focus our attention on the actual appearance of the formation under the particular angle of illumination and record the shapes of shadows, the distribution of lighter and darker tones, and any other details. More experienced observers sometimes attempt to mark tiny details directly onto a prepared enlarged photograph or map of the area under scrutiny. As a general rule, it is best to restrict our detailed drawings to small areas that can be depicted within a reasonable time, say 10–30 minutes. If the work takes too long, not only does the eye get tired but the appearance of the lunar landscape changes, particularly under oblique illumination. For example, it is impossible to draw accurately, in such a short time, a crater such as Copernicus, which is full of minute and complicated detail. The scale of a drawing depends on the magnification used. For example, the famous selenographer P. Fauth recommended a scale of about 1 : 2 000 000 (i.e. 1 mm on the drawing corresponds to 2 km on the Moon) using a telescopic magnification of 150×, or a scale of about 1 : 1 000 000 with a magnification of 300×, etc. To have any value, each drawing must be inscribed with the date and time of the observation, the type of telescope used, the magnifying power, and an evaluation of the atmospheric 'seeing' conditions.

The section 'Fifty views of the Moon' provides plenty of suggestions for regular observations of interesting formations and regions under varying angles of illumination. Often substantial changes in the appearance of lunar formations occur, sometimes leading to apparent invisibility and at other times emphasizing their conspicuousness. The period when the area under observation will be close to the terminator can be determined beforehand by referring to the co-longitude tables (p. 212) or by consulting an astronomical almanac. Every part of the lunar surface is crossed by a morning or evening terminator 25 times per year. Usually the illumination of a given formation is such that fine detail may be observed for perhaps two days or slightly longer, so this equates to a possible 50 observational opportunities per year. However, of these, about two consecutive thirds cannot be exploited because of bad weather, and about one-third of the remaining 16 nights are subject to acute atmospheric turbulence. Disregarding five nights of average, indifferent 'seeing', we are left with five opportunities per year to observe the selected area under excellent or at least good conditions! Perhaps it would be too pessimistic to mention that many years could elapse before an observer were able to repeat his or her observation of a given area under almost identical phase, libration and excellent seeing conditions; this is why all systematic observations are unique and valuable.

Up to now we have been considering visual observation only. In the historical study of the Moon, however, it is undoubtedly the application of photography that has played the decisive role in producing a precise and objective representation of the Moon's surface. The Moon has also become a popular target for the photographically inclined amateur astronomer. Simple everyday types of cameras can take impressive black-and-white or colour photographs of the Moon above a terrestrial landscape, but with the usual photographic lens (e.g. $f = 50$ mm) the size of the lunar image is only 0.5 mm, which is hardly sufficient to resolve the lunar phase. Far more useful are telephoto lenses, with their longer focal lengths. As a general rule, we can calculate the size of the lunar image, d, in the camera by dividing the image-forming focal length in centimetres by 115. For example, for a telephoto lens of focal length $f = 13.5$ cm, the image size $d = 0.12$ cm or 1.2 mm; with an even longer focal-length lens of, say, $f = 30$ cm, the image size $d = 2.6$ mm, which is sufficient to reveal the phase and the shapes of the lunar maria.

It is obvious from the above that something more than the usual type of photographic equipment has to be used in order to obtain detailed images of the Moon. The answer is to use the telescope's long-focus objective lens or mirror as a camera lens and to attach a camera, minus its lens, to the tube that normally houses the eyepiece (Fig. 15a). In this way the telescopic lunar image passes through the open camera shutter and falls onto the photographic emulsion. In the past, focusing was an awkward problem, but this has now been solved by using the modern 35 mm single-lens-reflex camera, which, without its lens, is usually attached by a special adaptor to the point where the eyepiece would otherwise be and which can be racked in and out until the best focus has been found. It also allows the lunar photographer to satisfy himself or herself that

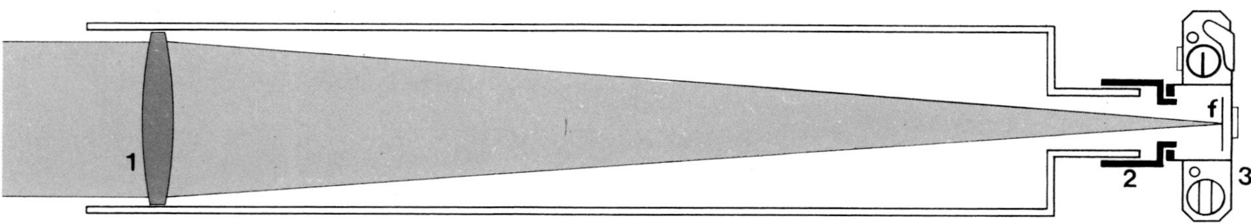

Fig. 15a. The scheme for photographing in the focal plane of the objective lens of a telescope. 1 = objective lens of telescope, 2 = tube for attaching the camera body, 3 = camera body (minus lens), f = focal image of the Moon.

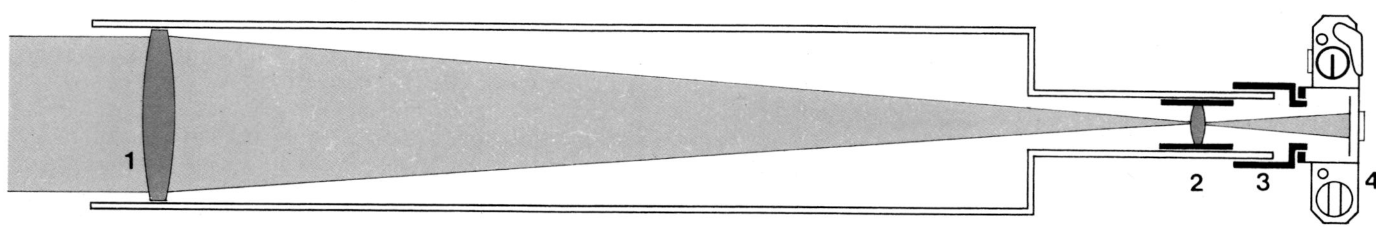

Fig. 15b. The scheme for photographing by eyepiece projection. 1 = objective lens of telescope, 2 = projection eyepiece (or Barlow lens), 3 = tube for attaching the camera body, 4 = camera body (minus lens).

the lunar image will actually fall onto the film when he or she presses the shutter release! If, for example, we have a telescope objective with a focal length of 120 cm, the projected image of the Moon will be approximately 1 cm in diameter on the film and will show the jagged edge of the terminator. However, the image will still not show fine detail and some of the detail will be lost in the grain of the photographic emulsion.

Larger images can be recorded by using a second method, which, in this case, employs the telescope's eyepiece to enlarge the primary image formed by the objective before it enters the camera (Fig. 15b). This may involve the use of special tubes or adaptors to support the eyepiece as well as the single-lens-reflex camera (without its own lens). Focusing is carried out as before. The greater the camera's distance behind the eyepiece, the greater will be the size of the lunar image: it can be several times larger than the image formed in the principal focal plane of the telescope objective. If need be, this can be verified by removing the camera and projecting the image onto a focusing screen or sheet of white paper held successively at, say, 5, 10 and 15 cm behind the eyepiece. Another way to enlarge the image is to employ a negative lens, the so-called 'Barlow lens', instead of the more conventional eyepiece.

Theoretically, there is no limit to how long the combined objective/eyepiece focal length can be, but practical considerations impose severe restrictions, for image size is only bought at the expense of speed or focal ratio. This involves progressively longer exposures and the problems that accompany them: telescope movement, atmospheric turbulence, following the Moon's motion in the sky, etc. Even the action of triggering the shutter can cause vibration, though this problem can be solved by setting the shutter to B or T and triggering it while the objective of the telescope is covered by a piece of black card held a few millimetres away from it. The exposure is made by removing and replacing the card, not forgetting to close the camera shutter! Ideally photographic exposures should be as brief as possible – there is no virtue in using a long exposure if a shorter one will achieve the desired result. With stationary apparatus, i.e. telescopes that are not driven to follow the motion of the Moon or stars, photography has to be done with very short exposures. Exposure times can be reduced by using 'faster' or more sensitive film, but, since the extra sensitivity is accompanied by increased graininess, it is best to choose films of medium sensitivity, i.e. of about ISO 100/21° rating. Exposure times depend on several factors, in particular the effective speed or sensitivity of the optical system, as well as the Moon's phase and altitude above the horizon, the transparency of the atmosphere, and the above-mentioned response of the film. As an illustration, consider the light-gathering power of an f/22 optical system with a film rated at ISO 100/21°. It turns out that suitable exposure times are 1/30 s for the Full Moon, 1/8 s for First and Last Quarter, and 1/2 s for the narrow crescents. In the case of modern single-lens-reflex cameras, which incorporate sensitive photometers, it is sometimes possible to obtain light readings that indicate exposure times as a function of film speed and focal ratio. With this type of equipment it is best to proceed by making an exposure as indicated by the photometer and then to back it up with others taken at haft and twice the recommended exposure. Differences in exposure times can be considerable, depending on which part of the Moon is being photographed: an exposure suitable for the terminator will overexpose the bright areas on the opposite side of the lunar disk, and vice versa.

In lunar photography the most serious problems are caused by atmospheric turbulence – which is the only problem that is beyond the control of the observer. With a motor-driven telescope, the image of the Moon can be seen in a secondary telescope or finder mounted on the larger instrument (with its attached camera). This enables the lunar photographer to choose the moments when the air is steadiest for making his or her exposures. But even this cannot assess the atmospheric conditions that prevail at the precise instant of the exposure. This is why it is preferable to make several exposures and to select the best photograph from the series. However, we have to live with the fact that even the most successful and sharpest photograph cannot record those elusive fine details that can be seen visually through the same telescope. During the period of observation the eye selects those fleeting moments when the steadiness of the atmosphere allows remarkable detail to be seen. In this respect the human eye is superior to the photographic emulsion, in which blurring naturally occurs, owing to the manner of its construction. An experienced observer is able to detect visually, with a telescope 10–15 cm in diameter, minute details that are capable of being photographed only with very much larger instruments. A 20–30 cm diameter telescope can reveal visually details that can be photographed only through the world's most powerful telescopes.

Now that the Moon has been mapped in detail, following the fantastic successes of the unmanned lunar probes and *Apollo* lunar missions, it might seem to the layman that there is little point in continuing to observe the Moon from the Earth. Nothing could be further from the truth, for much remains to be cleared up. As a rule, the photographs taken by the probes show the lunar surface under only one angle of illumination, while a whole succession of angles of illumination, both east and west, are really required to reveal the detail that is actually there. Systematic observation of the terminator has not been accomplished by the lunar probes, so we still have much to learn about the surface structure of the maria, including low wrinkle ridges, flat domes and other rounded convexities that show up only under grazing or oblique illumination. Such observations can be conducted from the Earth and they may result in new discoveries.

Since 1783 numerous lunar observers have occasionally reported that they have witnessed 'mysterious happenings' on the Moon. So far about 1200 sightings of what are now called 'transient lunar phenomena' (TLPs) have been recorded, but any observer wishing to search for such events will soon find that it is an exacting and problematical undertaking. Phenomena reported include local hazes, temporary colour variations (sometimes red), localized brightness changes (glows), temporary obscurations of surface formations, etc. The reliability of many of these reports is questionable, for the eye, working under extreme conditions, is easily deceived. Other TLP observations seem to be authentic, especially when they have been confirmed by other experienced observers. One of these was Kozyrev's 1958 observation of a gaseous emission from the central peak of the crater Alphonsus, for which he obtained spectrographic evidence. No TLPs were reported by the *Apollo* lunar astronauts. It appears that these events are confined to certain regions; for example, about 300 TLPs have been observed in the environs of the crater Aristarchus, over 70 near the crater Plato, and 25 in and around Alphonsus. Some phenomena have been seen in the peripheral areas of the maria. On-the-spot evidence comes from the results of an *Apollo* experiment in which sensitive instruments detected the emission of the radioactive gas radon in the neighbourhood of the crater Aristarchus and along the edges of some of the circular maria.

The causes of transient lunar phenomena are unknown, but although lunar volcanism ceased in the remote past (see p. 16) and the present seismic activity is negligible, one must conclude that the Moon is not a completely dead world. If parts of the Moon's interior are still in a molten state, it would seem reasonable to expect the occasional escape of gases, or mixtures of gas and dust, from fissures near the surface. Could these be the events that produce the transient obscurations of surface detail? Here, then, is a field of study that demands much time, perseverance, dedication, a thorough knowledge of the lunar surface, and, from the practical point of view, a large, well-mounted, high-quality telescope.

The direct exploration of the Moon by unmanned lunar probes and the *Apollo* lunar astronauts has led to the introduction of new methods of observation. The chemical composition, mechanics, structure and other properties of lunar surface materials have been analysed on the Moon and in terrestrial laboratories. The sites from which these materials were taken have been identified and can be compared visually with other areas that have not yet been explored. Thus we can extend our knowledge of the lunar surface. In recent decades, the remote exploration of the Earth from space, using optical and radio techniques, has unveiled much that could not be discovered from ground level. The same will be true of the Moon (already very precise laser measurements of the Moon's distance and dynamics have been made).

In the future, most telescopes directed from the Earth to the Moon will be used by amateur astronomers observing our closest celestial neighbour, which also represents the longest step taken by mankind. There is no doubt that Man will return to the Moon again, probably before the end of this century or early in the twenty-first century. By that time priority will have been given to the establishment of a manned lunar base, leading, perhaps, to permanent settlements for scientists and explorers, and who knows, it might be possible to see, with Earth-based telescopes, some of the results of their labours. Isn't this an exciting prospect for the future?

Tables for the calculation of co-longitude

Table I: year

Year	c°	M°
1989	179.32	0.9
1990	308.95	0.7
91	78.57	0.4
92	220.39	1.1
93	350.01	0.9
94	119.63	0.6
1995	249.26	0.4
96	31.07	1.1
97	160.69	0.8
98	290.32	0.6
99	59.94	0.3
2000	201.76	1.0
01	331.38	0.7
02	101.01	0.5
03	230.63	0.2
04	12.44	1.0
2005	142.07	0.7
06	271.69	0.5
07	41.31	0.2
08	183.13	0.9
09	312.75	0.7
2010	82.38	0.4
11	212.00	0.2
12	353.81	0.9
13	123.44	0.6
14	253.06	0.4
2015	22.68	0.1
16	164.50	0.8
17	294.12	0.6
18	63.75	0.3
19	193.37	0.1
2020	335.18	0.8

Table II: month

Month	c°	M°
January	0.00	356.0
(January)*	347.81	355.0
February	17.91	26.6
(February)*	5.72	25.6
March	359.25	54.2
April	17.17	84.7
May	22.89	114.3
June	40.80	144.8
July	46.53	174.4
August	64.44	204.9
September	82.35	235.5
October	88.07	265.1
November	105.99	295.6
December	111.71	325.2

* The figures for January and February given in parentheses are valid for leap-years.

Table III: day

Day	c°	M°
1	12.19	1.0
2	24.38	2.0
3	36.57	3.0
4	48.76	3.9
5	60.95	4.9
6	73.14	5.9
7	85.34	6.9
8	97.53	7.9
9	109.72	8.9
10	121.91	9.9
11	134.10	10.8
12	146.29	11.8
13	158.48	12.8
14	170.67	13.8
15	182.86	14.8
16	195.05	15.8
17	207.24	16.8
18	219.43	17.7
19	231.62	18.7
20	243.81	19.7
21	256.01	20.7
22	268.20	21.7
23	280.39	22.7
24	292.58	23.7
25	304.77	24.6
26	316.96	25.6
27	329.15	26.6
28	341.34	27.6
29	353.53	28.6
30	5.72	29.6
31	17.91	30.6

Table IV: hour

Hour	c°	M°
1	0.51	0.0
2	1.02	0.1
3	1.52	0.1
4	2.03	0.2
5	2.54	0.2
6	3.05	0.2
7	3.56	0.3
8	4.06	0.3
9	4.57	0.4
10	5.08	0.4
11	5.59	0.5
12	6.10	0.5
13	6.60	0.5
14	7.11	0.6
15	7.62	0.6
16	8.13	0.7
17	8.64	0.7
18	9.14	0.7
19	9.65	0.8
20	10.16	0.8
21	10.67	0.9
22	11.17	0.9
23	11.68	0.9
24	12.19	1.0

Table V: correction

M°	Correction	M°
0	0.0	360
10	−0.4+	350
20	−0.7+	340
30	−1.0+	330
40	−1.3+	320
50	−1.5+	310
60	−1.7+	300
70	−1.8+	290
80	−1.9+	280
90	−1.9+	270
100	−1.9+	260
110	−1.8+	250
120	−1.6+	240
130	−1.5+	230
140	−1.2+	220
150	−1.0+	210
160	−0.6+	200
170	−0.3+	190
180	0.0	180

Lunar eclipses

Lunar eclipses are one of the most interesting astronomical phenomena and their observation provides useful programmes for amateur astronomers.

The geometrical conditions leading to lunar eclipses are indicated in Fig. 16a. The participants of this celestial 'hide-and-seek' are the Sun, Earth and Moon. From the Sun's rays, S, coming from the left, we choose those that, for the present purpose, may be described as the 'external' common tangents to the Sun and Earth. These, which are denoted by *t,* form the limits of the convergent cone of the Earth's full shadow. Similarly, the rays that are called the 'internal' common tangents, *t',* define the limits of the divergent cone, in which the light of the Sun is reduced. Let us imagine a large projection plane, π, situated at the distance of the Moon and perpendicular to the axis of the shadow cones. When the shadow cones are projected onto this plane the full circular shadow, which is termed the *umbra, u,* surrounded by an annulus of less dense shadow, the *penumbra, p,* appears. It can be appreciated that when the Moon passes across this area an eclipse will occur. We may then ask what are the conditions that have to be fulfilled for this to happen.

First of all, the Moon has to line-up with the Sun and the Earth so that the centres of the three bodies are close to being in a straight line. Naturally, this can happen only at Full Moon for a lunar eclipse, but there is another constraint: the Moon must be sufficiently close to the ecliptic (the plane of the Earth's orbit around the Sun). The Moon's orbit, however, is inclined to the ecliptic by an angle of about 5°, and its orientation is continually changing in space. This is why the Full Moon does not pass through the Earth's shadow cone every month – it usually passes a little bit 'above' or 'below' or, more correctly, to the north or south of the shadow cone. Thus an eclipse cannot occur at every Full Moon.

Assuming that the geometrical conditions for lunar eclipses are as stated above, then there are three alternative configurations (Fig. 16b). In configuration 1, the Moon passes through only the penumbra, *p,* and a *penumbral lunar eclipse* occurs. Sometimes these are difficult to observe because the de-

Explanatory note

An observer of the Moon needs to know the position of the terminator at any given moment. For this purpose the value of the so-called *co-longitude (c)* is usually specified in astronomical almanacs: it is, in fact, the selenographic longitude of the morning terminator, measured westwards from the prime or central meridian, and its numerical value ranges between 0° and 360°. The selenographic longitude of the evening terminator is equal to the co-longitude plus 180°. The relationship between selenographic longitude, λ, and co-longitude, *c*, is as follows:

λ_E = selenographic longitude of morning terminator = 360° – *c*
(see note 1, p. 213)
(sunrise on the Moon)

λ_W = selenographic longitude of evening terminator = 180° – *c*
(sunset on the Moon)

To simplify matters let us suppose that the terminator coincides with one or other of the meridians (neglecting the selenographic *latitude* of the Sun).

If an astronomical almanac is not at hand, we can calculate the co-longitude by referring to the Tables, given on p. 212, which are derived from those of J. Meeus ('Terminatortabellen voor Maanwaarnemingen', *Hemel en Dampkring,* Vol. 61 (1963), No. 12). Universal Time (UT) is used throughout.

Example 1

What is the position of the terminator on 19 April 1991 at 21h UT?

Referring to Tables I, II, III and IV, we obtain the following figures:

Time	Table	c°	M°	(correction)
1991	I	78.57	0.4	
April	II	17.17	84.7	
19	III	231.62	18.7	
21h	IV	10.67	0.9	
		338.03	104.7	(totals)

From Table V we see that the nearest value for M° is 100° and that the appropriate correction is therefore –1.9° (notice the change in sign for values of *M* greater than 180°). Subtracting this correction from *c* (i.e. 338.03° – 1.9° = 336.1°), we get the true value of the co-longitude, *c.* Since the prime or central meridian of the Moon has a value of 0° or 360°, it can be seen that our derived value of *c* is not far away from it. In fact 360° – 336.1° = +23.9°, i.e. the selenographic longitude east, λ_E = 23.9°E. Since, however, the other terminator is 180° away, we can also subtract the above value of *c* from 180°: 180° – 336.1° = –156.1°, i.e. the selenographic longitude west, λ_W = 156.1°W.

From the above calculations we have found that the morning terminator occupies the position defined by the meridian 23.9°E; and from the map we can see that the Sun, at that instant, is rising over the craters Plinius, Lamont and the trio of Theophilus, Cyrillus and Catharina, and the fault Rupes Altai is coming into view. It goes without saying that the evening terminator is on the averted hemisphere.

Example 2

When does the Sun rise above the crater Bullialdus (λ = –22°) in April 1991?

By definition the selenographic longitude, λ_E, of the morning terminator is λ_E = 360° – *c;* therefore the co-longitude, *c* = 360° – λ_E. The date and hour in April 1991 will be that when the value of the co-longitude, *c,* is numerically equal to that of the selenographic longitude of the meridian of Bullialdus. Hence *c* = 360° – (–22°) = 382° = 22°.

From Tables I and II we obtain the following values:

Time	Table	c°	M°	(correction)
1991	I	78.57	0.4	
April	II	17.17	84.7	
		95.74	85.1	(totals)

Subtracting 95.74° from the required value of *c,* i.e. 382°, we get 382° – 95.74° = 286.26°. Referring now to Table III, we discover that the closest value to 286.26° is 280.39° (*M* = 22.7°) for the 23rd day (in April). By adding the three 'year, month, day' values of the co-longitude, *c,* for this date, we obtain: *c* = 78.57° + 17.17° + 280.39° = 376.13°. Subtracting 360°, *c* = 376.13° – 360° = 16.13°.

We now apply a correction by adding the three values in the *M* columns and by referring to Table V. In this case, *M* = 0.4° + 84.7° + 22.7° = 107.8°. From Table V the nearest correction value is – 1.8°. Therefore the true value of the co-longitude *c* = 16.13° – 1.8° = 14.33°.

Thus the co-longitude for 23 April 1991 at 0h UT is 14.33°. But this is still 7.7° away from the meridian of Bullialdus (λ_W = 22°). In Table IV 7.7° is equivalent to about 15 h UT, so this is the time that an observer would see the Sun rising over Bullialdus.

Note 1
When the difference 360° – *c* is greater than 180°, subtract 360°. Starting from the prime or central meridian, the selenographic longitude, λ, is measured *positively* to the east and *negatively* to the west, from 0° to 180° or 0° to –180°.

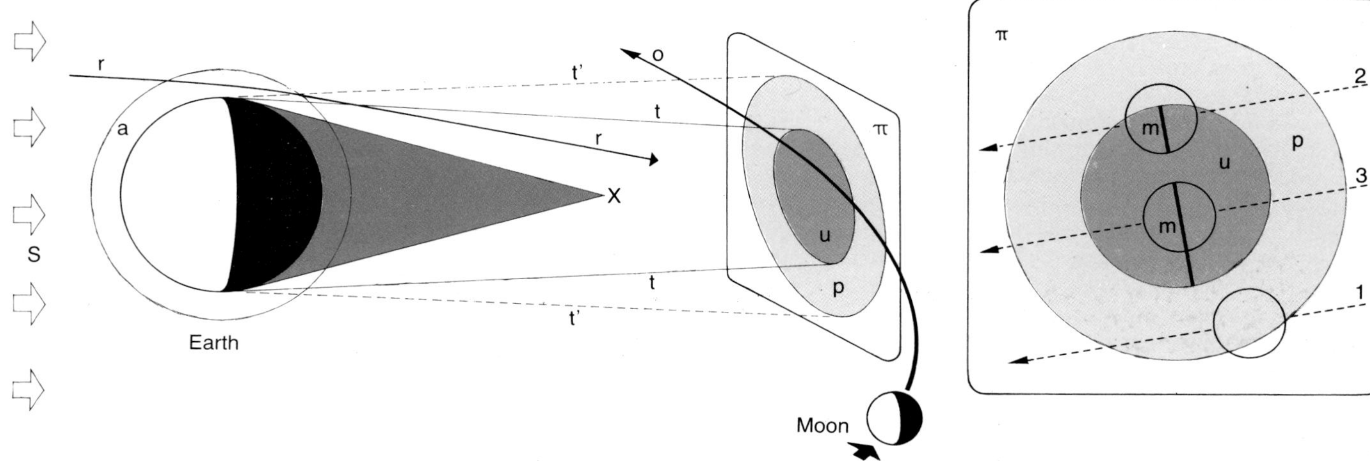

Fig. 16a. *Geometrical conditions for a lunar eclipse.*
Fig. 16b. *Penumbral (1), partial (2) and total (3) lunar eclipse.*

crease in the Moon's brightness is so slight. In configuration 2, the Moon, having entered the penumbra, passes through the edge of the umbra and observers witness a *partial lunar eclipse.* Finally, in configuration 3, the Moon fully enters the umbra and is in *total eclipse.* A lunar eclipse reaches its maximum when the Moon is closest to the centre of the umbra. At that moment, the magnitude, *m,* of the eclipse is specified in units of the Moon's diameter. If $0 < m < 1$, the eclipse is partial; if $m \geqslant 1$, the eclipse is total. The maximum possible magnitude, *m,* of a total eclipse is 1.888. The greater the magnitude, the longer the duration of the whole phenomenon; usually, the duration lasts from tens of minutes for a partial eclipse to several hours for a total eclipse.

The exact times and other data necessary for the observation of eclipses are published in astronomical almanacs. It is obviously essential to know in advance whether the phenomenon in question is going to be visible from a given geographical location. In general, it can be said that a lunar eclipse will be visible from any part of the terrestrial hemisphere that is turned towards the Moon. (If this seems obvious, remember that astronomers often have to travel to far-distant places to witness a total eclipse of the Sun!)

Ancient astronomers knew how to forecast lunar and solar eclipses; they even discovered the lengths of the recurrent periods that determine when eclipses occur. The fundamental period, called the *Saros,* is 6585.32 days (approximately 18 years 11 days), as can be verified from the accompanying table.

An apparent paradox is the fact that during a total lunar eclipse the Moon's disk is faintly visible and is usually reddish in colour. This is caused by the Earth's atmosphere (the depth of which, *a,* has been considerably exaggerated in Fig. 16a), which, like a glass prism, 'bends' the light passing through it, so that the dark umbral cone of shadow is diluted. In effect, this reduces the length of the umbral cone of shadow, resulting in the Moon never being near enough to the Earth to be totally immersed. Differences in the brightness of the eclipsed Moon arise, owing to the varying amounts of cloud, dust, etc. in the Earth's atmosphere, and may be estimated using the so-called Danjon classification:

Degree	Description
0	Very dark eclipse; Moon almost invisible, especially during the middle of totality.
1	Dark eclipse, grey or brownish colouring, details hardly recognizable.
2	Dark red or rusty red colouring with dark patch in the middle of the shadow; brighter edges.
3	Brick-red eclipse; sometimes shadow bordered with bright or yellowish margin.
4	Copper- or orange-red, very bright eclipse with a bluish and very bright marginal zone.

Small amateur telescopes and binoculars are most convenient for estimating the brightness and colour of a lunar eclipse according to the above scheme. It is recommended that the estimates be repeated several times during the course of totality, as well as during the partial phases. Other possible observations include the recording of the times selected craters enter and emerge from the eclipse shadow (see also page 25).

Total and partial lunar eclipses 1990–2011

Date	Maximum eclipse		Visible in
	Time: UT	Magnitude	
1990 February 9	19.11	1.07	Asia, Europe, Africa, Australia, East Atlantic
1990 August 6	14.10	0.69	Australia, Asia, Antarctica
1991 December 21	10.32	0.10	NE Asia, North America, Greenland, part of Australia
1992 June 15	04.56	0.68	North and South America, Antarctica, West Africa
1992 December 9	23.43	1.27	Africa, Europe, North America, West Asia
1993 June 4	13.00	1.56	Australia, East Asia, Antarctica
1993 November 29	06.24	1.09	North and South America, Western Europe, NE Asia, West Africa
1994 May 25	03.30	0.24	North and South America, Western Europe, Africa
1995 April 15	12.17	0.11	East Asia, Australia, Pacific Ocean, western part of North America
1996 April 4	00.09	1.38	Africa, Europe, South and eastern part of North America, West Asia
1996 September 27	02.53	1.24	North and South America, Africa, Europe
1997 March 24	04.39	0.92	North and South America, West Africa, Western Europe
1997 September 16	18.45	1.19	Asia, Europe, Africa, Australia
1999 July 28	11.32	0.40	Pacific Ocean, Australia, East Asia, NW North America
2000 January 21	04.43	1.33	North and South America, Europe, West Africa
2000 July 16	13.55	1.77	Australia, East Asia, Antarctica
2001 January 9	20.20	1.19	Europe, Asia, Africa, Arctic
2001 July 5	14.55	0.49	Australia, Antarctica, SE Asia, Pacific Ocean
2003 May 16	03.39	1.13	North and South America, Antarctica, Africa, Western Europe
2003 November 9	01.18	1.02	Europe, Africa, North and South America
2004 May 4	20.30	1.30	Africa, Europe, Asia, Australia, Antarctica
2004 October 28	03.03	1.31	North and South America, Europe, Africa
2005 October 17	12.02	0.06	Australia, East Asia, North America, Pacific Ocean
2006 September 7	18.50	0.18	Africa, Eastern Europe, Asia, Australia
2007 March 3	23.20	1.23	Africa, Europe, Asia, South America, eastern part of North America
2007 August 28	10.36	1.48	Pacific Ocean, Australia, North America, western part of South America
2008 February 21	03.25	1.11	North and South America, Africa, Europe
2008 August 16	21.09	0.81	Africa, Europe, Asia, Australia, Antarctica
2009 December 31	19.22	0.08	Asia, Europe, Africa, Australia
2010 June 26	11.38	0.54	Pacific Ocean, Australia, Antarctica
2010 December 21	08.16	1.26	North and South America, East Asia, Pacific Ocean
2011 June 15	20.11	1.70	Africa, Europe, Asia, Australia, Antarctica
2011 December 10	14.31	1.10	Asia, Australia, Pacific Ocean

After Meeus, J. & Mucke, H.: *Canon of Lunar Eclipses* – 2002 to + 2526, Astronomisches Büro, Vienna, 1979.

Glossary*

Age of Moon The period that has elapsed since the last New Moon.

Albedo A measure of surface reflectivity. The ratio of incident to reflected illumination. A good reflecting surface has a high albedo (e.g. clouds, 0.70); a poor reflector has a low albedo (e.g. granite, 0.31; the Moon, 0.12; lava, 0.04).

Altitude, absolute The height of any point on the lunar surface above the so-called 'reference sphere', which is defined as a perfect sphere of 3476 km diameter.

Altitude, relative The height of any point, e.g. a mountain peak, above its immediate surroundings.

Angular measure The division of a circle into 360° (degrees), each consisting of 60′ (minutes), each of which can be further subdivided into 60″ (seconds). The apparent diameter of a lunar or planetary disk, or of a lunar crater, is often expressed in angular measure.

Apogee The point on the orbit of the Moon (or of an artificial Earth satellite) where it is furthest from the Earth.

***Apollo* (programme)** US manned space programme, 1968–1972. The first soft-landing of man on the Moon was that of *Apollo 11* on 20 July 1969. Exploration of the Moon; collecting rock samples (a total of 382 kg were returned to Earth); installation of geophysical stations, etc.

Basalt Igneous rock, solidified lava or magma; fine-grained; black or dark grey in hue. Various kinds were found on the Moon and also exist on the other terrestrial planets (Mercury, Venus and Mars).

* Words *italicized* are defined elsewhere in the glossary. Words in **bold** are defined within the particular entry.

Caldera A large volcanic crater formed by an eruption of magmatic material, followed by the collapse of the interior as molten rock drains away into an underground cavity.

Cartographic representation A mathematical or geometrical method of representing the surface of a spherical (or ellipsoidal) body on the flat surface of a map. The **orthographic projection** is often used for lunar mapping because it depicts the Moon as it is seen from the Earth. The **conformal projection** is useful for preserving the shapes of formations that would otherwise be distorted. Examples of conformal projections include the stereographic and Mercator projections.

Celestial equator The prime circle of the celestial sphere. It is a projection of the Earth's equator onto the sky and the dividing line separating the northern and southern celestial hemispheres. Its declination is 0°.

Co-longitude The selenographic longitude (λ) of the morning terminator, measured from the central meridian westwards from 0° to 360°. Co-longitude values are used to determine the positions of lunar formations above which the Sun rises and sets.

Co-ordinates, astronomical Pairs of spherical co-ordinates that define the angular positions of celestial bodies. They are given in units of degrees, minutes and seconds of arc (°, ′, ″), which in some cases need to be converted to time-scales measured in hours, minutes and seconds (h, min, s). These systems are analogous to the well-known terrestrial co-ordinates of latitude and longitude, but despite these similarities there are many differences between the fundamental planes of reference and orientations, not forgetting the nomenclature. The Moon's selenographic co-ordinates in latitude, β, and longitude, λ, are defined by the lunar axis of rotation.

Crater A circular depression, usually with an elevated wall. These formations are found on cosmic bodies with solid surfaces, most notably planetary moons.

Crater, impact (also meteorite crater) A crater formed by the impact of a meteoritic body, or (in the case of **secondary craters**) by the ejecta from such impacts.

Crater, volcanic A crater formed by the escape of magma, gases, etc., from the interior of a cosmic body.

Declination A celestial co-ordinate similar to terrestrial latitude. It is the angular distance north or south of the *celestial equator* of a point on the celestial sphere. The declination of the celestial equator is 0°, of the north celestial pole +90°, and of the south celestial pole −90°. See also *right ascension.*

Diameter, apparent The diameter of a celestial object expressed in angular measure, e.g. the apparent diameter of the Moon is about 30′ = 0.5°. Also called the **angular diameter**.

Eccentricity One of the elements of an elliptical orbit of a celestial body. It is the elongation of the ellipse, obtained by dividing the distance between the two focal points by the length of the major axis. The eccentricity, *e*, = 0 for a circular orbit. For an elliptical orbit, *e* lies between 0 and 1.

Ecliptic 1. The apparent orbit of the Sun around the celestial sphere. The Sun appears to complete one circuit along the ecliptic in one year. 2. The plane of the Earth's orbit around the Sun.

Erosion Disintegration of a lunar or planetary surface caused by natural forces, e.g. water, wind, frost, etc. On celestial bodies without atmospheres, e.g. the Moon, erosion is caused mainly by the impacts of meteorites and micrometeorites.

Escape velocity The minimum velocity that a body has to attain to escape from the gravitational field of a more massive body, e.g. a planet. It is also called 'parabolic velocity' or the 'second cosmic velocity'.

Focal length The distance from the centre of a lens or mirror to its **focal point** or *focus,* where a sharply defined or focused image of a distant object is formed.

Focus The point to which a parallel beam of light entering an optical system, axially, is refracted or reflected to form a sharply defined or focused image. It lies on a focal plane that is perpendicular to the optical axis of the system. Also called the *focal point.*

Greek alphabet
On detailed lunar maps, isolated hills, mountain massifs and peaks of mountain ridges were denoted by lower-case Greek letters until the early 1970s. Nowadays, however, these identities are used only unofficially and to limited extent.

α	alpha	ι	iota	ϱ	rho
β	beta	ϰ	kappa	σ	sigma
γ	gamma	λ	lambda	τ	tau
δ	delta	μ	mu	υ	upsilon
ε	epsilon	ν	nu	ϕ	phi
ζ	zeta	ξ	xi	χ	chi
η	eta	ο	omicron	ψ	psi
ϑ	theta	π	pi	ω	omega

Libration The lunar librations in latitude and longitude are the apparent vertical and horizontal rocking motions of the Moon as it orbits around the Earth. Their combined effect enables us to observe rather more than 50% of the Moon's surface over a period of time.

Libration zones Peripheral areas of the lunar surface that are alternately turned towards and away from the Earth as a result of the lunar *librations.* The approximate longitudinal centres of these zones are the meridians 90°E and 90°W.

Light-gathering power The ratio of the diameter, *d,* of a telescope objective or mirror to its focal length, *f,* i.e. the ratio *d* : *f.* It is also called the **focal ratio** (or **speed**).

Line of nodes The line joining the two points where the Moon's orbit intersects the *ecliptic,* to which it is inclined by 5°. The point at which the Moon crosses the ecliptic from south to north is termed the **ascending node;** similarly, the point at which it crosses it moving from north to south is termed the **descending node**. The line of nodes continually changes its orientation in space, but returns to the same points of intersection with the ecliptic every 18.61 years (the **Saros**).

Luna A series of Soviet lunar probes and automatic stations that explored the Moon during the period 1959–1976. The following are of special importance:

Luna 2: First crash-landing of a man-made body on the Moon, 12 September 1959.
Luna 3: First photographs of the far-side of the Moon, 10 October 1959.
Luna 9: First soft-landing on the Moon, 3 February 1966.
Luna 10: First artificial satellite of the Moon, 3 April 1966.
Lunas 16, 20 and *24:* Automated sampling of lunar rocks, which were brought back to Earth, 1970, 1972 and 1976.
Lunas 17, 21: Automatic mobile laboratories *Lunokhod 1* and *2.* 1970–1971 and 1973.

Lunar Orbiter A series of five US lunar probes, all of which were successfully put into orbit around the Moon and undertook detailed photographic mapping of almost the entire lunar surface during 1966 and 1967.

Lunation The *synodic month,* which is the time taken by the Moon to complete one set of phases, e.g. from New Moon to New Moon: 29 days 12 hours 44 minutes 2.8 seconds. To facilitate computations, lunations are numbered, no. 1 commencing on 16 January 1923. Lunation no. 850 commences with the New Moon of 8 September 1991.

Magma Subterranean molten rock, which emerges as *lava* during volcanic eruptions.

Mascon Abbreviated form of the term 'mass concentration', which refers to denser material under the lunar mare basins that was manifest by local anomalies in the Moon's gravitational field. The mare basins, which were formed more than 4000 million years ago, contain solidified magma to a considerable depth and this is thought to be responsible for these anomalies.

Month, anomalistic The time that elapses between two successive orbital passages of the Moon through *perigee*: 27 days 13 hours 18 minutes 33.1 seconds. Since perigee is successively displaced in the direction of the Moon's motion, the Moon describes an angle of slightly more than 360° during an anomalistic month. This is why an anomalistic month is longer than a *sidereal month*.

Month, sidereal The time taken by the Moon to complete one revolution around the Earth with respect to the background stars: 27 days 7 hours 43 minutes 11.5 seconds.

Month, synodic The time taken by the Moon to complete one set of phases, e.g. from New Moon to New Moon. Also called a *lunation*, it is 29 days 12 hours 44 minutes 2.8 seconds.

Mounting for a telescope A telescope support comprising two mechanical bearings, set at right angles, which form the axes of rotation. There are two principal designs: **altazimuth** and **equatorial**. In the former, the movements conform to the horizon system of co-ordinates, i.e. are in altitude and azimuth. In the latter, the principal axis of rotation is parallel to the Earth's axis, so that when rotated about this axis the telescope can be made to follow the apparent path of a star across the sky.

Nodes The two points of intersection of the Moon's orbit with the plane of the *ecliptic*. See also 'line of nodes'.

Perigee The point on the orbit of the Moon (or of an artificial Earth satellite) where it is closest to the Earth.

Phase The amount of the illuminated disk of a dark body, e.g. the Moon, which shines by reflected sunlight, that is visible from the Earth. It depends on the ever-changing angle between the Sun, the dark body and the Earth. The main lunar phases are: New Moon, First Quarter, Full Moon and Last Quarter.

Phase angle The angle between a line connecting a dark body, e.g. the Moon, which shines by reflected sunlight, with the Earth, and another line connecting the Moon with the Sun. It determines the area of the illuminated hemisphere of the body that will be seen by an observer from the Earth.

Ranger A series of US lunar probes, of which *Rangers 7, 8* and *9*, which were launched to the Moon in 1964 and 1965, returned detailed photographs of the area where they were due to impact, showing surface details less than one metre across.

Regolith The incoherent surface layer of the Moon and many other bodies in the solar system that do not have a protective atmosphere. It consists of crushed and fragmented rocks, resulting largely from millions of years of meteorite impacts. The lunar regolith is said to have a depth of about 10–100 m.

Right ascension (RA) The equatorial co-ordinate which, together with *declination*, unambiguously defines the position of a point on the celestial sphere. It is analogous to terrestrial longitude, and is measured eastwards along the *celestial equator* from the **vernal equinox** (the point where the Sun appears to cross the celestial equator, moving from south to north around 21 March) to the declination circle passing through the given point. By convention it is measured in time, ranging from 0 to 24 hours.

Rotation, captured or synchronous The axial rotation of a satellite in a period equal to the time of its revolution around its primary, so that it always presents the same hemisphere (as in the case of our Moon). It is a consequence of the action of tidal forces between the two bodies.

Selenodesy The branch of astronomy that deals with the shape, size and gravitational field of the Moon. By using selenodetic methods, precise three-dimensional co-ordinates of selected points on the lunar surface can be determined, including their real or absolute altitude above the Moon's reference sphere. A net of such points allows the shape of the Moon to be found.

Selenography The mapping of the Moon, together with descriptions of its surface detail, the development of nomenclature, etc.

Sidereal period The time taken by a planet or satellite to complete one revolution around its primary with respect to the background stars. In the case of the Moon, it is the *sidereal month*.

Surveyor A series of US lunar probes, of which *Surveyors 1, 3, 5, 6* and *7* soft-landed on the lunar surface and carried out a programme of photography and chemical analysis of the lunar *regolith* between 1966 and 1968.

Synodic period The time taken by a planet or satellite to complete one revolution around its primary as observed from the (moving) Earth. In the case of the Moon, it is the time that elapses between two successive similar phases, i.e. the *synodic month*.

Temperature scale A graduated scheme for measurement of the quantity of heat energy possessed by a body. In astronomy temperatures are usually quoted in Kelvin (K) or degrees absolute (0 K = –273.16°C). For more mundane purposes the Celsius (or Centigrade) scale is replacing the Fahrenheit scale, which is, however, still widely used in some countries.

Water (H_2O)	°C	°F	K
Freezing point	0	32	273.16
Boiling point	100	212	373.16

Terminator The boundary between light and shadow, or between day and night, on a body that does not shine by its own light. In the case of the Moon, the morning terminator heralds sunrise; the evening terminator is the beginning of the 14-day long lunar night.

Tidal forces The mutual gravitational forces between two or more neighbouring cosmic bodies, which can deform their shapes, and, in extreme cases, can cause their disintegration. Familiar terrestrial phenomena are the tides caused by the attraction of the Sun and Moon.

Bibliography

Catalogues

1. Andersson, L.E. & Whitaker, E.A.: *NASA Catalogue of Lunar Nomenclature.* NASA, Washington, DC, 1982.
2. Arthur, D.W.G.: Consolidated Catalogue of Selenographic Positions. *Communications of the Lunar and Planetary Laboratory,* Vol. 1, No. 11. University of Arizona, Tucson, 1962.
3. Arthur, D.W.G. *et al.:* The System of Lunar Craters. Quadrants I; II; III; IV. *Communications of the Lunar and Planetary Laboratory,* Vol. 2, No. 30, 1963; Vol. 3, No. 40, 1964; Vol. 4, No. 50, 1965, Vol. 5, Part 1, No. 70, 1966. University of Arizona, Tucson.
4. Davies, M.E., Colvin, T.R. & Mayer, D.L.: A Unified Lunar Control Network – The Near Side. A RAND Note. RAND Corporation, Santa Monica, 1987.
5. International Astronomical Union: Annual Gazetteer of Planetary Nomenclature. US Geological Survey, Flagstaff, 1986.
6. Meeus, J. & Mucke, H.: *Canon of Lunar Eclipses* −2002 to +2526. Astronomisches Büro, Vienna, 1979.

Atlases

7. Gutschewski, G.L., Kinsler, D.C. & Whitaker, E.A.: *Atlas and Gazetteer of the Near Side of the Moon.* [Lunar Orbiter IV Photographs.] NASA SP-241, Washington, DC, 1971.
8. Kopal, Z., Klepešta, J. & Rackham, T.W.: *Photographic Atlas of the Moon.* Academic Press, New York and London, 1965.
9. Kopal, Z.: *A New Photographic Atlas of the Moon.* Robert Hale, London, 1971.
10. Kuiper, G.P. *et al.: Photographic Lunar Atlas* [Kuiper Atlas]. University of Chicago Press, 1960.
11. Kuiper, G.P., Arthur, D.W.G. & Whitaker, E.A.: *Orthographic Atlas of the Moon.* University of Arizona Press, Tucson, 1961.
12. Kuiper, G.P. *et al.: Consolidated Lunar Atlas.* Lunar and Planetary Laboratory, University of Arizona, Tucson, 1967.
13. Rükl, A.: *Moon, Mars and Venus.* Hamlyn/Artia, Prague, 1976.
14. Schwinge, W.: *Fotografischer Mondatlas.* J.A. Barth, Leipzig, 1983.
15. Voigt, A. & Giebler, H.: *Berliner Mondatlas.* Wilhelm-Foerster-Sternwarte, West Berlin, 1989.
16. Whitaker, E.A. *et al.: Rectified Lunar Atlas.* University of Arizona Press, Tucson 1963.

Maps

17. Arthur, D.W.G. & Agnieray, A.P.: *Lunar Designations and Positions, Quadrants I–IV. Revised Lunar Quadrants Maps.* University of Arizona, Tucson, 1969.
18. *Lunar Astronautical Charts* (LAC series), 1 : 1 000 000. USAF Aeronautical Chart and Information Center, NASA, Washington, DC, 1960–1967.
19. *Map Showing Relief and Surface Markings of the Lunar Far Side,* 1 : 5 000 000. US Geological Survey, Flagstaff, 1980.
20. *Map Showing Relief and Surface Markings of the Lunar Polar Regions,* 1 : 5 000 000. US Geological Survey, Flagstaff, 1980.
21. *NASA Lunar Chart* (LPC-1), 1 : 10 000 000. Defence Mapping Agency Aerospace Center, St. Louis, 1979.
22. *Polnaja Karta Luny,* 1 : 5 000 000. Nauka, Moscow, 1979.
23. *Polnaja Karta Luny,* 1 : 10 000 000. Gosudarstvennyi Astronomitsheskii Institut im. Sternberga, Moscow, 1985.
24. Rükl, A.: *Maps of Lunar Hemispheres,* 1 : 10 000 000. D. Reidel, Dordrecht, 1972.
25. Rükl, A.: *Skeleton Map of the Moon,* 1 : 6 000 000. [Appendix to atlas reference no. 8.] Central Institute of Geodesy and Cartography, Prague, 1965.
26. *The Earth's Moon,* 1 : 10 460 000. National Geographic Society, Washington, DC, 1976.
27. Wolf, H.: *Erdmond.* Vorderseite – Rückseite, 1 : 12 000 000. Kosmos, Stuttgart, 1985.

Monographs

28. Cherrington Jr, E.H.: *Exploring the Moon through Binoculars and Small Telescopes.* McGraw-Hill Inc., 1969; Dover Publications, 1984.
29. Fauth, P.: *Unser Mond – wie man ihn lesen sollte.* Breslau, 1936.
30. French, B.M.: *The Moon Book. Exploring the mysteries of the lunar world.* Penguin Books, Harmondsworth, England, 1977.
31. Goodacre, W.: *The Moon, with a description of its surface formations.* Privately published, Bournemouth, 1931.
32. Kopal, Z.: *An Introduction to the Study of the Moon.* D. Reidel, Dordrecht, 1966.
33. Kopal, Z. & Carder, R.W.: *Mapping of the Moon, past and present.* D. Reidel, Dordrecht, 1974.
34. Lipskij, J.N. & Rodionova, Z.F.: Kartometritsheskie issledovania vidimovo i obratnovo polusharia Luny. In *Atlas obratnoi storony Luny,* Part III. Nauka, Moscow, 1975.
35. Rackham, T.W.: *Moon in Focus.* Pergamon Press, Oxford, 1968.
36. Shevchenko, V.V.: *Sovremennaya selenografia.* Nauka, Moscow, 1980.
37. Taylor, S.R.: *Lunar Science: A Post-Apollo View.* Pergamon Press, New York, 1975.

SOURCES USED FOR COMPILATION OF THE MAP AND PICTORIAL SECTIONS

The main part of the present atlas, i.e. the detailed map of the near-side of the Moon in 76 sections, was compiled largely from catalogues, atlases and maps provided by the Lunar and Planetary Laboratory of the University of Arizona, USA. The positions of lunar formations were derived from references 2, 3, 11, 16 and 17. The lunar charts drawn by the author were based primarily on references 12 and 16. The *Consolidated Lunar Atlas* (reference 12), published in 1967, is still the best and most detailed Earth-based photographic atlas of the Moon; it shows areas under several angles of solar illumination, which is essential for the work of the cartographer. Fine surface details, such as narrow rilles, which are not resolved in photographs taken from the Earth, were based on reference 7 and other photographs taken by *Lunar Orbiters IV* and *V*.

The nomenclature of lunar formations, their approximate co-ordinates and their dimensions, which facilitate their identification on the map, were taken from references 1 and 5. Some later names, accepted by the General Assembly of the IAU in 1988, have been added from a communication from V. V. Shevchenko (President of the Task Group for Lunar Nomenclature). Brief explanations concerning names and bibliographical data have been extracted from the *Transactions of the IAU,* which are published every third year and list newly approved names given to formations on various bodies of the solar system. Maps reference nos 19 and 20, atlases reference nos 7 and 16, and photographs of the libration zones taken by *Lunar Orbiters IV* and *V* formed the basis for the maps of the libration zones (pp. 182–189).

The illustrations in the section 'Fifty views of the Moon' are based mainly on reference 12. For printing purposes, photographs of selected details of the lunar surface were either modified or completely redrawn using the air-brush technique. The drawings of the close-up views, presented in these pages, are based on a series of photographs showing each selected formation from references 7 and 12. In this respect, the author's intention was to offer the observer illustrations showing how each selected formation appears, so as to make identification easier. These illustrations should therefore be regarded as precise drawings or maps and not necessarily the views that would be obtained photographically. The same can be said about the representation of the phases of Moon on pp. 20–24.

ACKNOWLEDGEMENTS

The author's intention in producing this book was to create an up-to-date atlas of the observable part of the Moon, based on modern catalogues, atlases, maps and photographs. These sources are cited on pp. 218–219. Such a task would not have been possible without the valuable co-operation of prominent specialists, who readily gave advice and allowed access to essential materials. Foremost among these were: Dr M. E. Davies, RAND Corporation, USA; Mr. E. A. Whitaker, Lunar and Planetary Laboratory, University of Arizona, USA; Professor Z. Kopal, University of Manchester, England; Dr T. W. Rackham, Jodrell Bank, England; Professor H. Mucke, Astronomisches Büro, Vienna, Austria; Dr V. V. Shevchenko, Gosudarstvennyi Astronomitsheskii Institut im. Sternberga, USSR, and Dr L. D. Jaffe, Jet Propulsion Laboratory, USA. It is a pleasure for the author to express his gratitude to all of them for their friendly help.

For their useful comments on the textual content of the atlas, the author is obliged to Dr T. W. Rackham and to both reviewers, Dr Z. Pokorný CSc. and Ing. P. Příhoda, as well as to the Artia editor, Mrs E. Skřivanová and the Octopus Books copy-editor, Mr. Peter Gill. The author also extends his sincere thanks to the graphical designers, Mr. S. Seifert and Mr. V. Kopecký, to the experienced staff of the Aventinum Publishing House and to the Printing House TSNP Martin who produced the book in its final form. Last but not least, the author extends his thanks to his wife, Sonja, for the patient moral support she gave him while he was working on this atlas.

INDEX OF NAMED FORMATIONS

This index lists in alphabetical order within the appropriate category the names of lunar formations and indicates by a number adjacent to each name the appropriate section of the map of the near-side of the Moon in which a particular formation may be found. Where a formation is shown in the libration zone maps, roman numerals are used.

Craters

Abbot 37
Abel 69, IV
Abenezra 56
Abetti 24
Abulfeda 45
Acosta 49
Adams 69
Agatharchides 52
Agrippa 34
Airy 55, 56
Al-Bakri 35
Al-Biruni III
Al-Marrakushi 48
Albategnius 44, 45
Aldrin 35
Alexander 13
Alfraganus 46
Alhazen 27
Aliacensis 55, 65
Almanon 56
Alpetragius 55
Alphonsus 44
Ameghino 38
Ammonius 44
Amontons 48
Amundsen 73, 74, V
Anaxagoras 4
Anaximander 2
Anaximenes 3
Andersson VI
Anděl 45
Angström 19
Ansgarius 49, IV
Anuchin V
Anville 37
Apianus 56
Apollonius 38
Arago 35
Aratus 22
Archimedes 12, 22
Archytas 4
Argelander 56
Ariadaeus 35
Aristarchus 18
Aristillus 12
Aristoteles 5
Armstrong 35
Arnold 5
Arrhenius VI
Artemis 20
Artsimovich 19
Aryabhata 36
Arzachel 55
Asada 37
Asclepi 74
Aston 8, VIII
Atlas 15
Atwood 49
Autolycus 12
Auwers 24
Auzout 38
Avery 49, IV
Avicenna VIII
Azophi 56
Baade 61, VI, VII
Babbage 2
Babcock III
Back 38, III
Baco 74
Baillaud 5
Bailly 71, VI
Baily 6
Balboa 17, VIII
Ball 64
Balmer 60
Banachiewicz 38, III
Bancroft 21
Banting 23
Barkla 49
Barnard 60, IV

Barocius 66
Barrow 4
Bartels 17, VIII
Bayer 71
Beals 16, III
Beaumont 57, 58
Beer 21
Behaim 60, IV
Beketov 25
Běla 22
Belkovich 6, 7, II
Bell VIII
Bellot 48
Bernoulli 16
Berosus 16
Berzelius 15
Bessarion 19
Bessel 24
Bettinus 71
Bianchini 2, 10
Biela 75, 76
Bilharz 49
Billy 40
Biot 59
Birmingham 3, 4
Birt 54
Black 49, IV
Blagg 33
Blancanus 72
Blanchard VI
Blanchinus 55
Bobillier 23
Bode 33
Boethius 38
Boguslawsky 74
Bohnenberger 58
Bohr 28, VIII
Boltzmann VI
Bombelli 38
Bonpland 42
Boole 2, I
Borda 59
Borel 24
Born 49
Boscovich 34
Boss 16, II
Bouguer 2
Boussingault 74, 75
Bowen 23
Brackett 24
Bragg VIII
Brayley 19
Breislak 66
Brenner 68
Brewster 25
Brianchon 2, 3, I
Briggs 17
Brisbane 76, V
Brown 64
Bruce 33
Brunner IV
Buch 66
Bullialdus 53
Bunsen 8, VIII
Burckhardt 16
Burnham 45
Bürg 14
Büsching 66
Byrd 4, II
Byrgius 50, 51

C. Herschel 10
C. Mayer 5
Cabeus 73, VI
Cajal 36
Calippus 13
Cameron 37
Campanus 53
Cannizzaro I
Cannon 27, III
Capella 47
Capuanus 63
Cardanus 28
Carlini 10
Carmichael 25
Carpenter 2, 3
Carrel 35
Carrillo 49, IV
Carrington 15
Cartan 38
Casatus 72, VI
Cassini 12
Catalán 61, VI
Catharina 57
Cauchy 36
Cavalerius 28
Cavendish 51
Caventou 20
Cayley 34
Celsius 67
Censorinus 47
Cepheus 15
Chacornac 14, 25
Chadwick VI
Challis 4
Chamberlin V
Chapman I
Chevallier 15
Chladni 33
Ching-te 25
Cichus 63
Clairaut 66
Clausius 62
Clavius 72, 73
Cleomedes 26
Cleostratus 1, I
Clerke 25
Collins 35
Colombo 59
Compton II
Condon 38
Condorcet 38
Conon 22
Cook 59
Copernicus 31
Couder VII
Cremona I
Crile 37
Crozier 48
Crüger 50
Curie IV
Curtis 26
Curtius 73
Cusanus 6, II
Cuvier 74
Cyrillus 46
Cysatus 73

d'Arrest 34
da Vinci 37
Daguerre 47
Dale 49, IV
Dalton 17, VIII
Daly 38
Damoiseau 39
Daniell 14
Darney 42, 53
Darwin 50
Daubrée 23
Davy 43
Dawes 24
de Gasparis 51
de la Rue 6
de Morgan 34
de Roy VI
de Sitter 5, II
de Vico 50, 51
Debes 26
Dechen 8
Delambre 46
Delaunay 55
Delisle 9, 19
Delmotte 26
Deluc 73
Dembowski 34
Democritus 5
Demonax 74, V
Desargues 2, I
Descartes 45
Deseilligny 24
Deslandres 64, 65
Dionysius 35
Diophantus 19
Doerfel VI
Dollond 45
Donati 55
Donner IV
Doppelmayer 52
Dove 67
Draper 20
Drebbel 62
Dreyer III
Drude VII
Drygalski 72, VI
Dubiago 38
Dugan II
Dunthorne 62, 63
Dziewulski III

Eckert 26
Eddington 17
Edison III
Egede 5, 13
Eichstadt 50, VII
Eimmart 27
Einstein 17, VIII
Elger 63
Ellison I
Elmer 49, IV
Encke 30
Endymion 7
Epigenes 4
Epimenides 63
Eppinger 42
Eratosthenes 21, 32
Erro III
Esclangon 25
Euclides 41
Euctemon 5
Eudoxus 13
Euler 20

Fabbroni 25
Fabricius 68
Fabry II, III
Fahrenheit 38
Faraday 66
Fauth 31
Faye 55
Fedorov 19
Fényi VI, VII
Fermat 57
Fernelius 65
Feuillée 21
Finsch 24
Firmicus 38
Flammarion 44
Flamsteed 40
Focas VII
Fontana 51
Fontenelle 3
Foucault 2
Fourier 51, 61
Fox III
Fra Mauro 42, 43
Fracastorius 58
Franck 25
Franklin 15
Franz 25
Fraunhofer 69
Fredholm 26
Freud 18
Froelich I
Fryxel VII
Furnerius 69

G. Bond 15
Galen 22
Galilaei 28
Galle 5
Galvani 1, 8, I
Gambart 32
Ganswindt V
Gardner 25
Gärtner 6
Gassendi 52
Gaudibert 47
Gauricus 64
Gauss 16, III
Gay-Lussac 31
Geber 56
Geissler 49
Geminus 16
Gemma Frisius 66
Gerard 8, VIII
Gernsback IV
Gibbs 60, IV
Gilbert 49
Gill 75, V
Ginzel III
Gioja 4, I, II
Giordano Bruno III
Glaisher 37
Goclenius 48
Goddard 27, III
Godin 34
Goldschmidt 4
Golgi 18
Goodacre 66
Gould 53
Graff 61, VII
Greaves 37
Grimaldi 39
Grove 14
Gruemberger 73
Gruithuisen 9
Guericke 43
Gum 69, IV
Gutenberg 48
Guthnick VI
Gyldén 44

Hagecius 75
Hahn 16
Haidinger 63
Hainzel 63
Haldane 49, IV
Hale V
Hall 15
Halley 45
Hamilton 69, IV
Hanno 76, V
Hansen 38
Hanskiy IV
Hansteen 40
Harding 8
Hargreaves 49
Harkhebi III
Harpalus 2
Hartwig 39, VII
Hase 59
Hausen 71, VI
Hayn 6, II
Hecataeus 60, IV
Hedin 28
Heinrich 21
Heinsius 64
Heis 10
Helicon 10
Hell 64
Helmert 49, IV
Helmholtz 75, V
Henry 51
Henry Frères 51
Heraclitus 73
Hercules 14
Herigonius 41
Hermann 39
Hermite 4, I
Herodotus 18
Herschel 44
Hesiodus 54
Hevelius 28
Heyrovský VII
Hill 25
Hind 45
Hippalus 52, 53
Hipparchus 44, 45
Hirayama IV
Hohmann VII
Holden 60
Hommel 75
Hooke 15
Hornsby 23
Horrebow 2
Horrocks 45
Hortensius 30
Houtermans 49, IV
Hubble 27, III
Huggins 65
Humason 8
Humboldt 60, IV
Hume IV
Huxley 22
Hyginus 34
Hypatia 46

Ibn Battuta 48
Ibn Rushd 46
Ibn Yunus III
Ideler 74
Idelson V
Il'in VII
Inghirami 61, 62
Isidorus 47
Ivan 19

J. Herschel 2
Jacobi 74
Jansen 36
Jansky 38, III
Janssen 67, 68
Jeans V
Jehan 19
Jenkins 38
Jenner IV
Joliot III
Joy 23
Julius Caesar 34

Kaiser 65, 66
Kane 5
Kant 46
Kao 49, IV
Kapteyn 49
Kästner 49
Keldysh 6
Kepler 30
Kies 53
Kiess 49, IV
Kinau 74
Kirch 12
Kircher 71, 72
Kirchhoff 15
Klaproth 72
Klein 44
Knox-Shaw 38, III
Kopff 50, VII
Kovalskiy IV
König 53
Krafft 17
Kramarov VII
Krasnov 50, VII
Kreiken 49, IV
Krieger 19
Krishna 23
Krogh 38
Krusenstern 55, 56
Kugler V
Kuiper 42
Kundt 43
Kunowsky 30

L'allemand 39, VII
la Caille 55
la Condamine 2
la Perouse 49
Lacroix 61
Lade 45
Lagalla 63

Lagrange 61
Lalande 43
Lamarck 50
Lamb IV
Lambert 20
Lamé 49, 60
Lamèch 13
Lamont 35
Landsteiner 11
Langley 1, I
Langrenus 49
Lansberg 42
Lassell 54
Laue VIII
Lauritsen IV
Lavoisier 8, VIII
Lawrence 37
le Gentil 72, VI
le Monnier 24, 25
le Verrier 11
Leakey 47
Lebesgue 49, IV
Lee 62
Legendre 59, 60
Lehmann 61, 62
Lepaute 62
Letronne 40
Lexell 65
Liapunov 27, III
Licetus 65
Lichtenberg 8
Lick 37
Liebig 51
Lilius 73
Lindbergh 48
Lindblad I
Lindenau 67
Lindsay 45
Linné 23
Liouville 38
Lippershey 54
Littrow 25
Lockyer 67
Loewy 52
Lohrmann 39
Lohse 49
Lomonosov III
Longomontanus 64, 72
Lorentz VIII
Louise 19
Louville 9
Lovelace I
Lubbock 48
Lubiniezky 53
Lucian 25, 36
Ludwig IV
Luther 14
Lyell 36
Lyot 69, 76, V

Maclaurin 49
Maclear 35
Macmillan 21
Macrobius 26
Mädler 47
Maestlin 29
Magelhaens 48
Maginus 65, 73
Main 4
Mairan 9
Malapert 73, V
Mallet 68
Manilius 23, 34
Manners 35
Manzinus 74
Maraldi 25
Marco Polo 22
Marinus 69, IV
Marius 29
Markov 1
Marth 63
Maskelyne 36
Mason 14
Maunder VII
Maupertuis 2
Maurolycus 66
Maury 15
Maxwell III
McAdie III
McClure 59
McDonald 10
McLaughlin I
Mee 63
Mees VIII
Menelaus 23
Menzel 36
Mercator 53
Mercurius 15, 16
Merril I
Mersenius 51
Messala 16
Messier 48
Metius 68
Meton 4, 5
Milichius 30
Miller 65
Mitchell 5
Moigno 5
Moltke 46
Monge 59
Montanari 64
Moretus 73
Morley 49
Moseley VIII
Mösting 43
Mouchez 3, 4
Moulton V
Müller 44
Murchison 33
Mutus 74

Nansen II
Naonobu 49
Nasireddin 65
Nasmyth 70
Natasha 19
Naumann 8
Neander 68
Nearch 75
Neison 5
Neper 38, III
Nernst VIII
Neumayer 75, V
Newcomb 15, 25
Newton 73
Nicholson 50, VII
Nicolai 67
Nicollet 54
Nielsen 8
Niépce I
Nobili 38
Noether I
Nöggerath 62, 70
Nonius 65
Norman 41
Nunn III

Oenopides I
Oersted 15
Oken 69, IV
Olbers 28
Omar Khayyam I
Opelt 53
Oppolzer 44
Orontius 65

Palisa 42
Palitzsch 59
Pallas 33
Palmieri 51
Paneth I
Parkhurst IV
Parrot 44, 55
Parry 42, 43
Pascal 3, I
Peary 4, II
Peek 38, III
Peirce 26
Peirescius 68, 69
Pentland 73, 74
Petavius 59
Petermann 5, II
Peters 5
Petit 38
Petrov 76, V
Pettit 50, VII
Petzval VI
Phillips 60
Philolaus 3
Phocylides 70
Piazzi 61
Piazzi Smyth 12
Picard 26, 37
Piccolomini 58, 68
Pickering 45
Pictet 64, 65
Pilâtre 70, VI
Pingré 70, VI
Pitatus 54, 64
Pitiscus 75
Plana 14
Plaskett II
Plato 3, 4
Playfair 56
Plinius 24
Plutarch 27, III
Poczobutt I
Poisson 56, 66
Polybius 57
Pomortsev 38
Poncelet 3, I
Pons 57
Pontanus 56
Pontécoulant 76
Popov III
Porter 72
Posidonius 14
Priestley V
Prinz 19
Proclus 26
Proctor 65
Protagoras 4
Ptolemaeus 44
Puiseux 52
Pupin 21
Purbach 55
Purkyně IV
Pythagoras 2, I
Pytheas 20

Rabbi Levi 67
Raman 18
Ramsden 63
Rankine 49
Rayleigh 27, III
Réaumur 44
Regiomontanus 55
Regnault 1, I
Reichenbach 59, 69
Reimarus 68
Reiner 29
Reinhold 31
Repsold 1, I
Respighi 38
Rhaeticus 33, 44
Rheita 68
Riccioli 39
Riccius 67
Richardson III
Riemann 16, III
Ritchey 45
Rittenhouse V
Ritter 35
Ritz IV
Robinson 2
Rocca 39
Rocco 19
Rosenberger 75
Ross 35
Rosse 58
Rost 71
Rothmann 67
Rozhdestvenskiy I
Römer 25
Röntgen VIII
Runge 49, IV
Russell 17
Ruth 19
Rutherfurd 72
Rydberg VI
Rynin I

Sabatier 38
Sabine 35
Sacrobosco 56, 57
Sampson 21
Santbech 59
Santos-Dumont 22
Sarabhai 24
Sasserides 64
Saunder 45
Saussure 65
Scheele 41
Scheiner 72
Schiaparelli 18
Schickard 62
Schiller 71
Schlüter 39, VII, VIII
Schmidt 35
Schonberger 73, 74
Schorr 60, IV
Schönfeld VIII
Schrödinger V
Schröter 32
Schubert 38, III
Schumacher 16
Schwabe 6
Schwarzschild II
Scoresby 4
Scott 73, 74, V
Secchi 37
Seeliger 44
Segner 71
Seleucus 17
Seneca 27, III
Shaler 61, VII
Shapley 38
Sharp 9, 10
Sheepshanks 5
Shi Shen II
Short 73
Shuckburgh 15
Shuleykin VII
Sikorsky V
Silberschlag 34
Simpelius 73
Sinas 36
Sirsalis 39
Sklodowska IV
Slocum 49, IV
Smithson 37
Smoluchowski I
Snellius 59, 69
Somerville 49
Sömmering 32, 43
Sosigenes 35
South 2
Spallanzani 67
Spörer 44
Spurr 22
Stadius 32
Steinheil 68, 76
Stevinus 69
Stewart 38
Stiborius 67
Stokes 1, I
Stöfler 65, 66
Strabo 6
Street 64
Struve 17, VIII
Suess 29
Sulpicius Gallus 23
Sundman VIII
Swasey 49, IV
Swift 26
Sylvester 3, I
T. Mayer 19
Tacchini 38, III
Tacitus 57
Tacquet 24
Talbot 49, IV
Tannerus 74
Taruntius 37
Taylor 46
Tebbutt 37
Tempel 34
Thales 6
Theaetetus 12
Thebit 55
Theiler 38, III
Theon Junior 45, 46
Theon Senior 45, 46
Theophilus 46, 47
Theophrastus 25
Timaeus 4
Timocharis 21
Tisserand 26
Titius IV
Tolansky 42, 43
Torricelli 47
Toscannelli 18
Townley 38
Tralles 26
Triesnecker 33
Trouvelot 4
Tucker 49, IV
Turner 43
Tycho 64

Ukert 33
Ulugh Beigh 8, VIII
Urey 27, III

van Albada 38
Van Biesbroeck 19
Van Vleck 49
Vasco da Gama 28, VIII
Vashakidze III
Väisälä 18
Vega 68
Vendelinus 60
Vera 19
Verne 20
Very 24
Vestine III
Vieta 51, 61
Virchow 38, III
Vitello 62
Vitruvius 25
Vlacq 75
Vogel 56
Volta 1, I
von Behring 49
Voskresenskiy 17, VIII

W. Bond 4
Wallace 21
Wallach 36
Walter 65
Wargentin 70
Warner 49, IV
Watt 76
Watts 37
Webb 49
Weichert V
Weierstrass 49
Weigel 71
Weinek 58
Weiss 63, 64
Werner 55
Wexler V
Whewell 34
Wichmann 41
Widmanstätten 49, IV
Wildt 38
Wilhelm 63, 64
Wilkins 57, 67
Williams 14, 15
Wilson 72
Winthrop 40

Wolf 54
Wollaston 9
Wöhler 67
Wright 61, VII
Wrottesley 59
Wurzelbauer 64
Wyld IV
Xenophanes 1, I
Yakovkin 70, VI
Yangel' 22
Yerkes 26, 37
Young 68
Zach 73
Zagut 67
Zasyadko III
Zähringer 37
Zeeman VI
Zeno 16, II
Zinner 18
Zöllner 46
Zsigmondi I
Zucchius 71
Zupus 51

Catena – crater chain

Catena Abulfeda 56, 57
Catena Davy 43
Catena Dziewulski III
Catena Humboldt 60, IV
Catena Krafft 17, 28
Catena Littrow 24
Catena Sylvester I
Catena Timocharis 21

Dorsa – network or group of wrinkle ridges

Dorsa Aldrovandi 24
Dorsa Andrusov 49
Dorsa Argand 19
Dorsa Barlow 25, 36
Dorsa Burnet 18
Dorsa Cato 37
Dorsa Ewing 41
Dorsa Geikie 48
Dorsa Harker 27, 38
Dorsa Lister 24
Dorsa Mawson 48
Dorsa Rubey 40
Dorsa Smirnov 24
Dorsa Stille 20
Dorsa Tetyaev 27
Dorsa Whiston 8

Dorsum – wrinkle ridge

Dorsum Arduino 19
Dorsum Azara 24
Dorsum Bucher 9
Dorsum Buckland 23
Dorsum Cayeux 37
Dorsum Cushman 37
Dorsum Gast 23
Dorsum Grabau 11
Dorsum Guettard 42
Dorsum Heim 10
Dorsum Higazy 21
Dorsum Nicol 24
Dorsum Niggli 18
Dorsum Oppel 26
Dorsum Owen 23
Dorsum Scilla 8
Dorsum Termier 37, 38
Dorsum von Cotta 23
Dorsum Zirkel 10, 20

Lacus – 'lake'

Lacus Aestatis	Summer Lake 39, 50
Lacus Autumni	Autumn Lake 39, 50, VII
Lacus Bonitatis	Lake of Good 25, 26
Lacus Doloris	Lake of Suffering 23
Lacus Excellentiae	Lake of Excellence 62
Lacus Felicitatis	Lake of Happiness 22
Lacus Gaudii	Lake of Joy 23
Lacus Hiemalis	Winter Lake 23
Lacus Lenitatis	Lake of Tenderness 23, 34
Lacus Mortis	Lake of Death 14
Lacus Odii	Lake of Hate 23
Lacus Perseverantiae	Lake of Persistence 38
Lacus Somniorum	Lake of Dreams 14
Lacus Spei	Lake of Hope 16
Lacus Temporis	Lake of Time 15
Lacus Timoris	Lake of Fear 63
Lacus Veris	Spring Lake 39, 50, VII

Mare – 'sea'

Mare Anguis	Serpent Sea 27
Mare Australe	Southern Sea 76, IV, V
Mare Cognitum	Known Sea 42
Mare Crisium	Sea of Crises 26, 27, 37, 38
Mare Fecunditatis	Sea of Fertility 37, 48, 49, 59
Mare Frigoris	Sea of Cold 1–6
Mare Humboldtianum	Humboldt's Sea 7, II
Mare Humorum	Sea of Moisture 51, 52
Mare Imbrium	Sea of Rains 9–12, 19–21
Mare Insularum	Sea of Isles 30–32, 41, 42
Mare Marginis	Border Sea 27, 38, III
Mare Nectaris	Sea of Nectar 47, 58
Mare Nubium	Sea of Clouds 53, 54
Mare Orientale	Eastern Sea 50, VII
Mare Serenitatis	Sea of Serenity 13, 14, 23, 24
Mare Smythii	Smyth's Sea 38, 49, III, IV
Mare Spumans	Foaming Sea 38
Mare Tranquillitatis	Sea of Tranquillity 35, 36
Mare Undarum	Sea of Waves 38
Mare Vaporum	Sea of Vapours 22, 33, 34

Mons – mountain

Mons Ampère 22
Mons Argaeus 24, 25
Mons Blanc 12
Mons Bradley 22
Mons Delisle 19
Mons Gruithuisen Delta 9
Mons Gruithuisen Gamma 9
Mons Hadley 22
Mons Hadley Delta 22
Mons Hansteen 40
Mons Herodotus 18
Mons Huygens 22
Mons La Hire 20
Mons Maraldi 25
Mons Moro 42
Mons Penck 46
Mons Pico 11
Mons Piton 12
Mons Rümker 8
Mons Usov 38
Mons Vinogradov 19
Mons Vitruvius 25
Mons Wolff 21

Montes – mountain range or group of mountains

Montes Agricola	18	
Montes Alpes	12	Alps
Montes Apenninus	22	Apennines
Montes Archimedes	22	
Montes Carpatus	20, 31	Carpathian Mountains
Montes Caucasus	13	Caucasus Mountains
Montes Cordillera	39, 50, VII	Cordillera Mountains
Montes Haemus	23	(mountain range in Balkan)
Montes Harbinger	19	
Montes Jura	10	Jura Mountains
Montes Pyrenaeus	48, 58	Pyrenees
Montes Recti	11	Straight Range
Montes Riphaeus	41, 42	(nowadays Ural Mountains)
Montes Rook	50, VII	
Montes Secchi	37	
Montes Spitzbergen	12	Spitsbergen
Montes Taurus	25	
Montes Teneriffe	3, 11	

Oceanus – ocean

Oceanus Procellarum	Ocean of Storms 8, 9, 17–19, 28, 29, 39–41, VIII

Palus – marsh

Palus Epidemiarum	Marsh of Epidemics 63
Palus Putredinis	Marsh of Rot 22
Palus Somni	Marsh of Sleep 26, 37

Promontorium – promontory

Promontorium Agarum 38
Promontorium Agassiz 12
Promontorium Archerusia 24
Promontorium Deville 12
Promontorium Fresnel 22

Rima – rille

Rima Agatharchides 53
Rima Archytas 4
Rima Ariadaeus 34
Rima Artsimovich 19
Rima Billy 51
Rima Birt 54
Rima Bradley 22
Rima Brayley 19
Rima Calippus 13
Rima Cardanus 28
Rima Cauchy 36
Rima Cleomedes 26
Rima Conon 22
Rima Delisle 9
Rima Diophantus 19
Rima Flammarion 44
Rima Furnerius 69
Rima G. Bond 15
Rima Galilaei 28
Rima Gärtner 6
Rima Gay-Lussac 31
Rima Hadley 22
Rima Hansteen 40
Rima Hase 69
Rima Herigonius 41
Rima Hesiodus 53, 54, 63
Rima Hyginus 33, 34
Rima Jansen 25, 36
Rima Mairan 9
Rima Marius 18
Rima Messier 37
Rima Milichius 30
Rima Oppolzer 44
Rima Réaumur 44
Rima Schröter 32
Rima Sharp 9
Rima Sheepshanks 5
Rima Suess 29

Rimae – network or group of rilles

Rimae Alphonsus 44
Rimae Archimedes 22
Rimae Aristarchus 18
Rimae Arzachel 55
Rimae Atlas 15
Rimae Bode 33
Rimae Boscovich 34
Rimae Bürg 14
Rimae Chacornac 14
Rimae Daniell 14
Rimae Darwin 50
Rimae de Gasparis 51
Rimae Doppelmayer 52
Rimae Fresnel 22
Rimae Gassendi 52
Rimae Goclenius 48
Rimae Grimaldi 39
Rimae Gutenberg 47
Rimae Hevelius 28
Rimae Hippalus 52, 53
Rimae Hypatia 46
Rimae Janssen 67
Rimae Littrow 25
Rimae Maclear 35
Rimae Maestlin 29
Rimae Maupertuis 3
Rimae Menelaus 24
Rimae Mersenius 51
Rimae Opelt 42
Rimae Palmieri 51
Rimae Parry 42, 43
Rimae Petavius 59
Rimae Pitatus 54
Rimae Plato 4
Rimae Plinius 24
Rimae Posidonius 14
Rimae Prinz 19
Rimae Ramsden 63
Rimae Repsold 1, I
Rimae Riccioli 39
Rimae Ritter 35
Rimae Römer 25
Rimae Secchi 37
Rimae Sirsalis 39, 50
Rimae Sosigenes 35
Rimae Sulpicius Gallus 23
Rimae Theaetetus 12
Rimae Triesnecker 33, 34
Rimae Zupus 51

Rupes – scarp

Rupes Altai 57
Rupes Cauchy 36
Rupes Kelvin 52
Rupes Liebig 51
Rupes Mercator 53
Rupes Recta 54
Rupes Toscanelli 18

Sinus – 'bay'

Sinus Aestuum	Bay of Billows	32, 33
Sinus Amoris	Bay of Love	25
Sinus Asperitatis	Bay of Asperity	46, 47
Sinus Concordiae	Bay of Concord	37
Sinus Fidei	Bay of Faith	22
Sinus Honoris	Bay of Honour	35
Sinus Iridum	Bay of Rainbows	10
Sinus Lunicus	Bay of Lunicus	12
Sinus Medii	Central Bay	33, 44
Sinus Roris	Bay of Dew	1, 8, 9, I, VIII
Sinus Successus	Bay of Success	38

Vallis – valley

Vallis Alpes 4, 12
Vallis Baade 61, VI
Vallis Bohr VIII
Vallis Bouvard 61, VII
Vallis Capella 47
Vallis Inghirami 61
Vallis Palitzsch 59
Vallis Rheita 68
Vallis Schrödinger V
Vallis Schröteri 18
Vallis Snellius 59, 69

Substitute Names

In 1972 new names were substituted for the original ones listed in the second column. The second list (p. 224) contains an alphabetical list of original names in the first column and the new names in column 2.

New name	*Original name*	*Map no.*
Abbot	Apollonius K	37
Acosta	Langrenus C	49
Al-Bakri	Tacquet A	24, 35
Al-Marrakushi	Langrenus D	48
Ameghino	Apollonius C	38
Ammonius	Ptolemaeus A	44
Anville	Taruntius G	37
Artsimovich	Diophantus A	19
Aryabhata	Maskelyne E	36
Asada	Taruntius A	37
Atwood	Langrenus K	49
Avery	Gilbert U	49
Back	Schubert B	38
Bancroft	Archimedes A	21, 22
Banting	Linné E	23
Barkla	Langrenus A	49
Beals	Riemann A	16
Beketov	Jansen C	25
Bilharz	Langrenus F	49
Black	Kästner F	49
Blanchard	Arrhenius P	VI
Bobillier	Bessel E	23
Boethius	Dubiago U	38
Bombelli	Apollonius T	38
Borel	le Monnier C	24
Born	Maclaurin Y	49
Bowen	Manilius A	23
Brewster	Römer L	25
Cajal	Jansen F	36
Cameron	Taruntius C	37
Carmichael	Macrobius A	25
Carrel	Jansen B	35
Cartan	Apollonius D	38
Caventou	La Hire D	20
Chadwick	de Roy X	VI
Clerke	Littrow B	25
Condon	Webb R	38
Couder	Maunder Z	VII
Crile	Proclus F	37
Curtis	Picard Z	26
Daly	Apollonius P	38
Daubrée	Menelaus S	23
Eppinger	Euclides D	42
Esclangon	Macrobius L	25
Fabbroni	Vitruvius E	25
Fahrenheit	Picard X	38
Franck	Römer K	25
Fredholm	Macrobius D	26
Fryxell	Golitsyn B	VII
Galen	Aratus A	22
Gardner	Vitruvius A	25
Geissler	Gilbert D	49
Golgi	Schiaparelli D	18
Greaves	Lick D	37
Hargreaves	Maclaurin S	49
Heinrich	Timocharis A	21
Heyrovský	Drude S	VII
Hill	Macrobius B	25
Humason	Lichtenberg G	8
Huxley	Wallace B	22
Ibn Battuta	Goclenius A	48
Ibn Rushd	Cyrillus B	46
Il'in	Hohmann T	VII
Jenkins	Schubert Z	38
Joy	Hadley A	23
Keldysh	Hercules A	6
Knox-Shaw	Banachiewicz F	38
Kramarov	Lenz K	VII
Krogh	Auzout B	38
Kuiper	Bonpland E	42
Kundt	Guericke C	43
L'allemand	Kopff A	50
Landsteiner	Timocharis F	11
Lawrence	Taruntius M	37
Leakey	Censorinus F	47
Lindbergh	Messier G	48
Lindsay	Dollond C	45
Liouville	Dubiago S	38
Lucian	Maraldi B	25, 36
Macmillan	Archimedes F	21
McDonald	Carlini B	10
Morley	Maclaurin R	49
Naonobu	Langrenus B	49
Nielsen	Wollaston C	8
Nobili	Schubert Y	38
Norman	Euclides B	41
Petit	Apollonius W	38
Pilâtre	Hausen B	70, VI
Pomortsev	Dubiago P	38
Pupin	Timocharis K	21
Raman	Herodotus D	18
Respighi	Dubiago C	38
Santos-Dumont	Hadley B	22
Sarabhai	Bessel A	24
Scheele	Letronne D	41
Shapley	Picard H	38

New name	*Original name*	*Map no.*
Shuleykin	Pettit T	VII
Smithson	Taruntius N	37
Somerville	Langrenus J	49
Spurr	Archimedes K	22
Stewart	Dubiago Q	38
Swift	Peirce B	26
Tacchini	Neper K	38
Tebbutt	Picard G	37
Theophrastus	Maraldi M	25
Tolansky	Parry A	42, 43
Toscanelli	Aristarchus C	18
Townley	Apollonius G	38
Urey	Rayleigh A	27
Väisälä	Aristarchus A	18
van Albada	Auzout A	38
Van Biesbroeck	Krieger B	19
Van Vleck	Gilbert M	49
Very	le Monnier B	24
Virchow	Neper G	38
von Behring	Maclaurin F	49
Wallach	Maskelyne H	36
Watts	Taruntius D	37
Weierstrass	Gilbert N	49
Wildt	Condorcet K	38
Winthrop	Letronne P	40
Yakovkin	Pingré H	70
Yangel	Manilius F	22
Zähringer	Taruntius E	37
Zinner	Schiaparelli B	18

Original name	*New name*	*Map no.*
Apollonius C	Ameghino	38
Apollonius D	Cartan	38
Apollonius G	Townley	38
Apollonius K	Abbot	37
Apollonius P	Daly	38
Apollonius T	Bombelli	38
Apollonius W	Petit	38
Aratus A	Galen	22
Archimedes A	Bancroft	21, 22
Archimedes F	Macmillan	21
Archimedes K	Spurr	22
Aristarchus A	Väisälä	18
Aristarchus C	Toscanelli	18
Arrhenius P	Blanchard	VI
Auzout A	van Albada	38
Auzout B	Krogh	38
Banachiewicz F	Knox-Shaw	38
Bessel A	Sarabhai	24
Bessel E	Bobillier	23
Bonpland E	Kuiper	42
Carlini B	McDonald	10
Censorinus F	Leakey	47
Condorcet K	Wildt	38
Cyrillus B	Ibn Rushd	46
de Roy X	Chadwick	VI
Diophantus A	Artsimovich	19
Dollond C	Lindsay	45
Drude S	Heyrovský	VII
Dubiago C	Respighi	38
Dubiago P	Pomortsev	38
Dubiago Q	Stewart	38
Dubiago S	Liouville	38
Dubiago U	Boethius	38
Euclides B	Norman	41
Euclides D	Eppinger	42
Gilbert D	Geissler	49
Gilbert M	Van Vleck	49
Gilbert N	Weierstrass	49
Gilbert U	Avery	49
Goclenius A	Ibn Battuta	48
Golitsyn B	Fryxell	VII
Guericke C	Kundt	43
Hadley A	Joy	23
Hadley B	Santos-Dumont	22
Hausen B	Pilâtre	70, VI
Hercules A	Keldysh	6
Herodotus D	Raman	18
Hohmann T	Il'in	VII
Jansen B	Carrel	35
Jansen C	Beketov	25
Jansen F	Cajal	36
Kästner F	Black	49
Kopff A	L'allemand	50
Krieger B	Van Biesbroeck	19
La Hire D	Caventou	20
Langrenus A	Barkla	49
Langrenus B	Naonobu	49
Langrenus C	Acosta	49
Langrenus D	Al-Marrakushi	48
Langrenus F	Bilharz	49
Langrenus J	Somerville	49
Langrenus K	Atwood	49
le Monnier B	Very	24
le Monnier C	Borel	24
Lenz K	Kramarov	VII
Letronne D	Scheele	41
Letronne P	Winthrop	40
Lichtenberg G	Humason	8
Lick D	Greaves	37
Linné E	Banting	23
Littrow B	Clerke	25
Maclaurin F	von Behring	49
Maclaurin R	Morley	49
Maclaurin S	Hargreaves	49
Maclaurin Y	Born	49
Macrobius A	Carmichael	25
Macrobius B	Hill	25
Macrobius D	Fredholm	26
Macrobius L	Esclangon	25
Manilius A	Bowen	23
Manilius F	Yangel'	22
Maraldi B	Lucian	25, 36
Maraldi M	Theophrastus	25
Maskelyne E	Aryabhata	36
Maskelyne H	Wallach	36
Maunder Z	Couder	VII
Menelaus S	Daubrée	23
Messier G	Lindbergh	48
Neper G	Virchow	38
Neper K	Tacchini	38
Parry A	Tolansky	42, 43
Peirce B	Swift	26
Pettit T	Shuleykin	VII
Picard G	Tebbutt	37
Picard H	Shapley	38
Picard X	Fahrenheit	38
Picard Z	Curtis	26
Pingré H	Yakovkin	70
Proclus F	Crile	37
Ptolemaeus A	Ammonius	44
Rayleigh A	Urey	27
Riemann A	Beals	16
Römer K	Franck	25
Römer L	Brewster	25
Schiaparelli B	Zinner	18
Schiaparelli D	Golgi	18
Schubert B	Back	38
Schubert Y	Nobili	38
Schubert Z	Jenkins	38
Tacquet A	Al-Bakri	24, 35
Taruntius A	Asada	37
Taruntius C	Cameron	37
Taruntius D	Watts	37
Taruntius E	Zähringer	37
Taruntius G	Anville	37
Taruntius M	Lawrence	37
Taruntius N	Smithson	37
Timocharis A	Heinrich	21
Timocharis F	Landsteiner	11
Timocharis K	Pupin	21
Vitruvius A	Gardner	25
Vitruvius E	Fabbroni	25
Wallace B	Huxley	22
Webb R	Condon	38
Wollaston C	Nielsen	8

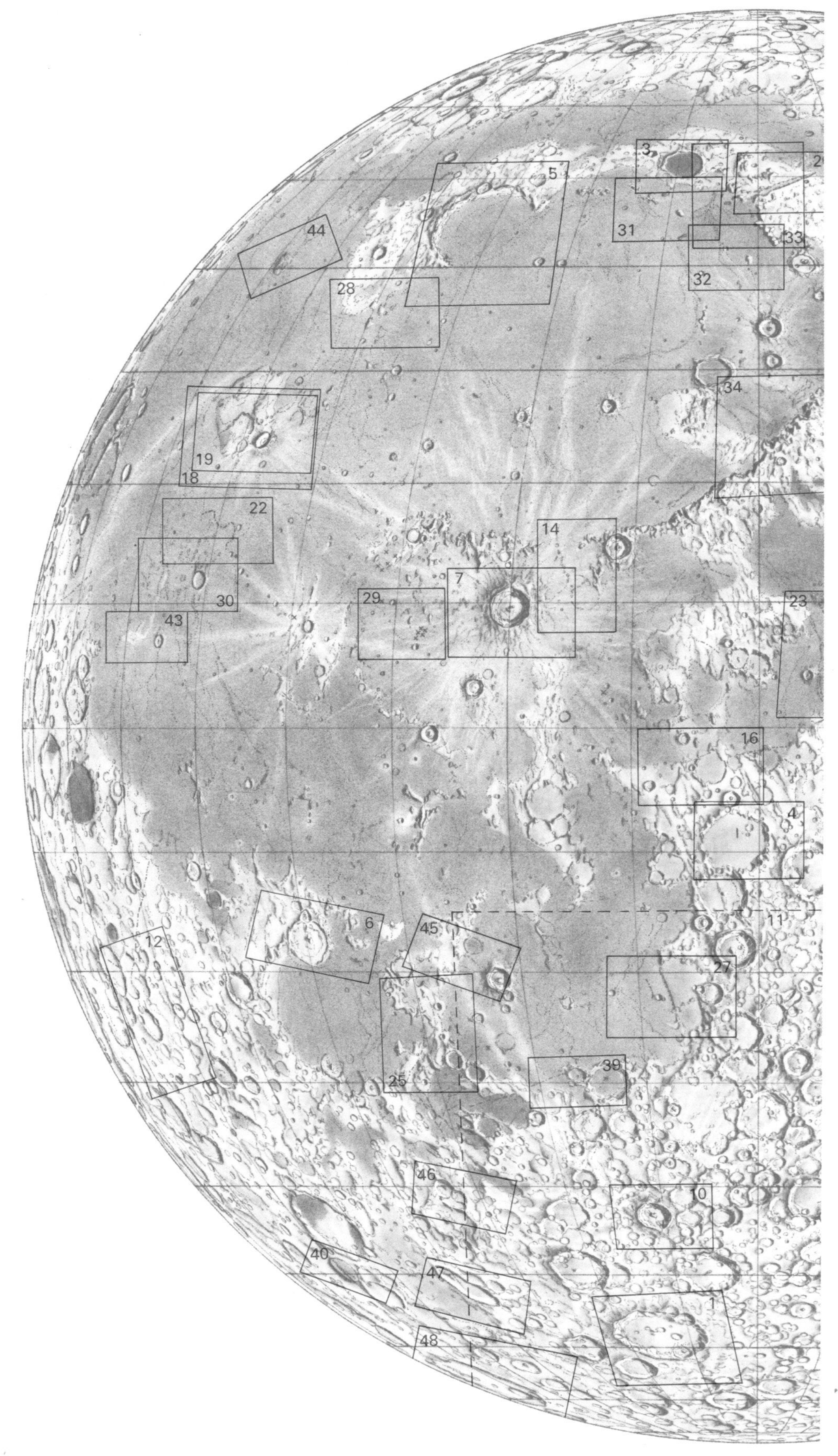
3
5
31
33
44
32
28
34
19
18
22
14
7
30
29
23
43
16
4
11
6
45
12
27
39
25
46
10
40
47
1
48